I0010903

Cloud Native Applications with Ballerina

A guide for programmers interested in developing
cloud native applications using Ballerina Swan Lake

Dhanushka Madushan

BIRMINGHAM—MUMBAI

Cloud Native Applications with Ballerina

Copyright © 2021 Packt Publishing

All rights reserved. No part of this book may be reproduced, stored in a retrieval system, or transmitted in any form or by any means, without the prior written permission of the publisher, except in the case of brief quotations embedded in critical articles or reviews.

Every effort has been made in the preparation of this book to ensure the accuracy of the information presented. However, the information contained in this book is sold without warranty, either express or implied. Neither the author, nor Packt Publishing or its dealers and distributors, will be held liable for any damages caused or alleged to have been caused directly or indirectly by this book.

Packt Publishing has endeavored to provide trademark information about all of the companies and products mentioned in this book by the appropriate use of capitals. However, Packt Publishing cannot guarantee the accuracy of this information.

Group Product Manager: Richa Tripathi
Publishing Product Manager: Sathyanarayanan Ellapulli
Senior Editor: Rohit Singh
Content Development Editor: Vaishali Ramkumar
Technical Editor: Karan Solanki
Copy Editor: Safis Editing
Project Coordinator: Deeksha Thakkar
Proofreader: Safis Editing
Indexer: Pratik Shirodkar
Production Designer: Shyam Sundar Korumilli

First published: September 2021

Production reference: 1210921

Published by Packt Publishing Ltd.
Livery Place
35 Livery Street
Birmingham
B3 2PB, UK.

ISBN 978-1-80020-063-0

www.packt.com

To my mother and father

– Dhanushka Madushan

Contributors

About the author

Dhanushka Madushan is a senior software engineer at WSO2 and has a bachelor of engineering qualification from the Department of Computer Science and Engineering, University of Moratuwa. He has 5+ years' experience in developing software solutions for cloud-based platforms in different business domains. He has worked on the WSO2 integration platform for 3+ years and is responsible for building and maintaining integration products. He often writes blogs about the latest cutting-edge, cloud native-related technologies using his experience of working on many open source projects, including Micro-integrator, WSO2 ESB, and Apache Synapse, as well as development of the Choreo iPaaS platform and the Zeptolytic SaaS data analytics platform. Dhanushka's extensive exposure to cloud native-related technologies, including Docker, Kubernetes, Jenkins, AWS, and multiple observability tools, is a key area of expertise.

First and foremost, my thanks go to my loving parents who have supported me throughout the long journey of writing this book. Next, my thanks extend to Sameera Jayasoma and all Ballerina team members for supporting me whenever I had questions. I especially need to thank Lakmal Warusawithana and Anjana Fernando for the support given in the initial drafting and code sample creation process. Also, I would like to thank the technical reviewers, Nadeeshaan Gunasinghe, Joy Rathnayake, Shiroshica Kulatilake, and Shenavi de Mel, for the amazing work they have done. I would also extend my gratitude to the Packt team who supported and guided me throughout the process of publishing this book. Finally, I would like to thank Dr. Sanjeewa Weerawarna for founding and this awesome programming language.

About the reviewers

Nadeeshaan Gunasinghe is a technical lead at WSO2 with over 6 years' experience in enterprise integration, programming languages, and developer tooling. Nadeeshaan leads the Ballerina Language Server team and is also a key contributor to Ballerina, which is an open source programming language and platform for the cloud era, as well as being an active contributor to the WSO2 Enterprise Service Bus. He is also passionate about sports, football and cricket in particular.

Joy Rathnayake is a solutions architect with over 16 years' industry experience and is part of the solution architecture team at WSO2, based in Colombo, Sri Lanka. He is primarily responsible for understanding customer requirements, identifying the products/technologies required, and defining the overall solution design/architecture.

Joy has been recognized as both a Microsoft **Most Valuable Professional** (**MVP**) and **Microsoft Certified Trainer** (**MCT**). He was the first to hold both MVP and MCT recognitions in Sri Lanka. He has contributed to developing content for Microsoft Certifications and has worked as a **Subject Matter Expert** (**SME**) for many Microsoft exam development projects. He has contributed a lot to the community by presenting at various events, including Tech-Ed Europe, Tech-Ed Southeast Asia, Tech-Ed Sri Lanka, Tech-Ed India, Tech-Ed Malaysia, Southeast Asia SharePoint Conference, and SharePoint Saturday. He enjoys traveling, speaking at public events/conferences, and reading.

Connect with him on LinkedIn at `https://www.linkedin.com/in/joyrathnayake/`.

Shiroshica Kulatilake is a solutions architect at WSO2 where she works with customers around the world to provide middleware solutions on the WSO2 stack for digital transformation projects. In her work, she is involved with real-world problems that organizations face in a rapidly changing digital world and gets the opportunity to help these organizations achieve what they require in their business from a technological standpoint. Her expertise lies in API Management, API Security, Integration, and EIPaaS. She started her career as a software engineer and is still passionate about the nitty-gritty aspects of building things, with her current focus being microservice architectures.

Shenavi de Mel is an experienced software solutions engineer with 7+ years' experience working in the computer software industry. She is passionate about building great customer relationships and enhancing her knowledge of the field. She has extensive hands-on experience in many development languages and technologies, including Java, Jaggery, Ballerina, JavaScript/jQuery, SQL, PHP, HTML, Docker, and Kubernetes. Currently, she is working as a lead solutions engineer as part of the solutions engineering team, assisting customers in implementing their solutions using the WSO2 platform. She is also very familiar with API management, integration, and identity protocols, having spent the majority of her career working at WSO2, one of the leading companies as regards middleware and open source technology.

Table of Contents

2
Getting Started with Ballerina

Section 2: Building Microservices with Ballerina

3
Building Cloud Native Applications with Ballerina

5

Accessing Data in Microservice Architecture

Section 3: Moving on with Cloud Native

6
Moving on to Serverless Architecture

7
Securing the Ballerina Cloud Platform

8
Monitoring Cloud Native Applications

9
Integrating Ballerina Cloud Native Applications

Preface

Ballerina is the latest general-purpose programming language that is specially written to build cloud native applications. Ballerina is a relatively new programming language initially released in 2019. Unlike other general-purpose programming languages, Ballerina provides built-in syntax to define services, native support for JSON and XML data types, and many more cloud native syntax styles. Kubernetes and Docker deployment support and remote functions are special built-in features provided by the Ballerina language.

This book has been written to help you to understand cloud native architectural design patterns and how to use the Ballerina language to implement them. We will discuss the availability, maintainability, deployability, resiliency, and security aspects of the Ballerina language with practical implementations.

We will discuss a sample order management system throughout the book to implement different cloud native architectural patterns. We will discuss architectural concepts that are essential in building cloud native systems. The examples given in each chapter start from simple applications to more advanced applications. We will use multiple free open source tools and libraries throughout the book to build more advanced cloud native systems.

In this book, we will focus on multiple different platforms you can use to deploy a Ballerina application. You can deploy on a stand-alone computer, cloud platform, Kubernetes, serverless platform, or Choreo **Integration Platform as a Service (iPaaS)** platform. You can select the best-fit platform for your system.

By the end of this book, you should be able to understand the basic concepts of cloud native and how it can be practically implemented with the Ballerina program in different deployment platforms.

Who this book is for

This book is intended for developers and software engineers who want to learn how to use the Ballerina language to develop cloud native applications. The samples given in this book cover a wide range of requirements for building cloud native applications. You should have a basic understanding of programming languages and programming to follow this book. Since we are discussing basic to advanced concepts of both cloud native architecture and the Ballerina language, both beginner and intermediate developers can easily follow the content.

What this book covers

Chapter 1, *Introduction to Cloud Native*, helps you to understand what cloud native is and important facts that you should keep in mind when you are building cloud native applications.

Chapter 2, *Getting Started with Ballerina*, explores Ballerina language syntaxes and setting up Ballerina on your local computer.

Chapter 3, *Building Cloud Native Applications with Ballerina*, helps you to understand the cloud native features provided by the Ballerina language and deploying a simple Ballerina service on a Kubernetes cluster.

Chapter 4, *Inter-Process Communication and Messaging*, teaches you about different types of communication protocols that can be used in a Ballerina-based cloud native application.

Chapter 5, *Accessing Data in Microservice Architecture*, covers using the database-per-service design pattern to build complex real-world applications with other microservice architecture design patterns.

Chapter 6, *Moving on to Serverless Architecture*, covers building a Ballerina application with the AWS and Azure serverless platforms.

Chapter 7, *Securing the Ballerina Cloud Platform*, helps you to understand security concepts and securing Ballerina applications with Ballerina security features.

Chapter 8, *Monitoring Cloud Native Applications*, covers monitoring Ballerina applications with logs, metrics, and traces.

Chapter 9, *Integrating Ballerina Cloud Native Applications*, covers exposing Ballerina services with a gateway and building integration with the Choreo platform.

Chapter 10, *Building a CI/CD Pipeline for Ballerina Applications*, teaches you how to build, test, and deploy a Ballerina program automatically with CI/CD pipeline tools.

To get the most out of this book

You don't need to have full knowledge of the Ballerina programming language to follow the samples given here. But you should have a basic understanding of at least one general-purpose programming language, such as Java, C, or Go, since Ballerina syntax is very similar to those.

Software/hardware covered in the book	Operating system requirements
Visual Studio Code	Windows, macOS, or Linux
Docker	Windows, macOS, or Linux
Kubernetes	Windows, macOS, or Linux
Minikube	Windows, macOS, or Linux
MySQL	Windows, macOS, or Linux

The samples given in this book are tested on macOS. These samples should also work with Linux and Windows operating systems as well. The tools and libraries used in this book are free and open source. You can try out all of these samples free of charge. But the samples given in iPaaS and **Software as a Service (SaaS)** platforms such as AWS, Azure, Choreo, and Snowflake may charge you based on usage of the resources.

If you are using the digital version of this book, we advise you to type the code yourself or access the code from the book's GitHub repository (a link is available in the next section). Doing so will help you avoid any potential errors related to the copying and pasting of code.

You can get the latest information on Ballerina and cloud native development by following the author on Twitter (`https://twitter.com/DhanushkaDEV`) or adding them as a connection on LinkedIn (`https://www.linkedin.com/in/dhanushkamadushan/`).

Download the example code files

You can download the example code files for this book from GitHub at `https://github.com/PacktPublishing/Cloud Native-Applications-with-Ballerina/`. If there's an update to the code, it will be updated in the GitHub repository.

We also have other code bundles from our rich catalog of books and videos available at `https://github.com/PacktPublishing/`. Check them out!

Code in Action

The Code in Action videos for this book can be viewed at `https://bit.ly/3l1Exb9`.

Download the color images

We also provide a PDF file that has color images of the screenshots and diagrams used in this book. You can download it here:

```
https://static.packt-cdn.com/downloads/9781800200630_
ColorImages.pdf.
```

Conventions used

There are a number of text conventions used throughout this book.

`Code in text`: Indicates code words in text, database table names, folder names, filenames, file extensions, pathnames, dummy URLs, user input, and Twitter handles. Here is an example: "The `sayHello` resource will simply send a `Hello, World!` response back to the caller."

A block of code is set as follows:

```
import ballerina/http;
service /hello on new http:Listener(9090) {
    resource function get sayHello(http:Caller caller,
        http:Request req) returns error? {
        check caller->respond("Hello, World!");
    }
}
```

When we wish to draw your attention to a particular part of a code block, the relevant lines or items are set in bold:

```
[dependencies]
"dhanushka/invoice_util" = "0.1.0"
```

Any command-line input or output is written as follows:

```
kubectl apply -f target/kubernetes/code_to_cloud
```

Bold: Indicates a new term, an important word, or words that you see onscreen. For instance, words in menus or dialog boxes appear in **bold**. Here is an example: "Then, you can log in to the Ballerina Central dashboard and see your package listed under the **My Packages** section."

> **Tips or important notes**
> Appear like this.

Get in touch

Feedback from our readers is always welcome.

General feedback: If you have questions about any aspect of this book, email us at customercare@packtpub.com and mention the book title in the subject of your message.

Errata: Although we have taken every care to ensure the accuracy of our content, mistakes do happen. If you have found a mistake in this book, we would be grateful if you would report this to us. Please visit www.packtpub.com/support/errata and fill in the form.

Piracy: If you come across any illegal copies of our works in any form on the internet, we would be grateful if you would provide us with the location address or website name. Please contact us at copyright@packt.com with a link to the material.

If you are interested in becoming an author: If there is a topic that you have expertise in and you are interested in either writing or contributing to a book, please visit authors.packtpub.com.

Share Your Thoughts

Once you've read *Cloud Native Applications with Ballerina*, we'd love to hear your thoughts! Scan the QR code below to go straight to the Amazon review page for this book and share your feedback.

https://packt.link/r/1800200633

Your review is important to us and the tech community and will help us make sure we're delivering excellent quality content.

Section 1:
The Basics

This first section focuses on the basics of cloud native technology concepts and the basic building blocks of the Ballerina language. This section is necessary to understand the more advanced concepts that we are going to discuss in later sections.

First, we will discuss what cloud native is, the history of cloud-based software architecture, the definition of cloud native, and transforming an organization to using cloud native technologies. Here, we will focus on the theoretical aspects of building a cloud native system.

Next, we will discuss the architecture of the Ballerina language, setting up a development environment, fundamental Ballerina syntaxes, the Ballerina type system, error handling, and controlling the program flow. Here, we will learn about the practical aspects of using the Ballerina language and fundamental concepts that are needed to build complex cloud native applications.

This section comprises the following chapters:

- *Chapter 1, Introduction to Cloud Native*
- *Chapter 2, Getting Started with Ballerina*

1
Introduction to Cloud Native

In this chapter, we will go through how developers came up with cloud native due to the problems that are attached to monolithic architecture. Here, we will discuss the old paradigms of programming, such as three-tier architecture, and what the weaknesses are. You will learn about the journey of shifting from an on-premises computation infrastructure model to a cloud-based computing architecture. Then we will discuss the microservice architecture and serverless architecture as cloud-based solutions.

Different organizations have different definitions of cloud native architecture. It is difficult to give a cloud native application a clear definition, but we will discuss the properties that cloud native applications should have in this chapter. You will see how the twelve-factor app plays a key role in building cloud native applications. When you are building a cloud native application, keep those twelve factors in mind.

Organizations such as Netflix and Uber are transforming the way applications are designed by replacing monolithic architecture with the microservice architecture. Later in this chapter, we will see how organizations are successful in their business by introducing cloud native concepts. It is not a simple task to switch to a cloud native architecture. We will address moving from a monolithic architecture to a cloud native architecture later in the chapter.

We will cover the following topics in this chapter:

- Evolution from the monolithic to the microservice architecture
- Understanding what the cloud native architecture is
- Building cloud native applications
- The impact on organizations when moving to cloud native

By the end of this chapter, you will have learned about the evolution of cloud native applications, what cloud native applications are, and the properties that cloud native applications should have.

Evolution from the monolithic to the microservice architecture

The monolithic architecture dictated software development methodologies until cloud native conquered the realm of developers as a much more scalable design pattern. Monolithic applications are designed to be developed as a single unit. The construction of monolithic applications is simple and straightforward. There are problems related to monolithic applications though, such as scalability, availability, and maintenance.

To address these problems, engineers came up with the microservice architecture, which can be scalable, resilient, and maintainable. The microservice architecture allows organizations to develop increasingly flexible. The microservice architecture is the next step up from the **Service-Oriented Architecture (SOA)**. Both these architectures use services for business use cases. In the next sections, we will follow the journey from the monolithic architecture to SOA to the microservice architecture. To start this journey, we will begin with the simplest form of software architecture, which is the N-tier architecture. In the next section, we will discuss what the N-tier architecture is and the different levels of the N-tier architecture.

The N-tier architecture in monolithic applications

The N-tier architecture allows developers to build applications on several levels. The simplest type of N-tier architecture is the one-tier architecture. In this type of architecture, all programming logic, interfaces, and databases reside in a single computer. As soon as developers understood the value of decoupling databases from an application, they invented the two-tier architecture, where databases were stored on a separate server. This allowed developers to build applications that allow multiple clients to use a single database and provide distributed services over a network.

Developers introduced the application layer to the two-tier architecture and formed the three-tier architecture. The three-tier architecture includes three layers, known as data, application, and presentation, as shown in the following diagram:

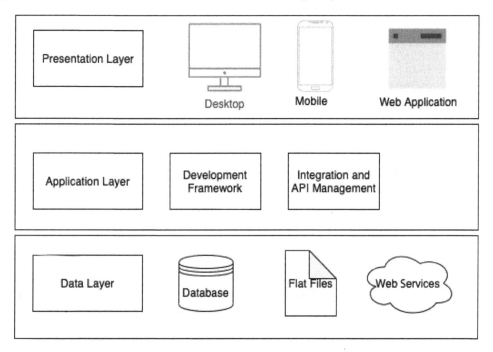

Figure 1.1 – Three-tier architecture

The topmost layer of the three-tier architecture is known as the **presentation layer**, which users directly interact with. This can be designed as a desktop application, a mobile application, or a web application. With the recent advancement of technology, desktop applications have been replaced with cloud applications. Computational power has moved from consumer devices onto the cloud platform with the recent growth of mobile and web technologies.

The three-tier architecture's middle layer is known as the **application layer**, in which all business logic falls. To implement this layer, general-purpose programming languages, along with supporting tools, are used. There are several programming languages with which you can implement business logic, such as Node.js, Java, and Python, along with several different libraries. These programming languages might be general-purpose programming languages such as Node.js and Java or domain-specific languages such as HTML, Apache Groovy, and Apache Synapse. Developers can use built-in tools such as API gateways, load balancers, and messaging brokers to develop an application, in addition to general-purpose programming languages.

The bottom layer is the **data layer**, which stores data that needs to be accessed by the application layer. This layer consists of databases, files, and third-party data storage services to read and write data. Databases usually consist of relational databases, which are used to store different entities in applications. There are multiple databases, such as MySQL, Oracle, MSSQL, and many more, used to build different applications. Other than using those databases, developers can select file-based and third-party storage services as well.

Developers need to be concerned about security, observability, delivery processes, deployability, and maintainability across all these layers over the entire life cycle of application development. With the three-tier architecture, it is easy and efficient to construct simple applications. Separating the application layer allows the three-tier architecture to be language-independent and scalable. Developers can distribute traffic between multiple application layer instances to allow horizontal scaling of the application. A load balancer sitting in front of the application layer spreads the load between the application instance replicas. Let's discuss monolithic application architecture in more detail and see how we can improve it in the next section.

Monolithic application architecture

The term "monolithic" comes from the Greek terms *monos* and *lithos*, together meaning a large stone block. The meaning in the context of IT for monolithic software architecture characterizes the uniformity, rigidity, and massiveness of the software architecture.

A monolithic code base framework is often written using a single programming language, and all business logic is contained in a single repository.

Typical monolithic applications consist of a single shared database, which can be accessed by different components. The various modules are used to solve each piece of business logic. But all business logic is wrapped up in a single API and is exposed to the frontend. The **user interface** (**UI**) of an application is used to access and preview backend data to the user. Here's a visual representation of the flow:

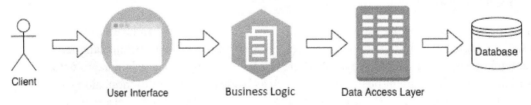

Figure 1.2 – A monolithic application

The scaling of monolithic applications is easy, as the developer can increase the processing and storage capacity of the host machine. Horizontal scalability can be accomplished by replicating data access layers and spreading the load of the client within each of these instances.

Since the monolithic architecture is simple and straightforward, an application with this paradigm can be easily implemented. Developers can start designing the application with a model entity view. This architecture can be easily mapped to the database design and applied to the application. It's also easy for developers to track, log, and monitor applications. Unlike the microservice architecture, which we will discuss later in this chapter, testing a monolithic application is also simple.

Even though it is simple to implement a monolithic application, there are lots of problems associated with maintaining it when it comes to building large, scalable systems:

- Monolithic applications are designed, built, and implemented in a single unit. Therefore, all the components of the architecture of the system should be closely connected. In most cases, point-to-point communication makes it more difficult to introduce a new feature component to the application later.

- As a monolithic application grows, it takes more time to start the entire application. Changing existing components or adding new features to the system may require stopping the whole system. This makes the deployment process slower and much more unstable.

- On the other hand, having an unstable service in the system means the entire system is fragile. Since the components of a monolithic application are tightly coupled with each other, all components should function as planned. Having a problem in one subsystem causes all the dependent components to fail.

- It is difficult to adopt modern technology because a monolithic application is built on homogeneous languages and frameworks. This makes it difficult to move forward with new technologies, and inevitably the whole system will become a legacy system.

Legacy systems have a business impact due to problems with the management of the system:

- Maintaining legacy systems is costly and ineffective over time. Even though developers continue to add features, the complexity of the system increases exponentially. This means that organizations spend money on increasing the complexity of the application rather than refactoring and updating the system.

- Security gets weaker over time as the dependent library components do not upgrade and become vulnerable to security threats. Migrating to new versions of libraries is difficult due to the tight coupling of components. The security features of these libraries are not up to date, making the system vulnerable to attacks.

- When enforcement regulations become tidal, it gets difficult to adjust the structure according to these criteria. With the **General Data Protection Act (GDPA)** enforcement, the system should be able to store well-regulated information.

- When technology evolves over time, due to compatibility problems, it is often difficult to integrate old systems with new systems. Developers need to add more adapters to the system to make it compliant with new systems. This makes the system a lot more complicated and bulkier.

Due to these obstacles, developers came up with a more modular design pattern, SOA, which lets developers build a system as a collection of different services. When it comes to SOA, a service is the smallest deployment unit used to implement application components. Each service is designed to solve problems in a specific business domain.

Services can be run on multiple servers and connected over a network. Service interfaces have loose coupling in that another service or client is able to access its features without understanding the internal architecture.

The core idea of SOA is to build loosely coupled, scalable, and reusable systems, where services work together to execute business logic. Services are the building block of SOA and are interconnected by protocols such as HTTP, JMS, TCP, and FTP. SOA commonly uses the XML format to communicate with other services. But interconnecting hundreds of services is a challenge if each service uses a point-to-point communication method. This makes it difficult to have thousands of links in interservice communication to maintain the system.

On the other hand, the system should be able to handle multiple different data formats, such as JSON, XML, Avro, and Thrift, which makes integration much more complicated. For engineers, observing, testing, migrating, and maintaining such a system is a nightmare. The **Enterprise Service Bus (ESB)** was introduced to SOA to simplify these complex messaging processes. In the next section, we will discuss what an ESB is and how it solves problems in SOA.

The ESB simplifies SOA

The ESB is located in the middle of the services, connecting all the services. The ESB provides a way of linking services by sitting in the center of services and offering various forms of transport protocols for communication. This addresses the issue of point-to-point connectivity issues where many services need to communicate with each other. If all of these services are directly linked to each other, communication is a mess and difficult to manage. The ESB decouples the service dependencies and offers a single bus where all of its services can be connected. Let's have a look at an SOA with point-to-point communication versus using an ESB for services to communicate:

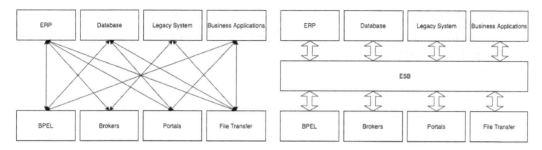

Figure 1.3 – ESB architecture

On incoming service requests, the ESB may perform simple operations and forward them to another service. This makes it easier for developers to migrate to SOA and expose existing legacy systems as services so that they can be easily handled instead of creating everything from scratch.

An ESB is capable of managing multiple endpoints with multiple protocols. For example, an ESB can use the following features to facilitate the integration of services in SOA:

- **Security**: An ESB handles security when connecting various services together. The ESB offers security features such as authentication, authorization, certificate management, and encryption to secure connected services.

- **Message routing**: Instead of directly calling services, an ESB provides the modularity of SOA by providing routing on the ESB itself. Since all other services call the ESB to route services, developers can substitute service components without modifying the service.

- **Central communication platform**: This prevents a point-to-point communication issue where each service does not need to know the address of the endpoint. The services blindly send requests to the ESB and the ESB routes requests as specified in the ESB routing logic. The ESB routes traffic between services and acts as smart pipes, and that makes service endpoints dumb.

- **Monitoring the whole message flow**: Because the ESB is located in the center of the services, this is the perfect location to track the entire application. Logging, tracing, and collecting metrics can be placed in the ESB to collect the statistics of the overall system. This data can be used along with an analytical tool to analyze bugs, performance bottlenecks, and failures.

- **Integration over different protocols**: The ESB ensures that services can be connected via different communication protocols, such as HTTP, TCP, FTP, JMS, and SMTP. It is also supported for various data interchange formats, such as JSON and XML.

- **Message conversion**: If a service or client application is required to access another service, the message format may be modified from one to another. In this case, the ESB offers support for the conversion of messages across various formats, such as XML and JSON. It also supports the use of transformation (XSLT) and the modification of the message structure.

- **Enterprise Integration Patterns (EIP)**: These are used as building blocks for the SOA messaging system. These patterns include channeling, routing, transformation, messaging, system, and management. This helps developers build scalable and reliable SOA platforms with EIP.

SOA was used as mainstream cloud architecture for a long time until the microservice architecture came along as a new paradigm for building cloud applications. Let's discuss the emergence of the microservice architecture and how it solves problems with SOA in the next section.

The emergence of microservices

SOA provides solutions to most of the issues that monolithic applications face. But developers still have concerns about creating a much more scalable and flexible system. It's easy to construct a monolithic structure. But as it expands over time, managing a large system becomes more and more difficult.

With the emergence of container technology, developers have been able to provide a simple way to build and manage large-scale software applications. Instead of building single indivisible units, the design of microservices focuses on building components separately and integrating them with language-independent APIs. Containers provide an infrastructure for applications to run independently. All the necessary dependencies are available within a container. This solves a lot of dependency problems that can impact the production environment. Unlike **virtual machines (VMs)**, containers are lightweight and easy to start and stop.

Each microservice in the system is designed to solve a particular business problem. Unlike monolithic architectures, microservices focus more on business logic than on technology-related layers and database designs. Even though microservices are small, determining how small they should be is a decision that should be taken in the design stage. The smaller the microservices, the higher the network communication overhead associated with them. Therefore, when developing microservices, choose the appropriate scope for each service based on the overall system design.

The architecture of microservices eliminates the concept of the ESB being the central component of SOA. Microservices prefer *smart endpoints and dumb pipes,* where the messaging protocol does not interfere with business logic. Messaging should be primitive in such a way that it only transports messages to the desired location. While the ESB was removed from the microservice architecture, the integration of services is still a requirement that should be addressed.

The following diagram shows a typical microservice architecture:

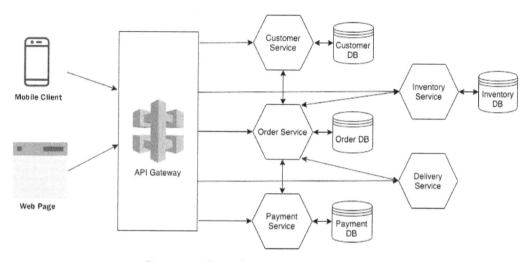

Figure 1.4 – Example microservice architecture

The general practice of the design of microservices is to provide a database for each service. The distributed system should be designed in such a way that disk space or memory is not shared between services. This is also known as shared-nothing architecture in distributed computing. Sharing resources creates a bottleneck for the system to scale up. Even if the number of processing instances increases, the overall system performance does not increase since accessing resources might become a bottleneck that slows down the whole process. As databases are shared resources for services, the microservice architecture forces the developer to provide separate databases for each service.

However, in traditional business applications, it is not feasible to split the schema into several mutually exclusive databases since different resources need to be shared by the same entities. This makes it important that services communicate with each other. Multiple protocols are available for interservice communication, which will be addressed in *Chapter 4, Inter-Process Communication and Messaging.* However, communication between services should also be treated with care, as consistency becomes another headache for the developer to solve.

There are multiple benefits of using microservices rather than monolithic architectures. One advantage of this type is the scalability of the system. When it comes to monolithic applications, the best way to scale is by vertical scaling, where more resources are allocated to the host computer. In contrast with monolithic applications, microservices can not only be scaled vertically but also horizontally. The stateless nature of microservices applications makes microservices more independent. These independent stateless services can be replicated, and traffic can be distributed over them.

Microservices enable developers to use multiple programming languages to implement services. In short, we refer to these as being **polyglot**, where each service is designed in a different language to increase the agility of the system. This provides freedom to choose the technology that is best suited to solve the problem.

As well as having advantages, the microservice architecture also has disadvantages.

The biggest disadvantage of the microservice architecture is that it increases complexity. Microservice developers need to plan carefully and have strong domain knowledge of the design of microservices. The following problems are also associated with the microservice architecture:

- **Handling consistency**: Because each service has its own database, sharing entities with other services becomes an overhead for the system compared to monolithic applications. Multiple design principles help to resolve this problem, which we will describe in *Chapter 5, Accessing Data in the Microservice Architecture*.

- **Security**: Unlike monolithic applications, developers need to develop new techniques to solve the security of distributed applications. Modern authentication approaches, such as JWT and OAuth protocols, help overcome these security issues. These methods will be explored in *Chapter 7, Securing the Ballerina Cloud Platform*.

- **Automated deployment**: Distributed system deployment is more jargon that needs to be grasped clearly. It is not very straightforward to write test cases for a distributed system due to consistency and availability. There are many techniques for testing a distributed system. These will be addressed in *Chapter 10, Building a CI/CD Pipeline for Ballerina Applications*.

Compared to an SOA, a microservice architecture offers system scalability, agility, and easy maintenance. But the complexity of building a microservice architecture is high due to its distributed nature. Programming paradigms also significantly change when moving from SOA to a microservice architecture. Here's a comparison between SOA and microservice architectures:

	SOA	Microservices
Communication	Communication via the ESB with smart pipes.	Pipes are dumb. Communication is through a well-defined API with lightweight protocols.
Data	Global data model on a shared database.	Independent data models on separate databases for each service.
Governance	The same governance rules across the whole application.	Different teams need to collaborate to govern the whole system.
Granularity	A large monolithic application with a modular design.	A small, fine-grained system to do some specific tasks.

Table 1.1 – SOA versus microservices

Having understood the difference between monolithic and microservice architectures, next, we will dive into the cloud native architecture.

Understanding what cloud native architecture is

Different organizations have different definitions of cloud native architecture. Almost all definitions emphasize creating scalable, resilient, and maintainable systems. Before we proceed to a cloud native architecture definition, we need to understand what cloud computing is about.

Cloud computing

The simplest introduction to the cloud is the on-demand delivery of infrastructure, storage, databases, and all kinds of applications through a network. Simply, the client outsources computation to a remote machine instead of doing it on a personal computer. For example, you can use Google Drive to store your files or share images with Twitter or Firebase applications to manage mobile application data. Different vendors offer services at different levels of abstraction. The companies providing these services are considered to be cloud providers. Cloud computing pricing depends mainly on the utilization of resources.

The cloud can be divided into three groups, depending on the deployment architecture:

- **Public cloud**: In the public cloud, the whole computing system is kept on the cloud provider's premises and is accessible via the internet to many organizations. Small organizations that need to save money on maintenance expenses can use this type of cloud service. The key issue with this type of cloud service is security.

- **Private cloud**: Compared to the public cloud, the private cloud commits private resources to a single enterprise. It offers a more controlled atmosphere with better security features. Usually, this type of cloud is costly and difficult to manage.

- **Hybrid cloud**: Hybrid clouds combine both types of cloud implementation to offer a far more cost-effective and stable cloud platform. For example, public clouds may be used to communicate with customers, while customer data is secured on a private network. This type of cloud platform provides considerable security, as well as being cheaper than a private cloud.

Cloud providers offer services to the end user in several ways. These services can rely on various levels of abstraction. This can be thought of as a pyramid, where the top layers have more specialized services than the ones below. The topmost services are more business-oriented, while the bottom services contain programming logic rather than business logic. Selecting the most suitable service architecture is a trade-off between the cost of developing and implementing the system versus the efficiency and capabilities of the system. The types of cloud architecture are as follows:

- **Software as a Service (SaaS)**
- **Platform as a Service (PaaS)**
- **Infrastructure as a Service (IaaS)**

We can generalize this to *X as a service*, where X is the abstraction level. Today, various companies have come up with a range of services that provide services over the internet. These services include **Function as a Service (FaaS)**, **Integration Platform as a Service (iPaaS)**, and **Database as a Service (DBaaS)**. See the following diagram, which visualizes the different types of cloud architecture in a pyramidal layered architecture:

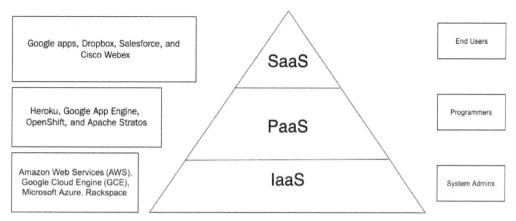

Figure 1.5 – Types of cloud platforms and popular vendors

Each layer in the cloud platform has different levels of abstractions. IaaS platforms provide OS-level abstraction that allows developers to create virtual machines and host a program. In PaaS platforms, developers are only concerned about building applications rather than infrastructure implementations. SaaS is a final product that the end user can directly work with. Check the following diagram, which shows the different abstraction levels of cloud platforms:

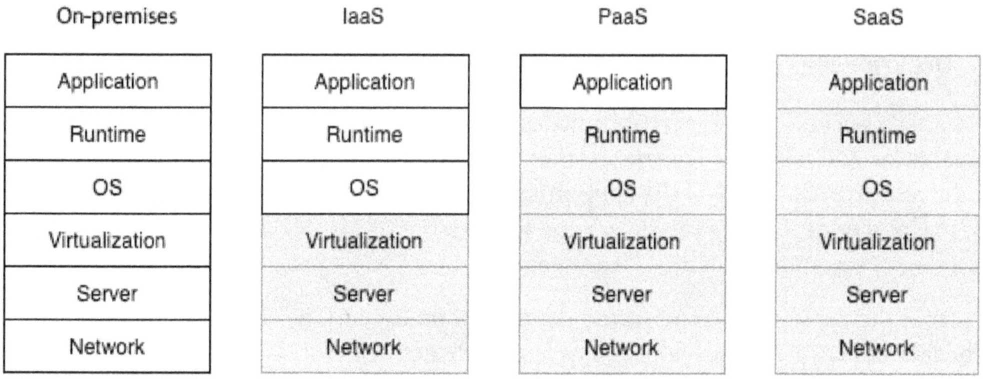

On-premises	IaaS	PaaS	SaaS
Application	Application	Application	Application
Runtime	Runtime	Runtime	Runtime
OS	OS	OS	OS
Virtualization	Virtualization	Virtualization	Virtualization
Server	Server	Server	Server
Network	Network	Network	Network

Figure 1.6 – Levels of abstraction provided by different types of platform

SaaS offers utilized platforms for the developer to work with where developers just need to take care of business logic. The majority of the SaaS platform is designed to run a web browser, along with a UI to work with. SaaS systems manage all programming logic and infrastructure-related maintenance, while end users just need to concentrate on the logic that needs to be applied.

In SaaS applications, users do not need to worry about installing, managing, or upgrading applications. Instead, they may use existing cloud resources to implement the business logic on the cloud. This also decreases expenses, as the number of individuals who need to operate the system is decreased. The properties of SaaS applications are better suited to an enterprise that is just starting out, where the system is small and manageable. Once it scales up, they need to decide whether to hold it in SaaS or switch to another cloud architecture.

On the PaaS cloud platform, vendors have an abstract environment where developers can run programs without worrying about the underlying infrastructures. The allocation of resources may be fixed or on demand. In this type of system, developers will concentrate on programming the business logic rather than the **OS**, software upgrades, infrastructure storage, and so on.

Here is a list of the advantages of building cloud applications using PaaS:

- Cost-effective development for organizations since they only need to focus on business logic.
- Reduces the number of lines of code that are additionally required to configure underlying infrastructure rather than business use cases.
- Maintenance is easy due to third-party support.
- Deployment is easy since the infrastructure is already managed by the platform.

IaaS is a type of cloud platform that provides the underlying infrastructure, such as VMs, storage space, and networking services, that is required to deploy applications. Users can pick the resource according to the application requirements and deploy the system on it. This is helpful as developers can determine what resources the application can have and allocate more resources dynamically as needed.

The cost of the deployment is primarily dependent on the number of resources allocated. The benefit of IaaS is that developers have complete control of the system, as the infrastructure is under the control of the system.

A list of the advantages of using IaaS is given here:

- It provides flexibility in selecting the most suitable infrastructure that the system supports.
- It provides the ability to automate the deployment of storage, networking, and processing power.
- It can be scaled up by adding more resources to the system.
- The developer has the authority to control the system at the infrastructure level.

Although these fundamental cloud architectures are present, there are additional architectures that will address some of the challenges of cloud architectures. FaaS is once such architecture. FaaS operates pretty much the same as the PaaS platform. On the FaaS platform, the programmer provides a function that needs to be evaluated and returns the result to the client. Developers do not need to worry about the underlying infrastructure or the OS.

Serverless architecture

There are a lot of components that developers need to handle in the design of microservices. The developer needs to create an installation script to containerize and deploy applications. Engineering a microservice architecture embraces these additional charges for managing the infrastructure layer functionality. The serverless architecture offloads server management to the cloud provider and only business logic programming is of concern to developers.

FaaS is a serverless architecture implementation. Common FaaS platform providers include AWS Lambda, Azure Functions, and Google Cloud Functions. Unlike in microservice architectures, functions are the smallest FaaS modules that can be deployed. Developers build separate functions to handle each request. Hardware provisioning and container management are taken care of by cloud providers. A serverless architecture is a single toolkit that can manage deployment, provisioning, and maintenance. Functions are event-driven in that an event can be triggered by the end user or by another function.

Features such as AWS Step Functions make it easier to build serverless systems. There are multiple advantages associated with using a serverless architecture instead of a microservice architecture.

The price of this type of platform depends on the number of requests that are processed and the duration of the execution. FaaS can scale up with incoming traffic loads. This eliminates servers that are always up and running. Instead, the functions are in an idle state if there are no requests. When requests flow in, they will be activated, and requests will be processed. A key issue associated with serverless architecture is cloud lock-in, where the system is closely bound to the cloud platform and its features. Also, you cannot run a long-running process on functions as most FaaS vendors restrict the execution time for a certain period of time. There are other concerns, such as security, multitenancy, and lack of monitoring tools in serverless architectures. However, it provides developers with an agile and rapid method of development to build applications more easily than in microservice architectures.

Definition of cloud native

In the developer community, cloud native has several definitions, but the underlying concept is the same. The **Cloud Native Computing Foundation (CNCF)** brings together cloud native developers from all over the world and offers a stage to create cloud native applications that are more flexible and robust. The cloud native definition from the CNCF can be found on their GitHub page.

According to the CNCF definition of cloud native, cloud native empowers organizations to build scalable applications on different cloud platforms, such as public, private, and hybrid clouds. Technologies such as containers, container orchestration tools, and configurable infrastructure make cloud native much more convenient.

Having a loosely coupled system is a key feature of cloud native applications that allows the building of a much more resilient, manageable, and observable system. **Continuous Integration and Continuous Deployment (CI/CD)** simplify and speed up the deployment process.

Other than the CNCF definition, pioneers in cloud native development have numerous definitions, and there are common features that cloud native applications should have across all the definitions. The key aim of being cloud native is to empower companies by offering a much more scalable, resilient, and maintainable application on cloud platforms.

By looking at the definition, we can see there are a few properties that cloud native applications should have:

- Cloud native systems should be loosely coupled; each service is capable of operating independently. This allows cloud native applications to be simple and scalable.
- Cloud native applications should be able to recover from failures.
- Application deployment and maintenance should be easy.
- Cloud native application system internals should be observable.

If we drill down a little further into cloud native applications, they all share the following common characteristics:

- **Statelessness**: Cloud systems do not preserve the status of running instances. All states that are necessary to construct business logic are kept within the database. All services are expected to read the data from the database, process data, and return the data where it is needed. This characteristic is critical when resilience and scalability come into play. Services are to be produced and destroyed, based on what the system administrator has said. If the service keeps its state on the running instance, it will be a problem to scale the system up and down. Simply put, all services should be disposable at any time.

- **Modular design**: Applications should be minimal and concentrate on solving a particular business problem. In the architecture of microservices, services are the smallest business unit that solves a particular business problem. Services may be exposed and managed as APIs where other modules do not require the internal implementation of each of the modules. Interservice communication protocols can be used to communicate with each provider and perform a task.

- **Automated delivery**: Cloud native systems should be able to be deployed automatically. As cloud native systems are intended to develop large applications, there could be several independent services running. If a new version is released for a specific module, the system should be able to adapt to changes with zero downtime. Maintaining the system should be achieved with less effort at a low cost.

- **Isolation from the server and OS dependencies**: Services run in an isolated system in which the host computer is not directly involved in services. This makes the services independent of the host computer and able to operate on any OS. Container technologies help to accomplish this by wrapping code with the container and offering OS-independent platforms to work with.

- **Multitenancy**: Multitenancy cloud applications offer users the ability to isolate user data from different tenants. Users can view their own information only. Multi-tenant architectures greatly increase the security of cloud applications and let multiple entities use the same system.

Why should you select a cloud native architecture?

The latest trend in the industry is cloud native applications, with businesses increasingly striving to move to the cloud due to the many benefits associated with it. The following are some of those benefits:

- Scalability

- Reliability

- Maintainability

- Cost-effectiveness

- Agile development

Let's talk about each in detail.

Scalability

As applications are stateless by nature, the system administrator can easily scale up or scale down the application by simply increasing or decreasing the number of services. If the traffic is heavy, the system can be scaled up and distribute the traffic. On the other hand, if the traffic is low, the system can be scaled down to avoid consuming resources.

Reliability

If one service goes down, the load can be distributed to another service and the work can be continued. There is no specificity about particular services because of the statelessness of a cloud native application. Services can easily be replaced in the event of failure by another new service. The stateless nature helps to achieve this benefit of building a reliable system. This ensures fault tolerance and reliability for the entire application.

Maintainability

The whole system can be automated by using automation tools. Whenever someone wants to modify the system, it's as simple as sending a pull request to Git. When it's merged, the system upgrades with a new version. Deployment is also simple as services are separate, and developers need to consider part of the system rather than the entire system. Developers can easily deploy changes to a development environment with CI/CD pipelines. Then, they can move on to the testing and production environment with a single click. The whole system can be automated by using automation tools.

Cost-effectiveness

Organizations can easily offload infrastructure management to third parties instead of working with on-site platforms that need to invest a lot of money in management and maintenance. This allows the system to scale based on the pay-as-you-go model. Organizations simply don't need to keep paying for idle servers.

Agile development

In cloud native applications, services are built as various independent components. Each team that develops the service will determine which technologies should be used for implementation, such as programming languages, frameworks, and libraries. For example, developers can select the Python language to create a machine learning service and the Go language to perform some calculations. The developer team will more regularly and efficiently deliver applications with the benefits of automated deployment.

Challenges of cloud native architecture

While there are benefits of cloud native architecture, there are a few challenges associated with it as well. We will cover them in the following sections.

Security and privacy

Even though cloud service providers provide security for your system, your application should still be implemented securely to protect data from vulnerabilities. As there are so many moving components in cloud native applications, the risk of a security breach is therefore greater. It also gets more complex as the application grows more and more. The design and modifications should always be done with security in mind. Always comply with security best practices and use security monitoring software for all releases to analyze security breaches. Use the security features of the language you use to implement services.

The complexity of the system

Cloud native is intended to develop large applications on cloud platforms. When applications get bigger and bigger, it's natural that they will get complicated as well. Cloud native applications can have a large number of components in them, unlike monolithic applications. These components need to communicate with each other, and this makes the whole system worse if it's not done correctly.

The complexity of cloud native applications is primarily due to communication between different services. The system should be built in a manner in which such network communications are well managed. Proper design and documentation make the system manageable and understandable. When developing a complex cloud native application that has a lot of business requirements, make sure to use a design pattern designed for a cloud native application such as API Gateway, a circuit breaker, CQRS, or Saga. These patterns significantly reduce the complexity of your system.

Cloud lock-in

Lock-in technology is not specific to cloud native architectures, where technology is constantly evolving. Cloud providers have ways of deploying and maintaining applications. For example, the deployment of infrastructures, messaging protocols, and transport protocols might vary from one vendor to another. Moving on to different vendors is also an issue. Therefore, while building and designing a system, ensure compliance with community-based standards rather than vendor-specific standards. When you are selecting messaging protocols and transport protocols, check the community support for them and make sure they are commonly used community-based standards.

Deploying cloud native applications

Cloud native systems involve a significant number of different deployments, unlike the deployment of a monolithic application. Cloud applications can be spread over multiple locations. Deployment tools should be able to handle the distributed nature of cloud applications. Compared to monolithic applications, you may need to think about infrastructure deployment as well as program development. This makes cloud native deployment even more complicated.

Deploying a new version of a service is also a problem that you need to focus on when building a distributed system. Make sure you have a proper plan to move from one version to another since, unlike monolithic applications, cloud native applications are designed to provide services continuously.

Design is complex and hard to debug

Each of the cloud native system's services is intended to address certain specific business use cases. But if there are hundreds of these processes interacting with each other to provide a solution, it is difficult to understand the overall behavior of the system.

Unlike debugging monolithic systems, because there is a replicated process, debugging a cloud native application often becomes more challenging.

With analytic tools, logging, tracing, and metrics make the debug process easy. Use monitoring tools to keep track of the system continuously. Automated tools are available that collect logs, traces, and metrics and provide a dashboard for system analysis.

Testing cloud native applications

Another challenge associated with delivering cloud native applications is that the testing of applications is difficult due to consistency issues. Cloud native applications are designed with a view to the eventual consistency of data. When doing integration testing, you still need to take care of the consistency of the data. There are several test patterns that you can use to prevent this kind of problem. In *Chapter 10, Building a CI/CD Pipeline for Ballerina Applications*, we will discuss automated testing and deployment further.

Placing Ballerina on cloud native architecture

The main goal of the Ballerina language is to provide a general-purpose programming language that strongly supports all cloud native aspects. Ballerina's built-in features let programmers create cloud native applications with less effort. In the coming chapters, you will both gain programming knowledge of Ballerina and become familiar with the underlying principles that you should know about to create cloud native applications.

Ballerina is a free, open source programming language. All of its features and tools are free to use. Even though Ballerina is new to the programming world, it has supported libraries that you can find from Ballerina Central. Ballerina provides built-in functionality for creating Docker images and deploying them in Kubernetes. These deployment artifacts can be kept along with the source code. Serverless deployment is also easy with Ballerina's built-in AWS Lambda and Azure support.

Ballerina provides standard libraries for both synchronous and asynchronous communication. It supports synchronous communication protocols such as HTTP, FTP, WebSocket, and GRPC, and asynchronous communication protocols such as JMS, Kafka, and NATS. By using these protocols, you can use both data-driven design and domain-driven design. These libraries are known as **connectors**, and they are used to communicate with remote endpoints. A Ballerina connector can be represented as a participant in the integration flow. A Ballerina database connector allows access to various forms of SQL and NoSQL databases.

Security is a must in cloud native architectures. With the Ballerina language, you can use different authentication and authorization protocols, including OAuth2 and Basic. You can easily connect with the LDAP store to authenticate and authorize users. Unlike other languages, certificate handling can be used easily to build a server; it is just a matter of providing certificates to the service description with configurations.

In each of the following chapters, we're going to discuss the principles of creating cloud native applications in the Ballerina language. Starting from creating Hello World applications, we'll explore how to create a more complex system with cloud native architecture design patterns. In each chapter, we will look at the aforementioned features offered by the Ballerina language. In the next section, let's discuss factors that you should follow when building cloud native applications.

Building cloud native applications

As previously mentioned, cloud native applications have associated challenges that need to be tackled to get them ready for production. Cloud native system designers should have a broad understanding of cloud native concepts. These concepts allow the developer to design and construct a better system. In this section, we will focus on software development methodologies used to build cloud native applications.

The twelve-factor app

Heroku engineers came up with twelve factors that should be adhered to for SaaS applications. The twelve-factor app methodology is specifically designed to create applications running on cloud platforms. The twelve-factor app allows the following to be carried out in cloud native applications:

- Minimize the time it takes to understand the system. Teams are agile and can be changed quickly. When new people join the team, they should be able to understand the system and set up the development environment locally.

- Provide complete portability between execution environments. This ensures that the system is decoupled from the underlying infrastructure and OS. The system is capable of being deployed independently from the OS.

- Suitable for use on modern cloud platforms. This reduces the maintenance costs associated with developing large cloud applications.

- Minimize the difference between the development and production environments. This makes CD straightforward, as the automated system can make any changes instantly with little effort.

- As the system is divided into several components that can perform independently, the system can scale with little effort. As these components are stateless, they can be scaled up or down easily.

Here are the twelve factors that you should consider when a cloud application is being developed:

- Code base
- Dependencies
- Config
- Backing services
- Build, release, run
- Processes
- Port binding
- Concurrency
- Disposability
- Dev/prod parity
- Logs
- Admin processes

The following subsections will discuss these factors in detail.

Code base

Version control systems (**VCSes**) are used for the code base in a central repository. The most common VCS is Git, which is faster and easier to use than **SubVersion** (**SVN**). It is not only important to have a VCS to manage the code in different versions, but it also helps to maintain the CI/CD pipeline. This allows for automated deployment where developers can send the pull request and automatically deploy the system with new improvements once it is merged.

The code base should be a single repository in the twelve-factor app, where all related programs are stored. This means you can't have several repositories for a single application. Having multiple repositories means that it is a distributed system rather than an app. You can split the code into several repositories and use it as a dependency when you need it in your application.

On the other hand, multiple deployments can be done in a single repository. These deployments may be for development, QA, or production purposes. But the code base should stay the same for all deployments. This means all deployments are generated by a single repository.

Dependencies

If you have dependencies that need to be added to a service, you should not copy the dependencies to the repository of the project code. Instead, dependencies are added via a dependency management tool. Manifest files are often used to add dependencies to a project.

Ballerina has a central repository where you can store dependencies and use them on a project by adding them to the dependency manifest. Ballerina uses the `Dependencies.toml` file as the dependency declaration file where you can add a dependency. During compilation, the Ballerina compiler pulls dependencies from the Ballerina Central repository if they are not available locally.

Another requirement of dependencies is that you specify the exact version of the dependencies used in the manifest file. Dependency isolation is the principle of ensuring that no implicit dependencies are leaked into a system. This makes code more reliable even when new developers start developing an application; they can easily set up a system without any problems. It's common to have a dependency conflict when creating an application. Finding this issue becomes a headache if the number of dependencies in the application is high. Always use the specific version that matches your application.

Ballerina offers built-in support for versioning and importing dependency modules along with the version. Ballerina pulls the latest dependencies from Ballerina Central at compile time. However, you can be specific with versioning when there are inter-module compatibility issues. This makes it simple when developers need to use a particular version, instead of always referring to the latest version, which may not be compatible.

Config

Applications certainly have configurations that need to be installed before a task is running. For example, think about using the following application configurations:

- Database connection string
- Backend service access credentials, such as username, password, and tokens
- Application environment data, such as IP, port, and DNS names

This factor requires that these configurations be kept out of the source code as the configurations can be updated more often than the source code. When the Ballerina program runs, it can load these configurations and configure the product out of the source code.

You can have a separate config file to keep all the required configurations. But twelve-factor app suggest using environmental variables to configure the system rather than using configuration files. Unlike config files, using an environment variable prevents adding configuration into the code repository accidentally as well.

On the other hand, it is a major security risk to maintain sensitive data such as passwords and authentication tokens in source code. Only at runtime should such data be needed. You can store this data in a secure vault or as an encrypted string that is decrypted at runtime.

Swan Lake Ballerina provides support for both file-based and environment-based configurations. The configs can be read directly from the environment variable by the Ballerina program. Otherwise, you can read configurations from the configuration file and overwrite it with an environment variable when the application starts up. This approach makes it simple for the config file to hold the default value and to override it with the environment variable.

Backing services

Backing services are the infrastructure or services that support an application service. These services include databases, message brokers, SMTP, and other API-accessible resources. In the twelve-factor app, these resources are loosely coupled so that if any of these resources fail, another resource can replace them. You should be able to change a resource without modifying the code.

There is no difference between local and third-party services for a twelve-factor app. Even though your database resides in your local machine or is managed by a third party, the application does not care about the location; rather, it cares about accessing the resource. For example, if your database is configured as local and later you need to switch to a third-party database provider, there should be no changes to the code repository. You should be able to change the database connection string by simply changing a system configuration.

Build, release, run

In the twelve-factor app, it is mandatory to split the deployment process into three distinct phases – build, release, and run:

- **Build**: In this phase, automated deployment tools use VCS code to build executable artifacts. The executable artifact is a set of JAR files for Ballerina. In this step, automated tests should be carried out to complete the task of building. If the tests fail, the entire deployment process should be stopped. When code changes are merged into the repository, the automated deployment tools will start building the application.

- **Release**: In the release stage, build the artifacts package together with the config to create the release package along with a unique release number. Each release has this unique number, which is mostly a timestamp-dependent ID. This specific ID can be used as an ID to switch back to the previous state. If the deployment fails, the system can automatically roll back to the last successful deployment.

- **Run**: This step is to run the application on the desired infrastructure. Deployment may take a few different stages, such as development, QA, stage, and production. The application is in the development pipeline to be deployed in production.

Processes

This factor forces developers to create stateless processes rather than stateful applications. The same process may be running in multiple instances of the system where scalability is necessary. Load balancers sitting in front of the processes distribute traffic between the processes. As there are several processes, we cannot guarantee that the same request dealt with earlier will obtain the same process again. Therefore, you need to make sure that these processes do not hold any state: it just reads data from the database, processes it, and stores it again.

Anything cached in memory to speed up the process is not worth it in the twelve-factor app. Since the next request will serve another process, caching is useless. Even for a single process, the process can die at any time and start again. There is also no great benefit in keeping the cache in memory or on disk.

Sticky sessions where a request is routed from the same visitor are not the best practice with the twelve-factor app. Instead, you can use scalable cache stores such as **Memcached** or **Redis** to store session information.

This applies to the filesystem as well. In certain cases, you may need to write data to a local filesystem. But keep in mind that the local filesystem is unreliable and, once the process is finished, keep it in a proper place.

Port binding

Some web applications, such as Java, run on the Tomcat web server, where the port is set across the entire application. The twelve-factor app does not recommend the use of runtime injections on a web server. Instead, apps should be self-contained; the port should read from the configuration and be set up by the application itself.

Compared to the Ballerina language, you can specify which ports should be exposed by specifying the port on the program itself. This is important where services are exposed as managed APIs for either the end user or another service. The principle of port binding is that the program should be self-contained.

Concurrency

Processes are first-class citizens in twelve-factor app. Instead of writing a single large application, divide it into smaller different processes where each process can separately start, terminate, and replicate independently. Don't depend on vertical scaling to add more resources to scale the application. Instead, increase the number of instances running the service. The load balancer may distribute traffic between these services. Increase the number of services when the load goes up, and vice versa.

Disposability

Twelve-factor app services should be able to start and stop within a minimal amount of time. Suppose, for instance, one service fails, and the system has to replace the failed system with a new instance clone. The new instance should be able to take responsibility as soon as possible to improve resilience. Ideally, an instance should be up and running within a few seconds after it is deployed. Since it takes less time to start the application, scaling up the system would also become more responsive.

On the other side, the process should be able to terminate as soon as possible. In the event of a system scale down or a process moving from one node to another node, the process should terminate quickly and gracefully with a `SIGTERM` signal from the process manager. In the event of the graceful shutdown of an HTTP server, no new incoming requests will be accepted. But long-running protocols, such as socket connections, should be handled properly by both the server and the client. When the server stops, the client should be able to connect to another service and continue to run.

In the event of sudden death due to infrastructure failure, the system should be designed in a way that it can handle client disconnection. A queuing backend can be used to get a job back in the queue when the job dies suddenly. Then another service will proceed with the task by reading the queue.

Dev/prod parity

This factor is to make sure that the gap is as minimal as possible between the development and production environments. A developer should be able to test the deployment in their local environment and push the code to the repository. Deployment tools take on these changes and deploy them in production. The goal of this factor is to make new improvements to production with minimal disruptions. The difference between dev and prod environments can be due to the following reasons:

- **The time gap**: After integrating new functionality into the system, the developer will take some time to deploy code into the production environment. The code can be quickly reflected in the deployment by reducing this time gap. This solves problems with code compatibility between multiple teams of developers. Since the application is built by different teams, it can cause tons of discrepancies that take time to integrate components.

- **The personal gap**: Another gap is that the code is created by developers and deployed by operations engineers. The twelve-factor app forces developers to also be involved in the deployment process and analyze how the deployment process is going.

- **The tool gap**: Developers cannot use tools that are not used for production. For example, an H2 database may be used by the developer to create a product, while a MySQL database may be used in production. Therefore, make sure that you use the same resources that have already been used for production in the development environment as well.

Containerized applications should be able to adjust the environment by simply modifying the configuration. For example, if the system is running on the production or development environment, the developer can change the configuration to the developer mode and test it on the local computer. The automated tool can change the mode to production mode and deploy it to the production environment.

Logs

Logs are output streams that are created by an application and are necessary for the system to debug when it fails. The popular way to log is to print it to your `stdout` or filesystem. The twelve-factor app is not concerned with routing or storing logs. Instead, the program logs are written directly to `stdout`. There are tools, such as log routers, that collect and store these logs from the application environment. None of this log routing or configuring is important to the code, and developers only concentrate on logging data into the console.

Ballerina includes a logging framework where you can quickly log in to different log levels. Logging can be easily incorporated with tools such as Elasticsearch, Logstash, and Kibana, where you can easily keep and visualize logs.

Admin processes

This factor enforces the one-off processes of running admin and management processes. Other than the long-running process that manages business logic, there could be other supporting processes, such as database migration and inspection processes. These one-off processes should also be carried out in the same environment as the business process.

Developers may attach one-off tasks to the deployment in such a way that it executes when it is deployed.

Twelve-factor app provide sets of guidelines to create cloud native applications that need to be followed. This allows developers to construct stateless systems that can be deployed with minimal effort in production. The Ballerina language offers support for all of these factors, and in the following chapters, we will look at how these principles can be implemented in practice.

Now let's look at another design approach, API-first design, which speeds up the development process by designing API interfaces first.

API-first design

As mentioned previously, services are designed to solve a particular business problem. Each service has its own functionality to execute. This leads to the problem of design first versus **code first**. The traditional approach, which is code first, is to build the API along with the development. In this case, the API documentation is created from the code.

On the other hand, design first forces the developer to concentrate more on the design of the API rather than on internal implementations. In API-first design, the design is the starting point of building the system. A common example for code first is that when you are building a service, you write the code and generate the WSDL definition automatically. Swagger also allows this type of development method by generating the relevant OpenAPI documentation from the source code. On the other hand, Swagger Codegen lets the developer create the source code based on the API specifications. This is known as the design-first approach, in which code is generated from the API.

The code-first approach is used when it comes to speed of delivery. Developers can create an application with the requirement documentation and generate the API later. The API is created from the code and can change over time. Developers can create an internal API using this approach, since an internal team can work together to build the system.

If the target customer is outside of your organization, the API is the way they communicate with your system. The API plays a key role in your application in this scenario. The API specification is a good starting point for developers to start developing applications. The API specification acts as a requirement document for an application.

Multiple teams can work in parallel because each service implementation is a matter for each team and the API determines the final output.

There are advantages and disadvantages associated with each of these design patterns. Nowadays, with the emergence of cloud-based architecture, API-first design plays a key role in software development. Multiple teams work collaboratively to incorporate each of the services in the design of microservices. REST is a common protocol that is widely used to communicate with services. The OpenAPI definition can be used to share the API specifications with others. Protobuf is another protocol that strongly supports API-first development, which is widely used in GRPC communication.

The 4+1 view model

Whatever programming languages you use, the first step when developing software applications is to gather all the specifications and design the system. Only then can you move on to the implementation. *Philippe Kruchten* came up with the idea of analyzing the architecture of software in five views called the **4+1 view model**. This model gives you a set of views where you can examine a system from different viewpoints:

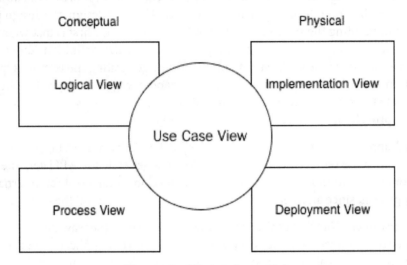

Figure 1.7 – The 4+1 view model

Let's talk about this briefly:

- **Logical View**: The logical view includes the components that make up the structure. This emphasizes the relationship between the class and the package. In this view, you can decompose the system into classes and packages and evaluate the system using the **Object-Oriented Programming (OOP)**. Classes, states, and object diagrams are widely used to represent this view.

- **Process View**: This explains the process inside the system and how it interacts with other processes. This view analyzes the process behavior, concurrency, and the flow of information between different processes. This view is crucial in explaining the overall system throughput, availability, and concurrency. Activity, sequence, and communication diagrams can be used to analyze and present this view.

- **Deployment View**: This view describes how the process is mapped to the host computer. This gives you a building block view of the system. Deployment diagrams can be used to visualize this kind of view.

- **Implementation View**: This view illustrates the output of the build system from the viewpoint of the programmer. This view emphasizes bundled code, components, and units that can be deployed. This also gives you a building block view of the system. Developers need to organize hierarchical layers, reuse libraries, and pick different tools and software management. Component diagrams are used to visualize this view.

- **Use Case View**: This is the +1 in the 4+1 view model, where all views are animated. This view uses goals and scenarios to visualize the system. Use case diagrams are usually used to display this view.

To demonstrate how a cloud native application is built with the Ballerina language, we will introduce an example system. We will build an order management system to demonstrate different aspects of cloud native technologies. We will discuss this example throughout this book. In the next section, let's gain an understanding of the basic requirements of this system and come up with an architecture that complies with cloud native architecture.

Building an order management system

The software design process begins with an overview of the requirements. During the initial event storming, developers, subject experts, and all other related roles work together to gather system requirements. Throughout this book, we're going to address developing cloud native systems by referring to a basic order management system.

You will gain an understanding of the requirements of this order management system in this section.

An example scenario of an order management system can be visualized with the following steps:

Figure 1.8 – Order management system workflow

The customer begins by placing an order. The customer may then add, remove, or update products from the order. The system will confirm the order with inventory once the order items are picked by the customer. If products are available in the inventory, the payment can be made by the customer. Once the payment has been completed, the delivery service will take the order and ship it to the desired location.

Use cases can be represented with a use case diagram, as follows for this system:

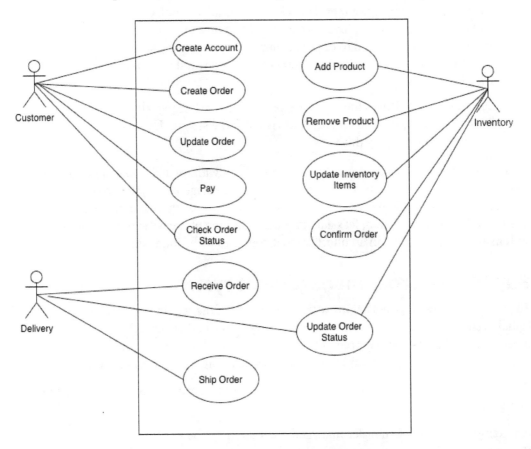

Figure 1.9 – Use case diagram for the order management system

Diagrams play a vital role in the design and analysis of a system. As defined in the 4+1 view, for each of these views, the designer may draw diagrams to visualize the different aspects of the system. Due to the distributed nature of cloud native applications, sequence diagrams play a key role in their development. Sequence diagrams display the association of objects, resources, or processes over a period of time. Let's have a look at a sample sequence diagram where the client calls a service, and the service responds after calling a remote backend endpoint:

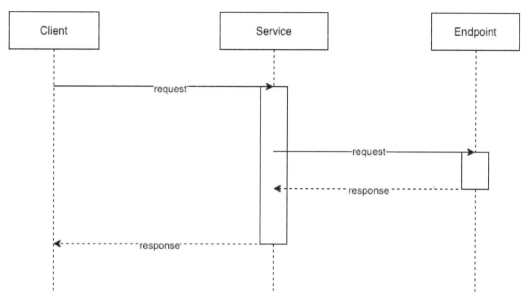

Figure 1.10 – Sequence diagram

Building order management systems to fulfill these requirements is easy and straightforward with monolithic applications. Based on the three-tier architecture, we can divide the system into three layers that manage the UI, business logic, and database. Interfaces can be web pages loaded in a web browser. The UI sends requests to the server, which has an API that handles the request. You can use a single database to store the application data. We can break down the application into several components to modularize it. Unlike a monolithic architecture, when we break this system into loosely coupled services, it allows developers to take a much more convenient approach to building and managing the system.

How small each service should be is a decision that should be made by looking at the overall design. Large services are harder to scale and maintain. Small services have too many dependencies and network communication overhead. Let's discuss breaking down the services of a large system with an example in the next section.

Breaking down services

When developing microservices, how small the microservices should be is a common problem. It's certainly not dependent on the number of code lines that each of the components has. The concept of the **Single-Responsibility Principle (SRP)** for object-oriented principles was developed by *Robert C. Martin*. SRP enforces that a class of a component should *have only one reason to change*. Services should have a single purpose and the service domain should be limited.

The order management system can be broken down into the following loosely coupled services, and each of the services has its own responsibilities:

- **Order service**: The order service is responsible for the placement and maintenance of orders. Customers should be able to generate orders and update them. Once customers have added products from an inventory, the payment will then be made.

- **Inventory service**: The inventory service maintains products that are available in the store. Sellers can add and remove items from the inventory. The customer can list products and add them to their order.

- **Payment service**: The payment service connects with a payment gateway from a third party and charges for the order. The system does not hold card details, as all of these can be managed by payment gateway systems. When the payment has been made, the system sends an email to the customer with payment details.

- **Delivery service**: The delivery service can add and remove delivery methods. The user can select a delivery method from the delivery service. The delivery service takes the order and ships the order to the location specified by the customer.

- **Notification service**: Details of the payment and order status are sent to the client via the email notification system. To send notifications, the notification service connects to an email gateway.

We need to come up with a proper way of communicating with each of these services once the single service is split into multiple services. The HTTP protocol is a good candidate for inter-process communication.

It is simple for scenarios such as making an order as the order service receives the request from the customer and adds a database record to the order table. Then several products are added to the same order by the client. The order service gets and stores these products in the database. The customer places the order for verification when they have finished selecting products. In this case, the order service must check with the inventory service for product availability. Since the inventory service is decoupled, the order service needs to send the inventory service a remote network call to confirm the order.

HTTP is known as a synchronous form of communication, where clients wait until the server responds. This might create a bottleneck in the order service as it waits to get the response from the inventory service. Asynchronous communication methods, such as JMS and AMQP, play an important role in developing distributed systems in this type of scenario. Brokers take the orders and dispatch them asynchronously to the inventory service.

Asynchronous communication protocol plays a key role in the architecture of microservices as it decouples services. Microservice architecture relies heavily on asynchronous communication to adopt event-driven architecture. Event-driven architecture allows programmers to create events-based programs rather than procedural commands. Message brokers serve as a layer of mediation for events. Message brokers offer the guaranteed delivery of messages with topics and queues. Events published on each channel are consumed and executed by other services. We will discuss event-driven architecture more in *Chapter 5, Accessing Data in Microservice Architecture.*

If the order has been verified, the customer may proceed to payment. The payment service uses a third-party payment gateway to complete the financial transaction. Once the transaction is completed, the order can be sent to the delivery service to deliver the product. Notification services send notifications such as payment and order information and delivery status to the customer.

Sample sequence diagrams can be generated as follows based on these service separations:

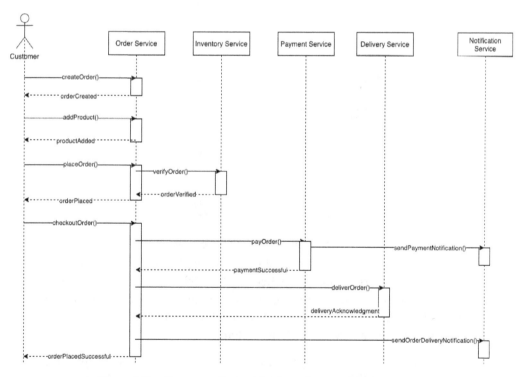

Figure 1.11 – Sequence diagram for the order management system

There are certain entities that interact with the system, such as customers, orders, products, inventory, and so on. Modeling these entities makes the architecture of the system more understandable. Next, we'll look at a class diagram to represent this system. Each entity in the system has a specific task associated with it. Each class in the diagram represents the functionality assigned to each object:

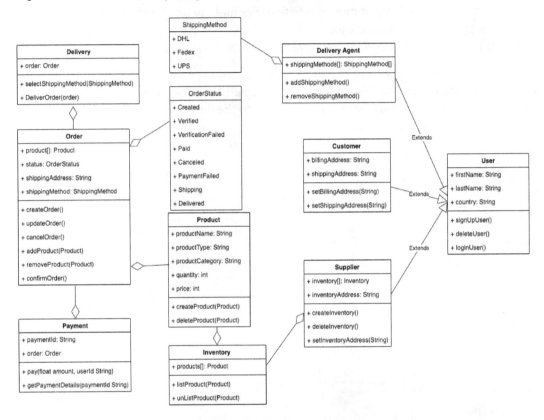

Figure 1.12 – Class diagram for the order management system

In data-driven design, a class diagram can easily be translated to a database scheme and the system design can be started from there. Because the architecture of microservices fits well with event-driven architecture, class diagrams do not explicitly infer system design, even though the class diagram plays a crucial role in defining and developing the system as it identifies the entities that interact with the system.

In the following chapters, we will review these initial specifications and come up with a design model to construct this order management system in a cloud native architecture using the Ballerina language.

So far, we have discussed what cloud native architecture is and the properties that it should have. But we always need to remember the fact that cloud native is not just all about technology. It is also associated with the humans that build, maintain, and use the system. The next section discusses how to adopt cloud native technologies in an organization.

Impact on organizations when moving to cloud native

There are hundreds of companies out there, such as Netflix, Uber, Airbnb, and WeChat, that use cloud native in their production. With cloud native applications, these organizations manage huge amounts of incoming traffic into the system. Some of these organizations have not only moved to cloud native architectures, but have also developed many open source tools to develop cloud native applications. Moving to cloud native is not easy, as there are a lot of obstacles to get over. Yet switching to cloud native makes a company even more agile and efficient in terms of organization management.

Challenges of moving to a cloud native architecture

Cloud native offers scalable, cost-effective architecture that can easily be managed. But without adequate system analysis, moving to a cloud native architecture is challenging. As technology evolves rapidly, switching from one software architecture to another is not that easy. With long-standing legacy software written in a monolithic architecture, migration is more difficult. Moving to cloud native is not just a technological transition but also a cultural change in the organization as a whole. The following are some of the difficulties of switching from a monolithic architecture to a cloud native architecture.

Outdated technologies

Outdated technologies have problems with rising costs, lack of competition, decreased productivity, lack of flexibility, and problems with security. Technologies used to construct the system may no longer be supported. Using the same technology stack without production support is challenging. Also, cloud native concepts might not suit these old technologies. When those technologies are no longer community-supported, things get worse. Often, certain components need to be designed from scratch. This takes a substantial amount of time and cost.

Operational and technology costs

It isn't an easy task to switch from one technology to another, and it takes both time and money. It is not only expensive to switch but also to support and maintain the new technology as it needs highly skilled engineers. Even though it comes at a cost, the advantages of cloud native make the end user experience much better.

Building cloud native delivery pipelines

Cloud native applications are expected to run on the cloud. The cloud may be public, private, or hybrid. Traditional apps can use an on-premises infrastructure to deploy the system. Moving from monolithic deployment to cloud native deployment is complicated, as there are many moving parts that need to be configured. A cloud native application cannot be deployed manually, as hundreds of different applications need to be deployed. Developers should build an automated deployment to automatically deploy the system. Building these delivery pipelines costs both time and money.

Interservice communication and data persistence

Unlike monolithic applications, the cloud native application database is divided into several databases where each service has its own database. Changing the database architecture can be a huge pain point for the developer, since the entire application architecture and systems may need to be updated. In addition, the architecture of microservices allows developers to use asynchronous communication methods rather than a synchronized communication pattern. If the previous system was designed with tightly coupled synchronized communication, there will be significant changes required to accommodate event-driven architecture.

In the next section, we will discuss Conway's law, which fits the cloud native architecture well. We can use this law to build more sophisticated cloud applications.

Conway's law

Moving to cloud native is not only an architectural change in technology but also a cultural change for the organization as a whole. Cloud native will deliver much more agile development than monolithic application development. *Melvin E. Conway* came up with Conway's law in 1967, stating that any organization that designs a system will design it with a structure that copies the organization's communication structure. After decades of microservices coming to the fore, Conway's law provides guidelines on how a team should be organized:

Organizations which design systems are constrained to produce designs
which are copies of the communication structures of these organizations.
— Melvin Conway (1967)

Simply put, the communication structure of the organization is often focused on the process of product development. For example, in the order management system, different teams in the order management system interact with each other as separate teams. The order service calls the inventory service to confirm whether the products are available. The order service calls the payment service to complete the transaction. The product we are creating has a communication structure similar to that of the organization.

> **Note**
>
> Conway's idea was initially not limited to software development. After *Harvard Business Review* rejected his article, Conway published it in a programming journal. Fredrick P. Brooks later quoted these ideas as Brooks's law in his book *The Mythical Man-Month*.

Next, we'll review Conway's laws.

The first law

The first of Conway's laws is about the size of teams in an organization. This law requires that a team be as small as possible. To understand how small it should be, consider how much of a personal relationship you have with each team member. If the relationships are weak, this means that the team size is big. Increasing the number of people to get things done quickly is not as straightforward as expected. Having bigger teams makes the coordination overhead bigger. Brooks proposes a method to calculate the cost of communication and complexity depending on the number of team members (n):

Communication cost $=n(n-2)/2$

Complexity $= O(n^2)$

According to the equation, you can clearly see that communication costs and complexity increase exponentially while increasing the number of team members. Therefore, to create a part of a product, small teams must be preferred as they perform much better.

The second law

The second law is all about there being no perfect systems. Human resources are limited in an organization, as is human intelligence. It is okay to neglect certain components when constructing a complex system. The focus of production is more on the delivery of the key components. There may be issues with a distributed system. You should develop a system that can withstand failures and recover from them instead of building a perfect failure-proof system. A system might fail at some point, but your system should be able to recover fast from a failure.

The third law

The third law discusses the homomorphism between the system and the design of the organization. As I mentioned earlier, the communication structure depends mostly on the process of product development. Team boundaries are also mapped to the system boundaries. Each team has its own business domain. Inter-team communication is often less frequent than intra-team communication. Make teams smaller and let team members concentrate on the task and do it in collaboration with other team members, only engaging with other teams if they must. This reduces the cost of communication between multiple teams.

The fourth law

The fourth law is about building organizational architecture with a divide and conquer approach. When a team gets bigger, we need to divide the team into smaller teams. Every team manager reports to a higher-level manager, with each manager in the company reporting to a manager in a higher layer in a pyramidal structure. The divide and conquer strategy helps to create and maintain a large organization.

By comparing the architecture of microservices and Conway's laws, we can clearly see similarities between these two approaches:

- Systems have distributed services/teams that solve a particular problem.
- Systems/organizations are separated by business lines.
- Services/teams focus on building the product rather than delivering the project.
- Believe that nothing works perfectly. Failures should be expected.

Next, we will look at an organization that has moved to cloud native technology and serves its products to millions of customers. Netflix is a video streaming company that migrated to cloud native architecture successfully, and also generates new open source tools for developers. Let's look at their journey to cloud native architecture in the next section.

Netflix's story of moving to cloud native architecture

Netflix was founded in 1997 by *Reed Hastings* and *Marc Randolph* in Scotts Valley, California. Initially, Netflix was a movie rental service where customers ordered movies on the Netflix website and received a DVD in the mail. As a result of the advancement of cloud technologies, Netflix announced in 2007 that it would launch a video streaming service and move away from selling DVDs.

In 2008, Netflix encountered major database corruption, which forced it to transfer all of its data to the AWS cloud. Netflix was deeply concerned about possible future failures since they were heavily focused on cloud streaming rather than renting DVDs. They started moving into the cloud by moving all customer databases to a highly distributed NoSQL database called Apache Cassandra.

Constructing a highly available, scalable, and high-performance infrastructure is a key aspect of moving into the cloud. It wasn't that easy for Netflix to move on-premises data centers to the public cloud. The migration process began in 2009 by transferring movie encodings and non-customer-facing applications. Netflix transferred account registration and movie selection by 2010 and completed cloud migration by 2011. The migration process was not simple since migration had to be carried out without any downtime.

They ensured that all on-premises servers and cloud servers were kept together to provide customers with continued operation. However, during the transition phase, they had to deal with lots of latency and performance problems. During this migration, Netflix implemented the following open source tools for the company to transition to cloud native architecture:

- **Eureka**: A service discovery tool
- **Archaius**: A configuration management tool
- **Genie**: A job orchestration tool
- **Zuul**: A layer 7 application gateway
- **Fenzo**: A scheduler Java library

You can find out about all of Netflix's cloud native tools in the official Netflix **Open Source Software** (**OSS**) Center in the GitHub repositories.

John Cianutti, Netflix VP of engineering, mentioned that they do not need to be concerned about upcoming traffic into the system. Since the platform is deployed on AWS, they can add resources as per requirements dynamically. Cloud native is about scaling the system when it is required. An organization can be scaled up on the basis of the traffic it receives. When an organization deploys services on the public cloud, it can quickly scale up with the dynamic resource allocation features provided by cloud providers. This means Netflix can focus on streaming services rather than maintaining data centers.

According to 2020 reports, Netflix serves more than 192 million subscribers worldwide. Netflix became a cloud native role model with how it converted on-premises monolithic applications into a scalable cloud native architecture. The tools it has developed are indeed very useful for constructing a cloud native application.

Summary

In this chapter, we discussed the monolithic architecture and its limitations, such as scalability and maintenance when the application gets bigger and bigger. Even though a simple application can adopt a monolithic architecture, for larger applications, it is hard to maintain a monolithic architecture. SOA provides a solution to break the system into multiple services and maintain those services separately. An ESB helps to create communication links between those services instead of using point-to-point communication. With the rise of container technology, these services became much more stateless and scalable. Microservices provide a way to build a more resilient and scalable system.

With the advancement of cloud computing, organizations have gradually switched to cloud-based technologies, rather than maintaining their own infrastructure. Cloud native is a new concept that addresses concerns related to old-school monolithic programming paradigms. The microservice architecture and the serverless architecture, together with cloud services, provide a convenient way to build more scalable systems. Various programming principles can be used to implement a cloud native microservice architecture.

The twelve-factor app makes cloud-based apps more cloud native.

Moving to cloud native may be difficult for an organization due to the associated costs. But the benefits associated with cloud native make it much more valuable than maintaining legacy systems. Theories such as Conway's law provide elegant guidelines on how organizations should design their microservice architecture.

Ballerina is a revolutionary programming language that supports cloud native. With built-in tools and network-oriented syntax styles, developers can easily build scalable, reliable, and maintainable systems. In the next chapter, we will discuss installing the Ballerina development kit and writing your first Ballerina code.

Questions

1. What are the characteristics of cloud computing?

2. Why is cloud native needed for fast-scaling businesses?

3. When should you consider moving to a microservice architecture?

4. How do you select a microservices or serverless architecture?

Further reading

You can refer to *Cloud Native Architectures: Design high-availability and cost-effective applications for the cloud*, by Tom Laszewski, Kamal Arora, Erik Farr, and Piyum Zonooz, available at `https://www.packtpub.com/product/cloud native-architectures/9781787280540`.

Answers

1. Cloud services are scalable, self-service provisioning, and billing is based on usage.

2. User base growth is unpredictable, and the system should be able to scale based on the requirement. The development process should also be quick so that developers can introduce new features into the system with minimal effort.

3. If you already have a monolithic application and it keeps on growing, and also scalability and maintainability are an issue, then it is better to move to a microservice architecture. Also, if you are building applications that will scale in the future and are looking for an efficient and fast development process, a microservice architecture will be a good solution.

4. Microservices are built with IaaS platforms such that developers think about the deployment process as well. Comparatively, a serverless platform is already maintained by cloud providers. In terms of cost, a serverless platform has some benefits over a microservices platform. On the other hand, a microservices platform offers more control over the infrastructure, and the developer can build a more sophisticated system.

2
Getting Started with Ballerina

Cloud native is a new concept that changes the way we think about building applications. It constitutes a whole new way of looking at programming compared to older monolithic architecture-based systems. There are several general-purpose programming languages available to develop cloud native applications. Ballerina is a new programming language that supports cloud native programming with a cloud-friendly programming style. It is a general-purpose programming language written with the integration of systems in mind.

We will start by introducing ourselves to Ballerina and its purpose, followed by performing some setup by downloading the installer and installing the Ballerina development kit. We will use **Visual Studio Code (VS Code)** and Ballerina VS Code plugins to develop Ballerina code. We will also explore the use of the Ballerina CLI tool to maintain Ballerina projects and the Ballerina type system, where we will cover Ballerina's basic types. We will also learn how to use Ballerina objects and classes to build object-oriented programs and handle errors in Ballerina programming language.

In this chapter, we will be learning the basic concepts of building applications with the Ballerina language. The topics covered are as follows:

- Introduction to the Ballerina language

- Setting up a Ballerina environment

- Understanding the Ballerina type system

- Using types to build applications with Ballerina

By the end of this chapter, you should be able to create a simple software application using the Ballerina language. You will learn concepts such as data types and flow control, which are the basic concepts of any programming language. Also, you will learn about advanced concepts, such as error handling and object-oriented programming.

Technical requirements

In this chapter, we will explain how to install Ballerina using the Ballerina installer, which is the easiest way to install Ballerina. If you install Ballerina from a ZIP file, you need to install the **Java Runtime Environment** (**JRE**) on your machine. On the other hand, you can install Ballerina by building the source code. For that, you will need to install JDK11 on your computer.

The command-line arguments given in this book are a reference to Unix architecture. Here we have used the Ballerina CLI tool to maintain the Ballerina projects. For any operating system, Ballerina CLI commands are common. If you are executing any other commands that are specific to an operating system, make sure to use the appropriate commands for each operating system.

The code files for this chapter can be found on GitHub at `https://github.com/ PacktPublishing/Cloud-Native-Applications-with-Ballerina/tree/ master/Chapter02`

The Code in Action video for the chapter can be found here: `https://bit.ly/3zWUfbZ`

Introduction to the Ballerina language

In this section, we will discuss how Ballerina came to be, the purpose of building this new language, and how it functions internally. We will also address how Ballerina manages its non-blocking threading model, which is an integral aspect of creating high-performance applications.

A glimpse of Ballerina

WSO2 has provided integrated solutions for over a decade. WSO2 has provided ESB solutions to integrate SOA-based services. WSO2 adapts Apache Synapse as the underlying mediation engine of the WSO2 ESB. Synapse offers a **Domain-Specific Language (DSL)** based on XML to describe message handling and logic for transformation. In SOA, messaging between services is mainly done with SOAP requests. XML message manipulation is mainly handled by using Xpath. However, creating complex message and transformation scenarios with a DSL can be difficult, and general programming languages such as Java are used to implement such scenarios.

The Ballerina language creator, **Dr. Sanjiva Weerawarna**, CEO and founder of WSO2, mentioned that WSO2 had been working on improving the ESB language's syntax for a long time. In its early stages, Ballerina was focused primarily on getting away from the Synapse DSL and offering a much more efficient way to write mediation logic.

On the other hand, the ESB is based on a centralized architecture in which it manages the data flow of the SOA-based system. This is an anti-microservice architecture pattern. Microservice architecture enforces the use of *smart endpoints and dumb pipes*, but the ESB behaves like a smart pipe and makes business decisions. Ballerina is a good candidate to implement microservices since it can be used to implement business logic and has strong integration support.

Ballerina makes integration developer-friendly and enables companies to be more agile when moving forward with integration. Ballerina is designed to bridge the gap between integration and general programming languages. With the support of connectors, Ballerina provides agility and ease of integration. Ballerina's designers have ensured that it complies strongly with cloud native principles. Built-in cloud native libraries allow developers to create complicated business logic simply. Its strong integration features let Ballerina interconnect with a variety of third-party services. Cloud native developers say that middleware is dead. The fact is that middleware is not dead, but it is no longer a component in a microservice architecture. The languages used to construct the system are responsible for integrating the components now.

Ballerina is a statically and strongly typed programming language; hence, it verifies and enforces type constraints at compile time, and the types of data should be predefined. Ballerina is strongly typed as it has strict typing rules. The following diagram compares the positioning of Ballerina against that of other programming languages:

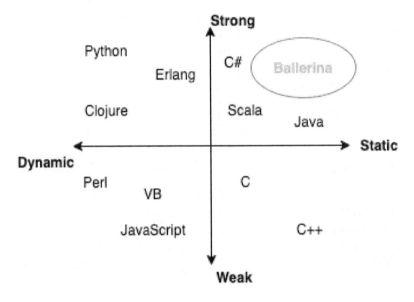

Figure 2.1 – Static versus dynamic typing

Strong and static typing systems resolve most type system issues at compile time instead of at runtime. Ballerina is flexibly typed and data-oriented, meaning you can use JSON and XML as built-in data types along with language-integrated queries for tabular data. Ballerina's wide and various type system allows the developer to work easily with several types of message structures. Ballerina treats network communication as a first-class concept. Cloud native applications are heavily dependent on network communication. Unlike other programming languages, Ballerina does not hide network communication using libraries. Instead, it shows network communication to programmers to make Ballerina even more expressive when writing network distributed systems.

Ballerina is not intended to be an object-oriented or functional programming language, but it supports object orientation and functional programming models. Even though **object-oriented programming (OOP)** is not strongly supported, you can still implement objects with OOP concepts. We will learn how to implement object-oriented principles later in this chapter.

Learning just the syntax does not make you adept when it comes to the Ballerina language. To learn more about Ballerina's capabilities, you need to dive into how Ballerina is implemented internally. In the next section, we will learn how the Ballerina compiler works.

The Ballerina compiler

Ballerina is a platform-independent programming language that runs on any operating system. Ballerina compiler compiles the source code into a JAR file that can be run on the **Java Virtual Machine (JVM)**. Ballerina currently uses the JVM as an interpreter to execute the JAR files that are generated. But Ballerina designers are planning to move away from the JVM and build their own runtime to run Ballerina applications. You can have a Java-independent Ballerina runtime in the upcoming version of Ballerina. Its compiler architecture looks as follows:

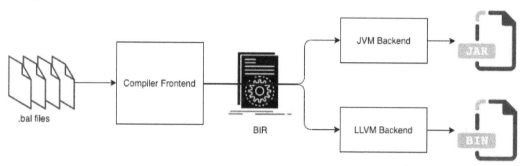

Figure 2.2 – Ballerina compiler architecture

The Ballerina compiler can be split into two main parts: the compiler frontend and the backend. The key responsibility of the frontend is to generate the **Ballerina Intermediate Representation (BIR)** from the `.bal` file. BIR is a platform-independent representation of a Ballerina program. BIR offers an elegant way to share packages with others rather than sharing source code or JAR files directly. In this case, BIR is important because it is platform and language-independent.

The Ballerina backend can use this BIR file to generate executable files. Ballerina currently supports the JVM runtime. Ballerina takes the BIR file and generates JAR files that can be executed on the JVM file. This backend compiler is known as **jBallerina**. Since Ballerina used BIR to represent the program, separate backends can be provided for the target executable format. For instance, the LLVM framework can be used to generate the machine code for the Ballerina program. Instead of running the Ballerina program on JVM, LLVM artifacts let Ballerina code run directly on the target platform. Programming languages such as C, C++, and Objective C use LLVM-based Clang compilers to generate executables. The LLVM Ballerina backend compiler is also known as **nBallerina**.

Since the Ballerina compiler is Java-based, it allows Ballerina to use Java libraries and built-in garbage collection functionality. Since Ballerina is a new programming language, this ensures that Ballerina has no shortage of features. Therefore, you can use Java libraries with Ballerina applications. Eventually, Ballerina will have a dedicated runtime instead of depending on the JVM runtime.

As we discussed earlier, the primary objective of the Ballerina frontend is to generate backend BIR files. The Ballerina frontend uses the `.bal` files and performs the following steps to create the BIR:

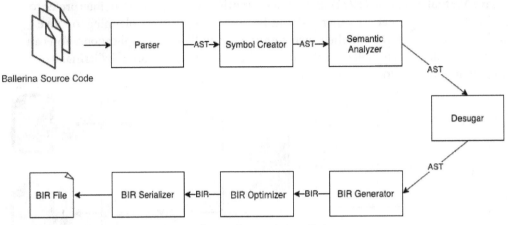

Figure 2.3 – Ballerina frontend compilation process

Let's discuss each stage in detail in the following subsections.

Parser

The Ballerina parser generates the **Abstract Syntax Tree (AST)** from the Ballerina source code. The AST represents the Ballerina code in a tree-like data structure, where the frontend can traverse the AST to validate and generate the BIR.

Symbol creator

The symbol creator phase creates a symbol for each entity. For example, the `int x = 5` statement creates a symbol to represent the x variable. This symbol includes the variable type, the location in the source code where the variable is defined, and more. When the x variable is assigned a new value, the compiler refers to the symbol by looking at its scope. Ballerina generates symbols for each type, which are used to hold information about each entity.

Semantic analyzer

In this stage, the compiler analyzes the AST statements. Statements can include assignment statements, assertion statements, return statements, and so on. For example, in assignment operations, compilers evaluate the left-hand side of the statement and check that the right-hand side can be assigned. The right-hand side returns an integer after some calculation. Then, the left-hand variable should be able to accept integers. For example, if you define an integer with the `int value` statement, you cannot assign a value with the `value = "hello"` statement since the types do not match.

Desugar

In the desugar phase, the AST nodes will be restructured to support the BIR file. It breaks down complex operations into basic sets of operations. For example, the desugar step converts *x += 1* into *x = x + 1*, which is a much simpler way of representing addition. The desugar phase simplifies all kinds of notations.

BIR generator

The BIR generator generates the BIR tree from the AST. The AST is a tree-like architecture that has nodes and links that represent the program. The AST is more about representing syntax, while the BIR represents program flow. Nodes and links in the BIR represent how the program should run. The BIR tree is constructed from simple blocks. Each basic block contains an associated instruction set that describes how the program should flow.

BIR optimizer

The BIR optimizer optimizes the BIR tree by eliminating redundant instructions and unnecessary variables. This phase eliminates the unnecessary code section of the Ballerina runtime. This step optimizes the performance of the code and simplifies the complexity of the program.

BIR serializer

At the final stage, the BIR tree is serialized into a byte array. This file type is also known as a BAL file. This is a platform-independent BIR file that is used to create Ballerina executable artifacts. The backend can deserialize the data and generate a JAR file that can be run on the JVM.

The serialized BIR file uses the Ballerina backend to create executable artifacts. You can generate artifacts for different platforms based on the backend compiler. The final executable artifacts will be a JAR file with jBallerina that can be executed in the JVM.

The Ballerina **threading model** is another thing that you need to understand regarding the Ballerina language. Ballerina has a non-blocking threading model. Learning about the threading model helps you understand how Ballerina runs on the JVM and how it runs in a concurrent environment. In the next section, we will talk more about the non-blocking architecture of the Ballerina runtime.

The Ballerina threading model

A process is a program that is under execution. It is built from multiple threads, and so processes are heavier than threads. There may be several cores in the processors of modern computers. Each of these cores can execute a single thread at a time. In certain cases, a thread might need to call the hard disk and wait for a response. These are called **Input/Output (I/O)** operations. It is a waste of computation power if the thread waits in the processor until it gets the result from an I/O operation. The **Operating System (OS)** provides context switching, which has a thread enter a blocked state if it has to wait for I/O operations.

If one thread gets blocked, the OS stores its state and releases the thread to another computation. This thread loads the state of another runnable thread and continues to execute. The switching between these threads is known as context switching.

Instead of threads, Ballerina has much smaller execution units called **strands**. Multiple strands can be used in a single thread to execute a Ballerina program. As described earlier, when a thread is blocked, the OS selects another runnable thread and executes it. If there are lots of I/O blocks, then the process needs to context switch frequently, which is an overhead for the overall performance.

The Ballerina threading model is based on a concurrent, non-blocking, asynchronous model where various workers are assigned to manage inbound and outbound requests. Ballerina contains a thread pool that are assigned to run strands. The Ballerina scheduler manages strand allocation for each thread in the thread pool. The Ballerina thread pool, by default, has twice as many threads as there are cores in the processor. Optionally, you can change the pool size by setting the `STRAND_THREAD_POOL_SIZE` value under `ballerina.lang.runtime` in the `Config.toml` file of the project home directory (you need to create this file manually since it is not available by default). Keeping the thread pool size similar to the number of CPU cores reduces context switching and improves the performance of your application.

When Ballerina code is executed, it creates a strand and allocates it to a thread. If the program has a blocking operation, it releases the thread to the Ballerina scheduler and keeps the status of the strand until a response comes back from the I/O operation. Then, another runnable strand is assigned to the thread and continues executing the strand with the allocated thread. When the blocked strand receives the desired response, the Ballerina scheduler submits that strand to the runnable strand list. Because of the strand model, Ballerina can perform better than OS-level thread scheduling with its own language runtimes.

You do not need to think about any of these threading models when you are building an application with Ballerina. Libraries that have blocking operations, such as network calls, database operations, and file access, are managed automatically by libraries. If you need to access the strand level and control the application, Ballerina gives you a library to handle strands.

That's everything that you need to know about the internals of Ballerina. This knowledge will help you understand how Ballerina functions and what its capabilities are. Now we are going to move on to the next section, which explains how to install Ballerina and how to build a basic "hello world" application.

Setting up a Ballerina environment

Setting up Ballerina is as easy as installing and running the Ballerina installer, which can be downloaded from the Ballerina website. In this section, we will discuss how to install the Ballerina development environment and setting up VS Code as the code editor.

Downloading and installing Ballerina

Ballerina is supported by the Windows, Linux, and macOS operating systems. You can download the installer from the official Ballerina download page (`ballerina.io`):

Figure 2.4 – Ballerina download page

Installation from the installer is as easy as double-clicking on the installer and following the instructions. If you install Ballerina using the installer, there are no pre-requisites to install the Ballerina compiler on your computer.

If you are installing Ballerina from the source code, you must first install the **Java Development Kit (JDK)** on your computer. You can download OpenJDK 11 from the download page at `https://openjdk.java.net/`.

After you have installed Ballerina, you can check the installation by running the `bal version` Ballerina CLI command on the terminal. This will give you details about the current version of Ballerina that you are using, the date of the language specification, and the Ballerina update tool version that is used for new updates.

Ballerina offers a CLI tool that allows developers to manage Ballerina applications. This CLI tool allows developers to create, build, run, and test Ballerina projects. Once you have installed Ballerina on your machine, the CLI tool will also be installed.

You can view all the supported CLI commands by running the `bal help` command in the terminal. The Ballerina CLI tool commands can be categorized into four groups as follows:

- Core commands
- Package commands
- Update commands
- Other remaining commands

Next, we will discuss the commands provided under each of these categories.

Core commands

These commands let Ballerina developers build and run a Ballerina program:

- `build`: Compile the Ballerina program and generate executable artifacts.
- `run`: Build the program and run it.
- `test`: Run the test cases.
- `doc`: Generate API documentation.
- `clean`: Clean the artifacts generated by the `build` command.
- `format`: Format the Ballerina source code.

Package commands

Package commands are used to work with Ballerina packages. These commands allow you to create a new Ballerina project, import packages from Ballerina Central, and use them in your project. You can also publish your own packages in Ballerina Central as well. The following are the commands to use to manage packages in Ballerina:

- `pull`: Pull packages from Ballerina Central.
- `push`: Push packages to Ballerina Central.
- `search`: Search packages in Ballerina Central.
- `new`: Create a new Ballerina project.
- `add`: Add a new Ballerina module to the project.
- `init`: Initialize a new Ballerina package in the current directory.

Update commands

Update commands let developers update Ballerina tools and switch between different Ballerina versions:

- `dist`: Manage the Ballerina distribution.

- `update`: Update the Ballerina platform.

Other commands

The following available commands provide additional functionalities for a Ballerina application:

- `grpc`: This command generates a gRPC protocol stub implementation for protobuf.

- `openapi`: This command generates Ballerina source code from a given OpenAPI definition.

- `version`: Prints the Ballerina version.

- `bindgen`: This command generates bindings for a Java API.

As opposed to other programming languages, Ballerina offers a built-in tool to upgrade the Ballerina language itself. The Ballerina CLI provides a distribution management tool to upgrade to the latest version of the Ballerina language:

```
bal dist <command>
```

This command includes a collection of functionalities to change the version of the Ballerina compiler. By executing the `bal help dist` command, you can list all supported `dist` commands. You can find commands such as `update`, `pull`, `use`, `list`, and `remove`.

Updating Ballerina to the latest version is as simple as executing the `bal dist update` command. This will update the Ballerina compiler to the latest version.

Furthermore, you can switch between Ballerina versions with the `bal dist` command in the Ballerina CLI. The Ballerina compiler can keep multiple Ballerina compilers on the same computer. You can specify the version that you need to compile with the `bal dist list` command, listing all Ballerina versions that are local and remote to you. If you do not have the desired Ballerina distribution on your local machine, you can pull it from a remote machine by executing the `bal dist pull <distribution-name>` command. Then, you can verify whether the distribution has been pulled to your local machine by executing the `bal dist list` command. This command lists pulled distributions under the `local` section. To switch between locally available versions, you can use the `bal dist use <distribution-name>` command. If you have distributions that you are no longer using, you can remove those distributions with the `bal dist remove <distribution-name>` command.

Setting up VS Code

VS Code is free and open source software published under the permissive Expat License by Microsoft. The features of VS Code include debugging, the completion of code, syntax highlighting, snippets, and refactoring code. VS Code has built-in Git support that allows developers to manage source code versioning in the IDE itself. The features for version control include code difference review, viewing stage data, and push and pull to and from remote repositories. VS Code is built on the Electron framework, and this means VS Code can run on any operating system.

Language capabilities can be extended with extensions, available in the central repository. Ballerina extensions are available in Visual Studio Marketplace; you can get set up with the following steps:

1. Go to the official download page of VS Code at `https://code.visualstudio.com/` and download the installer that corresponds to your operating system:

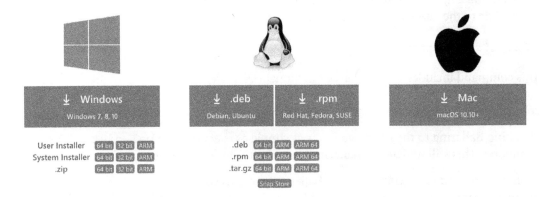

Figure 2.5 – VS Code download page

2. Click on the installer and follow the instructions to install VS Code on your computer.

3. Next open the VS Code to install Ballerina language plugin. Click on the **Extensions** icon in the left activity bar and search for Ballerina as shown:

Figure 2.6 – Installing the Ballerina extension

4. Once you've picked Ballerina from the list, you will see the **Install** button on the right-hand side of the panel. Click on it and wait a few seconds for the installation to complete.

More information about the Ballerina VS Code plugin is available at `https://ballerina.io/learn/tooling-guide/visual-studio-code-extension/quick-start/` page.

You can then start using the VS Code editor to create a Ballerina program. You can find code navigation, search, version control, and extension features in the VS Code activity bar. You can type Ballerina code in the editor and visualize it in real time as a sequence diagram. The sequence diagram view can be displayed by clicking the **Ballerina Diagram** button at the top right of the editor view. Here's a screenshot of the VS Code layout:

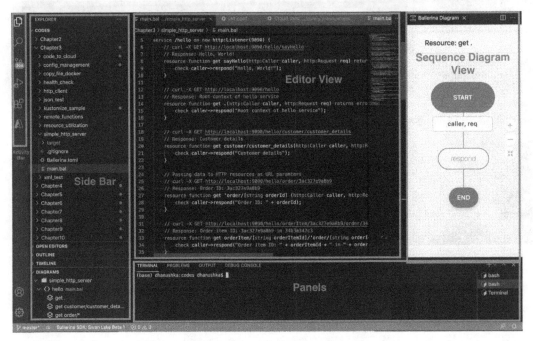

Figure 2.7 – VS Code layout

You will find a terminal on the panel where you can run commands in VS Code. By default, it sets the current directory as the project location. The debug console can be used to debug the Ballerina application. You can also find the status bar in the bottom that previews details and the status of the project.

Using VS Code to develop a Ballerina application

The VS Code editor offers many features that allow developers to easily create applications, such as these:

- **Intellisense**: While you are writing the Ballerina application, VS Code gives auto-complete suggestions.

- **Formatting**: You can use features such as code formatting to format Ballerina code.

- **Code navigation**: VS Code also simplifies navigation through source code.

- **Code actions**: VS Code provides support to run and debug Ballerina applications inside the IDE. You can use debug pointers to pause the execution of a Ballerina program and analyze the stack. VS Code lets you navigate through the execution of the Ballerina program, which helps programmers get a better understanding of variables and other language symbols.

- **Ballerina sequence diagram**: The Ballerina sequence diagram lets programmers view the application's communication and workflow. Sequence diagrams play a key role in building distributed systems. As sequence diagrams help you understand the flow of software, Ballerina uses them to explain real-time code.

Always use version control when developing applications with any programming language. Git is the **Version Control System (VCS)** that is most widely used in the industry. You can download Git from the official Git website and use it along with a remote Git repository such as GitHub or Bitbucket. Git is not only about maintaining checkpoints or commits on your program but also about allowing developers to maintain their platforms using delivery pipelines.

There are many Git remote repositories available, such as GitHub, GitLab, and Bitbucket. You can keep a clone of your work in remote repositories and share it with others. Central code repositories allow multiple developers to work with a single code base. In *Chapter 10*, *Building a CI/CD Pipelines for Ballerina Applications*, we will explore the use of Git repositories to automatically deploy Ballerina code to desired platforms.

Writing a simple "hello world" program

To start creating a Ballerina project, we need to initialize the project with the Ballerina CLI tool. To run the CLI tool, first you need to open the terminal view by clicking on **View | Terminal** in the menu bar, and then you execute this command:

```
bal new <project-path>
```

This is the command to build a new Ballerina project in the specified project path. If the given path does not exist, the directories will be generated and the Ballerina project will be initialized in the specified directory.

To create a new Ballerina project, execute `bal new HelloWorld` in the terminal (change your directory path according to the location where you are going to create the project). This command generates a Ballerina project in the current directory. Executing this command generates a `Ballerina.toml` file and a `main.bal` file inside the project directory.

You can open this project on VS Code by clicking on **File| Open...** in the menu bar. Select the directory that we have just created and open the project. This will reload the IDE and import the newly created project into the workspace.

> **Important note**
> You cannot run `bal new <project-path>` inside another Ballerina project. If you try to execute this command, it will throw an error.

The `main.bal` file is the main access point to the Ballerina application. Ballerina uses `.bal` as its file extension. If you open the `main.bal` file, its content will be as follows:

```
import ballerina/io;
public function main() {
    io:println("Hello World!");
}
```

If you are already familiar with the Java, C, or Go languages, you will instantly realize that this program will print `"Hello World!"` to `stdout`. As in those programming languages, the `main` function is the main entry point for the Ballerina application. Ballerina uses the `function` keyword to define a function. The `public` keyword represents that the `main` function is accessible from anywhere.

To print the `"Hello World!"` string to `stdout`, we need to import the `ballerina/io` package to use the `println` command. Here, `ballerina` is the organization name and `io` is the package name. The following is the syntax to import a Ballerina package into the current Ballerina application:

```
import [<org-name>]/<package-name> [as <identifier>];
```

Optionally, you can set an explicit identifier instead of using the package name as is. The `as` keyword creates an alias with the identifier name. Since we did not use an identifier, we can simply invoke the `println` function using the `io:println` (`"Hello World!"`) statement.

Ballerina has an integrated collection of libraries that can be used in most programming scenarios. Standard libraries contain different libraries, such as data structure, cloud deployment, calculation, file access, and others. Ballerina reserves the `ballerina` organization name for these libraries. If you are creating your own library, make sure to not use `ballerina` as your organization name.

The `lang` library contains language-specific libraries such as basic data types, including `int`, `arrays`, `maps`, `string`, and so on. These libraries are released together with the Ballerina language. Other Ballerina libraries, such as `http`, `file`, `log`, and `io`, are released separately from the Ballerina language release. The Ballerina organization includes these libraries as standard libraries and you can import them with the `ballerina` organization. Additionally, Ballerina provides support to work with MySQL, JDBC, Azure Functions, and more. These supportive libraries are released with `ballerinax` as the organization.

To run the application, you can use the `run` command as follows:

```
bal run [<options>] <ballerina_file_path> | <ballerina_package_
path> | <executable-jar> [--] [<arguments>]
```

Ballerina lets you run the project by specifying either a package or a generated JAR file. Since we have created a package, we can simply execute the `bal run HelloWorld` command to run the application. This command prints `Hello World!` in the terminal.

Building a Ballerina project

The Ballerina CLI tool provides commands to build and maintain Ballerina applications. In the `run` commands that we discussed earlier, the building and running of the Ballerina application happen together. `run` commands first generate the artifacts and then run the application. The `build` command does not execute the program and only generates the artifacts. To build this Ballerina project, you can use the Ballerina `build` command as follows:

```
bal build [<options>] <ballerina_file_path> | <ballerina_
package_path>
```

For example, you can use `bal build HelloWorld` to build a Ballerina package that has the package name `HelloWorld`. This command builds an executable JAR file that can run on the JVM. While building the project, tests also run. If any of the tests fail, it will not generate any build artifacts and show you an error message. If you do not need to run tests, you can avoid running test cases with the `--skip-test` option. Then, your `build` command without running test cases would be `bal build --skip-test HelloWorld`.

If you check the directory structure, you'll see there is a directory named `target` created inside the `project` directory. Inside this directory, there are another two directories named `bin` and `cache`. These are the build artifacts generated by the Ballerina compiler:

- The `bin` directory contains the executable JAR file named `<package_name>.jar`.
- The `cache` directory contains a compiler cache to speed up the compilation process. You can clear the cache with the following command:

```
bal clean
```

Since we already generated JAR artifacts with the `build` command, we can point to that JAR file and execute the Ballerina program with the `run` command. For this example, the command would be `bal run HelloWorld/target/bin/HelloWorld.jar`. This will print `Hello World!` in the terminal.

Until now, it's all been about setting up and starting your Ballerina environment. The program is a collection of data and control statements. In the next section, we will discuss the Ballerina type system and how we can use it to build our applications.

Understanding the Ballerina type system

The type system is the most important part of learning any programming language. The Ballerina type system lets developers define variables with different structures. Starting from basic types, a developer can build much more complicated data structures. Ballerina type systems include data types as well as behavioral types such as functions and errors. Ballerina types can be separated into five categories, as follows:

- **Simple types**: These are the primitive data types of the Ballerina language, such as `boolean`, `float`, and so on. These types are immutable by nature.
- **Structured types**: These are simple types organized in structures, such as arrays and maps.
- **Sequence types**: This is a combination of simple and structured types.
- **Behavioral types**: Rather than representing data, these types are used to handle behaviors of the Ballerina language. These types include errors, functions, objects, and so on.
- All other types.

The following sections will discuss each type category in detail.

Simple types

Simple types are the basic building blocks of all data types. All other structured types and sequence types are built with simple types. In this section, we will learn about each of the simple types that are available in the Ballerina type system.

Nil

Nil represents the absence of a value. Nil is represented as `()` in Ballerina. If you need to define that a value can be nil, you can use `()` in your variable definition. See the following example of assigning nil to a nil type variable:

```
() nilValue = ();
```

Boolean

`Boolean` represents a single bit value and it can be either `true` or `false`. Booleans are important in logical operations to determine the outcome. A Boolean can be defined as follows:

```
boolean a = true;
boolean b = false;
boolean aAndb = a && b; // Result is false
boolean aOrb = a || b; // Result is true
boolean negetion = !a; // Result is false
```

Boolean operations include the OR operation, which is represented by `||`, and the AND operation, which is represented by `&&`. You can represent negation with the `!` sign.

Integer

Ballerina's integer is 64-bit and it can be used to represent signed integers. Integers can represent any number between -9,223,372,036,854,775,808 and 9,223,372,036,854,775,807. They can be defined as follows:

```
int a = 7;
int b = 5;
int aPlusb = a + b; // Result is 12
int aMinusb = a - b; // Result is 2
int aMultib = a * b; // Result is 35
int aDivideb = a / b; // Result is 1
```

Operations between integers also produce integers. For example, if you add, subtract, multiply, or divide two integers, it always produces an integer. Ballerina performs integer division for the divide operation. Additionally, you can also perform bitwise operations such as AND (&), OR (|), XOR (^), left shift (<<), and right shift (>>) as well.

Float

Ballerina's float is 64-bit and is used to represent signed, floating-point numbers. The float type supports addition, subtraction, multiplication, and division, just as an integer does. Unlike integer division, float division generates floating-point numbers:

```
float valueFloat = 2.5;
```

Decimal

The decimal type is for 128-bit floating-point numbers used to represent decimal values. Decimal can be used to represent much larger values than float. This is important since some third-party applications, such as databases, have this kind of large decimal value representation. You can use the same operations as with float, such as addition, subtraction, multiplication, and division, to work with decimal values as well. The following is an example definition of decimal data:

```
decimal valueDecimal = 5.3;
```

Structured types

Ballerina structured types provide a way to implement much more complicated data structures in the Ballerina language. You can use structured data types along with basic data types to create complex data structures. Here we are going to go through each of these structured data types.

Array

An array is an iterable item whose index starts with zero. Any data type can be used with an array. Arrays can be accessed by their index. In line 2 of the following example code, the value is accessed using index 2; this is the definition of an integer array:

```
int[5] valueArray = [2,3,4,5,1];
io:println(valueArray[2]);
```

Arrays have built-in operations that can be performed on an array. You can add a new item to an array with valueArray.push(4). If you need to remove the last item, you can use valueArray.pop(). You can reverse an array with the valueArray.reverse() statement. The length can be returned with the valueArray.length() statement.

Tuple

A tuple is a set of different types. Each member of a tuple can have different types. You can use tuples to return multiple types from a function. We will learn about using tuples with functions in a later section. See the following example of defining a tuple. In this tuple, we have integer and string values. Tuple initialization syntax is very similar to that of arrays:

```
[int, string] valueTuple = [3, "Hello"];
```

Map

A map is a key-value pair where the key is always a string. The value type is defined in the variable declaration as follows:

```
map<string> user = {
    name: "Tom",
    address: "New York"
};
```

Here we create a map of `string` with the variable name `user`. As you may notice, `user` has two fields, and both fields are strings. Variable elements can be accessed with the key as follows. This is similar to the array, but instead of accessing value using an index, here we access with the key name:

```
io:println(user["name"]);
```

Record

A record is a generalized version of a map, where you can define multiple types in a single record. In the previous example, we defined a map using the string data type. Here, you can have multiple attributes that have different data types. See the following record definition, which initializes `string`, `int`, and `boolean` types:

```
record {
    string name;
    int age;
    boolean married;
} userRecord = {
    name: "Tom",
    age: 21,
    married: false
};
```

Instead of defining the record along with the variable declaration, we can use the record type to define a new type such that we can reuse the record. We can use the `type` keyword and define a new type as follows:

```
type User record {
    string name;
    int age;
    boolean married;
};
```

The `User` record type that we have just created can be initialized as follows:

```
User = {
    name: "Tom",
    age: 21,
    married: false
};
```

Records can be accessed the same way that maps are. To assign `age` to a variable, you can use the `int newAge = user.age;` statement. You can add custom attributes to the variable at runtime. For example, you can add a new attribute to the `user` variable by using the `user["address"] = "New York";` statement. This statement adds a new `address` attribute to the `user` variable and assigns `New York` as the value of the attribute. These types of records are known as open records. Open records can be changed dynamically. If you need to restrict adding a new attribute to the record, you can use a closed record. See the following record definition with a closed record:

```
type User record {|
    string name;
    int age;
    boolean married;
|};
```

This restricts adding new types in the runtime. If you try to execute `user["address"] = "New York"`, then the compiler generates an error.

You can also define an optional field for record initialization. For example, right now, if you initialize a user record without an `age` field, the compiler throws an error saying that you missed defining the `age` attribute. If `age` is optional, then you can mark the field as an optional field:

```
type User record {
    string name;
    int age?;
    boolean married;
};
```

In this example, `age` is an optional field that you can ignore when the `user` record gets initialized. When you try to access the optional field, it may or may not be initialized. Ballerina does not allow accessing uninitialized variables. Therefore, to access optional fields in a record, you need to use the **optional field access** operator, `?.`, instead of the **field access** dot (`.`) operator. This operator returns either the value assigned to the attribute or nil if it is not initialized. To find out what exactly the return type is, we can use a type guard. Your code should be as follows to read the age from the `user` variable with a type guard:

```
int? age = user?.age;
if age is int {
    io:println("Age is " + age.toString());
} else {
    io:println("Age is nil");
}
```

The `age` variable can be either integer or nil. If we need to access the integer value, first we need to do a type check and then use the value. Line 2 checks whether `age` is an integer. If it is an integer, it prints the value. Else, it prints the value as nil.

You can also create a more complex record by defining a record inside a record, which is known as using nested records. For example, the address information of the user might be another record. See the following code example where a User record has a nested Address record:

```
type User record {
    string name;
    int age;
    boolean married;
    Address;
};
type Address record {
    string lane1;
    string city;
    string country;
    string zipCode;
};
public function main() returns error? {
    User = {
        name: "Tom",
        age: 21,
        married: false,
        address: {
            lane1: "Water St.",
            city: "New York",
            country: "USA",
            zipCode: "10001"
        }
    };
}
```

Here, the User record contains an Address record. You can initialize the complete User record, which looks similar to a JSON object.

Table

Table lets developers store data in a table-like format. It is a collection of records where each record represents a row of a table. You can define an index for the table and use it to access tables. See the following example for columns with name, age, and married fields for the User type that we defined earlier. Here we will define a new type by setting the indexing key as the name field:

```
type User record {
    readonly string name;
    int age;
    boolean married;
};
type UserTable table<User> key(name);
```

The readonly keyword at the beginning of the name attribute is to mark the name field as only for reading. Since we select name as the key in this table, you need to mark this field as read only. Since it is read only, you cannot change it at runtime. Now we can use this new type to initialize a new variable and access its content with the get(key) function provided by the table:

```
UserTable users = table [
    {name: "Tom", age: 25, married: false},
    {name: "Alice", age: 23, married: true},
    {name: "Bob", age: 34, married: true}
];
io:println("User name", users.get("Tom"));
```

Sequence types

Sequence types include string and XML data types. These types are the combination of basic and structured data types.

String

A string is a sequence of Unicode characters. A string can be used to represent a plain text value. Each character in the string can be accessed in the same way as arrays:

```
string valueString = "Hello";
io:println("First character is " + valueString[0]);
```

Strings have built-in features to work with strings. For example, if you need to convert a string to uppercase, then you can use `valueString.toUpperAscii()`. In the same way, you can use the `toLowerAscii()` function to convert a string to lowercase. There are other functions, such as `substring` to extract a string segment from the given offset, `length()` to get string size, `join()` to join two strings with a separator, `trim()` to remove white spaces, and so on.

You can use a string template expression to define strings. See the following example, which defines strings using template variables:

```
string firstName = "Tom";
string lastName = "Wilson";
string fullName = string 'Mr.${firstName} ${lastName}';
```

With this syntax, you can avoid using concatenation functions and addition operators to join multiple strings Template variables are surrounded with the ${} syntax. Inside the brackets, you can reference a variable of a function that returns a string.

XML

The XML type is used to represent parsed XML objects. XML is natively supported by Ballerina, and we will discuss XML more in later chapters. For now, this is an example definition of an XML value:

```
xml valueXml = xml '<user>
<name>Tom</name>
<age>45</age>
</user>';
```

Behavioral types

Behavioral types are not data types. These types play a key role in the flow of a Ballerina application. For example, the `error` type handles errors and the `future` type handles multi-threading. In this section, we will go through each of these behavioral types and see how to use them in the Ballerina language.

Error

Errors can be generated in the event of failure and passed along to the application for resolution. The following code shows how to create an error. Ballerina treats errors as a type that can be handled similarly to data types. You can pass and return errors to functions in the same way as with data types:

```
error valueError = error("Error name");
```

Function

Functions provide a way to organize your data by separating instructions into multiple blocks. Functions are important when there are repeated instruction sets to execute. Functions take input, perform operations over the input, and return an output. For example, the following function takes two values as input and returns the sum as the output:

```
function sum(int a, int b) returns int {
    return a + b;
}
```

Future

The future type is used to reference a worker that runs and returns a result in the future. Future plays a key role in building thread and asynchronous calls. See the following example, which executes the sum function in an asynchronous way. Here the sum function executes separately from the main thread and a future reference is used to reference the worker:

```
future<int> futureType = start sum(5, 6);
```

Object

Objects are a lot like records. Unlike records, objects can have functions as well. Ballerina provides object-oriented support with the Object type. We will discuss objects more in the *Ballerina objects* section later in this chapter.

Stream

Ballerina streams are lists of iterable variables that can be used to represent a stream of data. A stream can be built with basic and sequential data types. The next () function can be used to access and retrieve data bit by bit. Note that a stream can be iterated over only once. You cannot iterate over the same stream twice. Streams are widely used to retrieve data from a database as an iterable list. See the following example, which creates a stream of users:

```
User[] users = [
    {name: "Tom", age: 25, married: false},
    {name: "Alice", age: 23, married: true},
    {name: "Bob", age: 34, married: true}
];
stream<User> userStream = users.toStream();
```

You can use the `filter` function with streams as follows to filter a list of married users from `userStream`:

```
stream<User> marriedUserStream =
    userStream.filter(function (User user) returns boolean {
        return user.married;
    });
```

A stream can be used to perform the same operation over a list of users. Always keep in mind that streams can iterate only once. Once iterated, the values in a stream vanish.

Other types

These are the remaining types that cannot be categorized into the aforementioned types.

Union

The union type lets variables have multiple types. The value can have any of the types defined in the variable declaration. In the following example, variables can be either integers or strings. You can assign both string and integer values to a variable:

```
int|string value = "this is a text";
value = 5;
```

Any

Any can represent any type other than the `error` type. The following code is possible with the any type descriptor as any data type can be accepted:

```
any someValue = 5;
someValue = "This is a text";
```

Optional

This means that the variable can either be the desired type or nil. The optional type can be defined by adding a question mark (?) at the end of the type definition. This is a shorthand representation of | (). See the following example of defining an `index` variable's type as either `int` or `nil`:

```
string statement = "this is a text";
int|() index = statement.indexOf("is");
```

`statement.indexOf()` returns either `int` or `nil`. This can be written in shorthand with an optional operator as follows:

```
int? index = statement.indexOf("is");
```

Anydata

Anydata is the same as the `any` type, but unlike the `any` type, `anydata` only lets you define plain data. You cannot assign behavioral types such as functions and objects to the `anydata` type. Anydata is equivalent to the following union:

```
()|boolean|int|float|decimal|string|(anydata|error)
[]|map<anydata|error>|xml|table
```

Byte

Byte is 8-bit and lets you assign values from 0 to 255. Byte is usually used to represent binary values. A binary value can be represented as an array of bytes in the Ballerina language:

```
byte byteValue= 255;
```

An example of defining a `byte` array to represent a binary value follows. Here we convert the base64-encoded string `Hello World` into a `byte` array and assign it to a variable. If you check the length of the array after assigning it, the length should be `11`:

```
byte[] byteValue= base64 'SGVsbG8gV29ybGQ=';
```

Working with types

Here we will discuss a set of functionalities that are provided to work with types. This set of keywords is important when it comes to variable definition.

var type

The `var` type lets the compiler decide the type of the variable. The variable type is inferred by evaluating the right-hand side of the assignment. Do not confuse `var` with the `any` data type. Any is a type that is a combination of all the types except the `error` type. The `var` type is not a combination of all the types.

With the `any` type, the following code is possible. Since the `value` variable accepts all types, it can have any value:

```
any value = "This is a string";
value = 6;
```

But the following code does not compile since var assigns the string type and then tries to assign an integer:

```
var value = "This is a string";
value = 6;
```

Ballerina forces developers to be specific when selecting the type for a variable. This helps the readability of code and prevents human error. Therefore, if you know the data type to use, use that specific data type instead of using var to find the type.

Ballerina constant

Constant and final are used to define values that will not change in the future. Constant values are initialized at compile time and final values are initialized at runtime. When a constant or final value is initialized, it cannot be changed at runtime. See the following example of using a constant:

```
const float PI = 3.141592;
```

The constant should define a top-level/module-level construct. You cannot define constants inside functions or objects.

Optionally, you can ignore the type and Ballerina can infer the type of the constant by evaluating the right-hand side of the assignment. You can simply write the previous statement as follows:

```
const PI = 3.141592;
```

Constants can be used in the same way as variables. See the following example of using the previous constant to calculate the circumference of a circle:

```
float circumference = 2 * PI * radius;
```

Enum

An enum represents a collection of constants. It is a shorthand for the combination of a string and a constant. You can define a set of constants that belong to a group with the enum keyword. See the following example, which creates a set of constants as an enum:

```
enum OrderStates {
    OrderCreated,
    OrderVerified,
    OrderDeliverted,
    OrderCanceled
}
```

Enum can be accessed the same way as constants. For example, the following statement assigns `OrderCreated` as a string to the `orderStatus` variable:

```
string orderStatus = OrderCreated;
```

The let expression

The `let` expression lets developers assign a value to a variable as a result of the evaluated expression. It has the following syntax:

```
<variable-type> <variable-name> = let <expression-variable-
definitions> in <expression-to-evaluate>;
```

Consider the following example of calculating the sum of 5 and 6 and attaching it as the value of a variable. In this example, `sum` has `11` as its value by evaluating the `let` expression:

```
int sum = let int a = 5, int b = 6 in a + b;
```

As you can see, the `let` expression contains two parts: a `let` part and an `in` part. The `let` part of the expression defines the variables that need to be evaluated and the `in` part contains the expression that needs to be evaluated to get the final result. You can also use function calls and variables that are available for the current scope inside the `let` expression. But you cannot directly access anything outside the scope from the `in` part. In the `in` part, the variable scope includes only the variables that are defined in the `let` part.

See the following example where a `let` expression reads values from outside of the scope. The code uses the `getFive` function and the already-initialized `c` variable to calculate the sum of 7 and 5. If you try to use the `getFive` function inside the `in` part, the compiler will throw a compilation error:

```
public function main() returns error? {
    int c = 7;
    int result = let int a = getFive(), int b = c in a + b;
    io:println(result);
}
function getFive() returns int {
    return 5;
}
```

So far, we have discussed the Ballerina type system, which enables developers to work with different types. In the next section, we will discuss using the Ballerina type system further to create a program by adding functionalities and behaviors to types.

Using types to build applications with Ballerina

As you have already learned, Ballerina has a diverse type system that you can use to represent complex data structures and behaviors. To build a functioning application, you need to use these types. In this section, we will focus on handling program flow by using conditions and loops, using object concepts to build an object-oriented system, error handling to handle errors.

Controlling program flow

Although types provide a way to use data in the Ballerina programming language, program flow controllers provide syntax to handle the data flow in an application. Ballerina flow control is inspired by the best of other programming languages. Ballerina contains program flow controllers such as `if else`, `while`, `foreach`, and `match`. The syntaxes of these are similar to those in programming languages such as Java, C, and Go.

if else

The `if else` syntax is pretty much the same as it is in other programming languages that let programs make decisions based on a condition. The `if else` syntax in the Ballerina language is as follows:

```
if <condition> {
    // statement
} else if <condition> {
    // statement
} else {
    // statement
}
```

Optionally, you can use parentheses to wrap the condition if there are complex conditions. The following example compares two variables and prints whichever one is bigger:

```
import ballerina/io;
public function main() {
    int a = 35;
    int b = 25;
    if a == b {            // Line 5
        io:println("a equal b");
```

```
    } else if a < b {      // Line 7
        io:println("a less than b");
    } else {                // Line 9
        io:println("a larger than b");
    }
}
```

The `if` conditions check for the equality of values a and b in line 5. `else` conditions get executed if an `if` condition or any of the `else if` conditions are not fulfilled. If the `else if` condition is not fulfilled on line 7, then the code continues to execute the `else` condition on line 9.

Ballerina provides the following four types of equality checks to compare two variables:

- **Deep value equality**: This is represented by == and checks whether variables' values are equal.

- **Deep value inequality**: This is represented by != and checks for the inequality of two variables.

- **Reference value equality**: This is represented by === and checks whether two variables are exactly the same, along with doing a type check.

- **Reference value inequality**: This is represented by !== and is the opposite of reference equality.

To understand the difference between deep value equality and reference value equality, consider comparing two float values, 1.0 and 1.00. Now, 1.0 == 1.00 is a deep value check and the result would be true. On the other hand, the 1.0 === 1.00 reference equality check would be false. In the first method, we only check for the represented value. In the second expression, we are checking for exact similarities, meaning the type should also be the same.

Logical expressions can be used to combine multiple Boolean values and evaluate the result. For example, AND can be represented by && in the Ballerina language. OR can be represented using ||. Negation can be represented by placing an exclamation mark (!) in front of a variable:

```
int age = 25;
boolean isEmployed = true;
if age>18 && isEmployed {
    io:println("Employed adult");
} else if age>18 && !isEmployed {
```

```
        io:println("Unemployed adult");
    } else {
        io:println("Not an adult");
    }
```

The while loop

while loops are used to perform repetitive tasks until a given condition is satisfied. The Ballerina syntax for the while loop is as follows:

```
while <condition> {
    // statement
}
```

Here is an example of using a while loop to print all the integers from 0 to 10:

```
int index = 0;
while index <= 10 {
    io:print(index.toString() + " ");
    index = index + 1;
}
```

A while loop can be broken with the break statement. You can stop running the nearest enclosed while loop and run the rest of the program by adding break inside the while block. Ballerina also supports the continue statement, which lets you stop running the nearest enclosed while loop and resume running while in the next iteration.

The previous example can be modified as follows by adding a break statement to the program. If index becomes 10, the while loop is exited and will continue running the rest of the program:

```
int index = 0;
while true {
    io:print(index.toString() + " ");
    if index == 10 {
        break;
    }
    index = index + 1;
}
```

foreach

foreach lets you loop through iterable Ballerina types such as arrays, map, JSON, XML, and tables. The syntax for the foreach statement is as follows:

```
foreach <element> in <iterable-item> {
    // statement
}
```

Let's look at an example of how to use foreach to calculate the sum of values in a products table:

```
public type Product record {     // Line 1
    string productName;
    int quantity;
    float unitPrice;
};                               // Line 5
public function main() {
    table<Product> products = table [     // Line 7
            {productName: "SD card reader", quantity: 1,
                unitPrice: 20},
            {productName: "USB cable", quantity: 2, unitPrice: 1}
        ];                           // Line 10
    float totalPrice = 0;
    foreach Product in products { // Line 12
        totalPrice =  totalPrice + <float>product.quantity *
            product.unitPrice;
    }
    io:println("Total price is: " + totalPrice.toString());
}
```

The data structure that we will use to construct the table as a record will be defined from lines 1 to 5. In the main function, from line 7 to line 10, the list of products is initialized to a variable called products. In line 12, a foreach statement is applied and iterated over the list of products one by one. This will calculate the total value in the products table variable.

match

match lets you execute a block of code based on a value. match is the same as a series of if else conditions. Unlike the if else syntax, match considers a single variable and matches different values against it. Different code blocks get executed for each value that matches. Other programming languages can also have a similar syntax, such as switch cases. Unlike other programming languages, Ballerina does not need the break statement to stop executing. Ballerina keeps trying to match values until it finds a match. Once it is found, it leaves the match flow and continues with the rest of the code. See the following example of using match to print the day of the week depending on the day value:

```
public function main() {
    int day = 2;
    match day {
        1 => {
            io:println("Monday");
        }
        2 => {
            io:println("Tuesday");
        }
        3 => {
            io:println("Wednesday");
        }
        4 => {
            io:println("Thursday");
        }
        5 => {
            io:println("Friday");
        }
        6 => {
            io:println("Saturday");
        }
        7 => {
            io:println("Sunday");
        }
    }
}
```

Ballerina functions

Ballerina lets programmers split a single execution program into multiple blocks with functions. Like most other programming languages, the Ballerina functions take input as arguments and return a value. The Ballerina syntax defines a function as follows:

```
<access-modifier> function <function-name>(<input-arguments?>)
   returns <return-types?> {
      // Function content
}
```

Ballerina supports public and module-level access modifiers for functions. These access types allow programmers to restrict access to a variable. The following are the access modifiers that are supported by the Ballerina language:

- **Public**: Public modifiers are visible everywhere.
- **Module-level**: The absence of an access modifier signifies that it is a module-level access modifier that only gives access to current modules. Other modules cannot access module-level variables.

You may or may not have input arguments and return types for a function. A function can have multiple input arguments separated by commas. See the following function example, which takes two arguments and calculates the sum of two values:

```
function sum(int a, int b) returns int {
    return a + b;
}
```

Here the two input arguments are a and b. The return type of this function is defined as an integer since it returns the sum of two integer values. The `return` statement in line 2 is used to give the result back to the caller function. If the function has been defined, the following syntax can be used to access the function:

```
<type_descriptor> <variable_name> = <function_name>(<input_
arguments>);
```

In cases where the function does not return anything, you can leave out using a variable to hold the return value from the function. If the function returns a value, you cannot ignore the result and it should be allocated to a variable.

For this example, we are calling the sum function. Here the return type is integer and is assigned to a variable called valueSum. The input arguments for the function contain two integer inputs:

```
int valueSum = sum(5, 6);
```

Ballerina lets you use function arguments with the default values. The default values for the input function are specified in the function definition. See the following example of setting a default time. This function prints the default time as the current timestamp if the function does not get an input argument:

```
function printDay(time:Time date = time:currentTime()) {
    io:println("Time is " + date.toString());
}
```

It is possible to call this function with or without input arguments. This method might give conflicting results when there are multiple function arguments with multiple default values on the function definition. Ballerina lets the caller use named arguments to solve any conflicts with arguments of the same type:

```
function sumABC(int a = 4, int b = 5, int c = 6) returns int {
    return a + b + c;
}
```

In this example, the functions take three optional function arguments. If you send only two values as an argument, Ballerina cannot recognize the corresponding arguments for each variable. To avoid such argument mapping problems, you can call this function with named arguments as follows:

```
int valueSum = sumABC(a = 5, c = 6);
```

Another advantage of using named arguments is that you do not need to provide arguments in the same order as in the function definition. You can directly pass arguments by name in the function definition.

If you do not know the number of arguments in a Ballerina function, you can use rest parameters in the function definition. Rest parameters let you take a list of arguments of the same type. See the following example where we describe a function that has a rest parameter:

```
function printUser(string name, int age, string... details) {
    io:println("Name " + name);
    io:println("Age " + age.toString());
    foreach string detail in details {
        io:println(detail);
    }
}
```

Here, the `details` variable is an array of string values. In lines 4-6, we loop through the `details` variable and print the content to the terminal. We can call this function with multiple `details` variables as follows:

```
printUser("Alice", 24, "Lives in New York", "Drink Coffee");
```

Ballerina does not have a built-in way to return multiple variables. Combining the return function with the tuple type makes for an elegant way to return multiple results to a single function call. See the following example:

```
function getUserDetails(string name, int age) returns [string,
    string] {
    string nameDetails = "User's name is " + name;
    string ageDetails = "User's age is " + age.toString();
    return [nameDetails, ageDetails];
}
```

Calling this function as follows is more readable than using other types such as records and maps to return multiple variables from a function:

```
[string, string] [nameDetail, ageDetail] =
    getUserDetails("Alice", 24);
```

In some scenarios, programmers may need to ignore the return values of a function if they are not important. Unlike other programming languages, in Ballerina, you cannot simply ignore return values and leave them empty. Ballerina forces you to allocate the function result to a variable. In this case, an underscore (_) can be used to assign unnecessary return values. Consider the same example we discussed earlier involving the return of string tuples. You may use an underscore as follows to skip the return values:

```
_ = getUserDetails("Alice", 24);
```

But always note that if the function returns an error along with the return type, you will have to manage the error. Since Ballerina forces you to handle errors, you cannot just ignore errors with an underscore.

Treating functions as variables

As we discussed earlier, Ballerina treats functions as variables. Functions may be assigned to and used as variables. See the following example of a function being used as a variable:

```
public function main() {
    function (int, int) returns int sumFunction = getSum;// Line 2
    io:println(sumFunction(5, 3)); // Line 3
}
function getSum(int a, int b) returns int { // Line 5
    return a + b;
}  // Line 7
```

In line 2, a function named sumFunction is created, of the function type, which takes two arguments with a single return value. The function definition is defined separately from lines 5-7, outside the main function. In line 2, sumFunction is assigned to the getSum function implementation and is called in line 3. Here you can see that a function can be assigned the same variables.

Instead of defining functions separately with a function name, Ballerina also lets programmers define functions without names. The preceding example can be modified as follows to define the getSum function as an anonymous function on the right side of the sumFunction variable assignment:

```
public function main() {
    function (int, int) returns int sumFunction =
        function(int a, int b) returns int {
        return a + b;
    };
    io:println(sumFunction(5, 3));
}
```

If the function only contains a single statement such as getSum() that adds two values and returns, you can define the expression-bodied function to define those functions. See the following example of defining the getSum() expression-bodied function:

```
function getSum(int a, int b) returns int => a + b;
```

Anonymous functions can also be defined as expression-bodied. The sumFunction()
function that we defined in the previous example can be written as expression-bodied
as follows:

```
var sumFunction = function (int a, int b) returns int => a + b;
```

For simplicity, I defined the function name with var, which assigns an appropriate type
based on the contents of the right-hand side. Expression-bodied functions are is helpful in
cases where functions have only a single expression.

Treating functions as variables allows programmers to pass functions as variables to other
functions. We can extend the previous program to handle multiple functions by specifying
multiple functions and passing these functions as input parameters for another function:

```
public function main() {
    function (int, int) returns int sumFunction =
      function(int a, int b) returns int {    // Line 2
        return a + b;
    };
    function (int, int) returns int multiplyFunction =
      function(int a, int b) returns int {    // Line 5
        return a * b;
    };    // Line 7
    io:println(performOperation(sumFunction, 5, 3)); // Line 8
    io:println(performOperation(multiplyFunction, 5, 3)); // Line 9
}
function performOperation(function (int, int) returns int
  operation, int a, int b) returns int {
    return operation(a, b);
}
```

Lines 2-4 define a function named sumFunction that adds two values, and lines 5-7
define another function called multiplyFunction that multiplies two values. The
performOperation function takes a function that has two input arguments and one
return type as its argument along with two integer values. In this function, it blindly calls
the operation function with given arguments and returns the result. In lines 8 and 9,
the performOperation function gets called with the target calculation function.

The Ballerina main function

In Ballerina, there are two starting points for a program: either the main function or a service. As opposed to the main function, services are meant to run long-running processes such as server and schedule tasks. The main function gives the starting point for a Ballerina application to run from.

Just like regular functions, main functions also have function inputs and outputs. The input may be a terminal input. The main function access modifier should always be public. At startup, Ballerina will take input arguments from the terminal. Consider the following instances, which use name and age as input arguments:

```
public function main(string name, int age) {
    io:println(name);
    io:println(age);
}
```

At execution time, you need to provide input arguments in the terminal. For example, if the package name is hello_world, then the terminal command to execute this program would be bal run hello_world -- Alice 24. This program casts Alice to a string and 24 to an integer and passes them into the main function.

Functions are the main concept of building functional programming architecture. Object-oriented programming architecture uses objects to design programs. In the next section, we will discuss building Ballerina applications with Ballerina objects.

Working with Ballerina classes

As discussed previously, Ballerina is not intended for object-oriented programming or functional programming. But you can use object-oriented programming concepts to create Ballerina programs. The object-oriented programming syntax of Ballerina is very similar to the syntax of the Python language. Here is an example of building a simple class:

```
class User {
    string firstName;
    string lastName;
    string address;
    function init(string firstName, string lastName, string
        address) {
        self.firstName = firstName;
        self.lastName = lastName;
```

```
        self.address = address;
    }
    function getUserDetails() returns string {
        return string '${self.firstName} ${self.lastName} from
          ${self.address}';
    }
}
```

A class is a prototype that can be used to create an object. The object definition begins with the `class` keyword. Here the class name is `User`, and `firstName`, `lastName`, and `address` are attributes in the `User` class. The `init` method is the constructor of the object and it takes attribute details and assigns them. The `self` keyword is used to access variables within the object. `getUserDetails` is used to return a string with details of the user. Here we used Ballerina string template literals to construct the string instead of using the concatenation (+) operator. You can initialize this object as follows:

```
User = new("Sherlock", "Holmes", "221b Baker St, Marylebone,
    London");
```

The `new` keyword is used to create an object from a class definition. The object will be initialized with the parameters provided through the `new` inputs. Once the object is initialized, object methods can be accessed with the dot (.) operator as follows:

```
io:println(user.getUserDetails());
```

This prints `Sherlock Holmes from 221b Baker St, Marylebone, London` in the terminal.

You can also access object attributes with the dot (.) operator as follows:

```
io:println("First name is " + user.firstName);
```

In object-oriented programming, there are the following four main concepts:

- Encapsulation
- Inheritance
- Abstraction
- Polymorphism

In the previous example, you may have noticed that we did not use any access modifiers in the class definition. This means that the default access modifier is module-level for the attributes and methods in the object. If you need to limit the access level of the object attribute and methods, you can use `private` as the access modifier. On the other hand, variables and methods can be exposed to other modules as well, with public modifiers. If we modified the previous example to have private modifiers for attributes and public modifiers for methods, the code would be as follows:

```
class User {
    private string firstName;
    private string lastName;
    private string address;
    public function init(string firstName, string lastName,
        string address) {
        self.firstName = firstName;
        self.lastName = lastName;
        self.address = address;
    }
    public function getUserDetails() returns string {
        return string '${self.firstName} ${self.lastName} from
            ${self.address}';
    }
}
```

Since attribute access levels are restricted, you cannot directly access attributes with the dot operator. To access these attributes, you need to add getters and setters. Depending on your requirements, you can add getters and setters for each attribute.

Ballerina has the concept of `final` to avoid setting values for already initialized objects. By adding the `final` keyword, Ballerina restricts the setting of object attributes. There are two ways to set a value as final. One way to define a final value is to provide it as a default value. Otherwise, you can initialize the value upon object creation using the `init` method. See the following class definition:

```
class User {
    final string firstName = "Tom";
    final string lastName;
    string address;
    public function init(string firstName, string lastName,
        string address) {
```

```
            self.lastName = lastName;
            self.address = address;
        }
    }
```

In this case, `firstName` and `lastName` are defined as final values. `firstName` is initialized with the default value and `lastName` is initialized in the `init` method. If you tried to change these values, the compiler would throw an error message.

Ballerina objects

Ballerina objects are much the same as Ballerina classes. Unlike Ballerina classes, Ballerina objects can have attribute definitions and method definitions only. Therefore, Ballerina objects are most similar to interfaces in the Java and C languages. Also, objects can have public-and module-level access control only:

```
type User object {   // Line 1
    string firstName;
    string lastName;
    string address;
    public function getUserDetails() returns string;
};      // Line 6
public function main() returns error? {
    User = object User {    // Line 8
      public function init() {
            self.firstName = "Sherlock";
            self.lastName = "Holmes";
            self.address = "221b Baker St, Marylebone, London";
      }
      public function getUserDetails() returns string{// Line 14
            return string '${self.firstName} ${self.lastName}
              from ${self.address}';
      }  // Line 16
    };   // Line 17
    io:println(user.getUserDetails());
}
```

In this example, an object named User is defined from lines 1 to 6. It includes a method definition as well. Objects are initialized from lines 8 to 17 with object constructor expressions. Here the init method adds default values to the object, and lines 14 to 16 add the function body. The final output of the program is the same as in the previous example.

We can extend Ballerina objects with the Ballerina class to implement inheritance. Object type referencing lets developers copy a member into another class or object. Instead of duplicating the same object, object references let you reuse the same object again and again. See the following example, which reuses the User object again with the customer and supplier classes.

The User object contains attribute and method definitions common to both the Customer and Supplier classes:

```
type User object {
    string firstName;
    string lastName;
    public function getUserDetails() returns string;
};
```

The Customer and Supplier classes separately have their own attributes, such as shippingAddress and inventoryAddress. The getUserDetails function also has different function definitions for each class type:

```
class Customer {
    *User;
    string shippingAddress;
    public function init(string firstName, string lastName,
        string shippingAddress) {
        self.firstName = firstName;
        self.lastName = lastName;
        self.shippingAddress = shippingAddress;
    }
    public function getUserDetails() returns string {
            return string '${self.firstName} customer from
                ${self.shippingAddress}';
    }
}
class Supplier {
```

```
    *User;
    string inventoryAddress;
    public function init(string firstName, string lastName,
        string inventoryAddress) {
        self.firstName = firstName;
        self.lastName = lastName;
        self.inventoryAddress = inventoryAddress;
    }
    public function getUserDetails() returns string {
            return string '${self.firstName} supplier from
                ${self.inventoryAddress}';
    }
}
```

In the main function, we can initialize these two different objects into the relevant class type. The following code sample creates two variables as customer and supplier and there is a getUserDatails method that returns a string:

```
public function main() returns error? {
    Customer = new Customer("Sherlock", "Holmes",
        "221b Baker St, London");
    Supplier = new Supplier("Tom", "Cruise", "Syracuse,
        New York");
    io:println(customer.getUserDetails());
    io:println(supplier.getUserDetails());
}
```

The same example can be extended to achieve polymorphism by using the User object to hold both the Customer and Supplier objects. Polymorphism refers to the ability to appear in many forms. A single User object can be used in many forms as follows:

```
    User customer = new Customer("Sherlock", "Holmes",
        "221b Baker St, London");
    User supplier = new Supplier("Tom", "Cruise", "Syracuse,
        New York");
```

As previously, you can use the customer and supplier variables to access the getUserDetails() function.

Error handling

In every programming language, error handling is a vital part of creating reliable software that does not have hidden functionality issues. All major programming languages have different ways to manage errors. In the case of Java, error management is achieved by exception handling. Exceptions can be created at runtime and thrown to the top of the stack to be handled. Ballerina's error handling is pretty similar to the Go language's error handling. Next, we will see how to handle an error with the Ballerina time library example.

In the following example, you can see how the Ballerina time parser generated an error and how the error was handled:

```
import ballerina/io;
import ballerina/time;
public function main() {
    time:Time|error decodedTime = time:parse("2020-10-06",
        "yyyy-MM-dd");   // Line 4
    if decodedTime is time:Time {  // Line 5
        io:println("The parsed time is ", decodedTime);
    } else {
        io:println("Error while parsing time", decodedTime);
    }
}
```

In line 4, we are trying to parse a date with a specified time format. On the left side of the statement, you can see the union type between time: Time and the error type. This means that the time:parse() function returns either a Time object or an error. If you execute this program, the terminal will print the parsed time correctly as if the condition on line 5 became true.

If you change time:parse("2020-10-06", "yyyy-MM-dd") to an invalid time string such as time:parse("2020-ab-cd", "yyyy-MM-dd"), the terminal prints an error message. Since the decodeTime variable is assigned as an error type, the program executes the else path.

In some scenarios, you do not need to handle errors inside the function where the error occurred. You can simply return the error to the parent function and handle it there. You can do this with the Ballerina check keyword. See the following example:

```
public function main() returns error?{
    error? result = readTime();
    if (result is error) {
        io:println("Error occured while getting time");
    }
}
public function readTime() returns error? {
    time:Utc decodedTime = check time:utcFromString("2007-12-
        03T10:15:30.00Z"); // Line 8
    io:println("The parsed time is ", decodedTime);
}
```

In this code segment, in line 8, you can see that the time:utcFromString() function generates time:Utc, and on the left-hand side, the decodedTime variable assigns its return value. If the time:utcFromString() function returns an error, the check keyword returns the error to the main function. In the readTime function definition, you can see the return type is set as error?. This means that the readTime function returns either an error or nil value. The main function gets the value and decides the return value from the readTime function with a type guard. We have to define the nil type in the readTime function since we are not returning anything.

If you run this application, it will print the decoded time in the terminal. As before, if you change the date to an invalid value, an error message will be printed to the terminal.

On the other hand, you can use the checkpanic keyword to check errors and stop running the program. In critical situations where an error is received, there may be no point in continuing to run the application. In these cases, we can use the checkpanic keyword instead of check to terminate the execution.

All of these errors are system-generated errors. If we can create our own errors, then we can define different types of errors that can occur in the application and return them when there is an error. In the next section, we will discuss generating custom errors and how to handle them.

Using custom errors

Ballerina lets you define custom errors that can be passed throughout the application and handled. Consider the following example of validating user input.

First, we create a record for `Customer` as follows. The `Customer` record has `firstName`, `lastName`, `address`, `mobileNumber`, `email`, and `birthDay` as its attributes:

```
type Customer record {
    string firstName;
    string lastName;
    string address;
    string mobileNumber;
    string email;
    string birthDay;
};
```

Next, we will create an `ErrorDetail` record, which can store the cause of the error and the error code. `cause` is an optional parameter that can be used to represent the cause of error. `code` is used as a unique identifier for each error type. The | signs surrounding the `ErrorDetail` record signifies that this record structure does not change dynamically. Put simply, you cannot add more attributes into the `ErrorDetail` record dynamically:

```
type ErrorDetail record {|
    error cause?;
    int code;
|};
```

In this example, we are going to validate the user's mobile number, email, and birthday. For each validation, we can create separate errors that will be generated when validation fails. `UserValidationError` is the abstract error type that defines all other errors:

```
type UserValidationError distinct error<ErrorDetail>;
type MobileNumberValidationError distinct UserValidationError;
type EmailValidationError distinct UserValidationError;
type BirthdayValidationError distinct UserValidationError;
```

Next, we look for the mobile number validation function. Here the mobile number validation is done by checking the mobile number using a regular expression. If no matches are found, the mobile number is not valid and we return the custom error that we just created. In the function definition, you can see that the function can return either the error that we have created or a nil value. If validation is successful, a nil value is returned:

```
function checkCustomerMobileNumber(Customer customer)
  returns UserValidationError? {
boolean matched = regex:matches(customer.mobileNumber,
  "^[(]?\\d{3}[)]?[(\\s)?.-]\\d{3}[\\s.-]\\d{4}$");
if !matched {
    return error MobileNumberValidationError("Invalid mobile
      number", code = 22331);
}}
```

Email validation is also done by checking the email address using a regular expression. If the regular expression does not find a match, a custom error is returned:

```
function checkCustomerEmail(Customer customer) returns
  UserValidationError? {
    boolean matched = regex:matches(customer.
      email,"^[\\w-\\.]+@([\\w-]+\\.)+[\\w-]{2,4}$");
    if !matched {
        return error EmailValidationError("Invalid email
          address", code = 22332);
    }}
```

Birthday validation is done by parsing the birthday with a given time format. Otherwise, you can use a regular expression for this validation as well. The time parser generates an error if it is unable to parse the given string correctly:

```
function checkCustomerBirthday(Customer customer) returns
    UserValidationError? {
      time:Utc|error birthday = time:utcFromString(customer.
        birthDay);
        if birthday is error {
            return error BirthdayValidationError("Invalid
              birthday format", code = 22333, cause = birthday);
        }
}
```

The main function for demonstrating the functionality of this application follows. Here we initialize the new customer variable with the Customer record. It will then go through a series of functions to validate the customer attributes. Here we used the do on fail syntax to catch errors. This block works similarly to the try catch syntax in Java. When the check keyword returns an error if an error is found, the on fail keyword catches all these errors with the type UserValidationError. Here UserValidationError is an abstract definition of all other error types:

```
public function main() {
    Customer = {
        firstName: "Tom",
        lastName: "Cruise",
        address: "Syracuse, New York",
        mobileNumber: "(123)-456-7890",
        email: "tomcruise@mail.com",
        birthDay: "1962-07-02"

    };
    do {   // Line 11
        _ = check checkCustomerMobileNumber(customer);
        _ = check checkCustomerEmail(customer);
        _ = check checkCustomerBirthday(customer);
```

```
    } on fail UserValidationError e {
        io:println("Error occurred while validating the
          customer", e);
    }  // Line 17
}
```

Once the error is caught in an `on fail` block, it gets printed to the terminal screen. If no errors are caught, the run continues without executing the `on fail` block. Other than using `do`, you can also use this `on fail` clause with `transaction`, `retry`, `lock`, `match`, `foreach`, and `while`.

If you need to catch errors separately, you can use a type guard to do so. See the following snippet, which can replace lines 11-17 of our code. This implementation goes through a series of `if` conditions to match a given error with a specific error:

```
UserValidationError? validationError;
validationError = checkCustomerMobileNumber(customer);
if validationError is MobileNumberValidationError{
    io:println("Error while validating mobile number",
      validationError);
    return;
}
validationError = checkCustomerEmail(customer);
if validationError is EmailValidationError{
    io:println("Error while validating email address",
      validationError);
    return;
}
validationError = checkCustomerMobileNumber(customer);
if validationError is BirthdayValidationError{
    io:println("Error while validating birthday",
      validationError);
    return;
}
io:println("Customer validated");
```

Summary

Ballerina is a cloud native programming language that provides a cloud native syntax style to enable programmers to easily implement cloud native applications. You can start building a Ballerina application by downloading the Ballerina development kit. In this chapter, we used VS Code as the IDE. We created a simple "hello world" application and discussed using the CLI tool to maintain a Ballerina project.

The Ballerina type system includes types to represent both data and behavior. The data types include simple data types, such as integers, Booleans, floats, and so on. We can build more complex data types by using arrays, maps, records, and tables. The Ballerina type system contains types such as functions, futures, and errors that are essential to building applications. Furthermore, we discussed control flow instructions such as loops and conditions.

Functions give you a way to organize your code and reuse it. The main function is the main entry point to the application, and you can provide arguments to the main function as you do with other functions. Errors are also treated as types that you can send and return from a function. You can generate your own errors to return when an error occurs and handle errors when they should be handled.

In this chapter, we learned about the basic functionalities provided by the Ballerina language. Ballerina provides its own set of functionalities to create cloud native applications. In the next chapter, we will talk more about cloud native features, such as cloud deployments, built-in JSON, XML support, remote functions, and Ballerina services.

Questions

1. You might already be familiar with printing star patterns using different programming languages. To familiarize yourself more with the Ballerina flow control syntax, write a Ballerina program to print the following star pattern:

Figure 2.8 – Symmetric star pattern

2. Write a function to sort an integer array. For example, an array with elements [5, 3, 9, 2, 4] should be returned as [2, 3, 4, 5, 9].

    Hint: You can simply use an insertion sort with two loops to implement a sorting algorithm. The first `for` loop should loop through the whole element and the second loop should insert the elements in the correct location. You can also use a merge sort with a recursive call to build a much more efficient sorting function.

Answers

1. This star pattern can be printed by using the following three `foreach` loops:

```
int rows = 10;
foreach int i in 1..< rows {
    foreach int j in i..< rows {
        io:print(" ");
    }
    foreach int k in 1..< 2*i-1 {
        io:print("*");
    }
    io:println("");
}
```

2. An insertion sort can be implemented in Ballerina using the following code:

```
int[] values = [5, 3, 9, 2, 4];
int element = 0;
int j = 0;
foreach int i in 1..< values.length() {
    element = values[i]; j = i - 1;
    while (j >= 0 && values[j] > element) {
        values[j + 1] = values[j];
        j = j - 1;
    }
    values[j + 1] = element;
}
```

Section 2: Building Microservices with Ballerina

This second section focuses on building microservices with the Ballerina language. In this section, we will go through the architectural concepts of microservice architecture and how we can implement a real-world scenario with the Ballerina language.

First, we will discuss the cloud native features provided by the Ballerina language, including Ballerina service syntaxes, remote methods, JSON/XML data type support, and the deployment of Ballerina artifacts on microservice architecture. We will discuss Docker and Kubernetes to deploy and orchestrate a Ballerina program as containers.

Next, we will discuss inter-process communication and messaging between services in a microservice architecture. We will discuss both synchronous and asynchronous communication protocols and how we can implement those with the Ballerina language.

Finally, we will understand using databases with the Ballerina language in a microservice environment. Here, we will discuss the database-per-service design pattern and how we can implement it with the Ballerina language with event-driven architecture.

This section comprises the following chapters:

- *Chapter 3, Building Cloud Native Applications with Ballerina*
- *Chapter 4, Inter-Process Communication and Messaging*
- *Chapter 5, Accessing Data in Microservice Architecture*

3
Building Cloud Native Applications with Ballerina

Cloud native is a new concept for developing scalable, reliable, and maintainable cloud applications. There are hundreds of programming languages available out there for building cloud native applications with supported libraries. Ballerina, on the other hand, is explicitly designed to simplify the building of cloud native applications using its own syntax style and lots of different connectors.

In this chapter, we will discuss the unique features provided by Ballerina to build cloud native applications. We will focus on building microservices applications with a popular containerizing application, which is Docker. Also in this chapter, Kubernetes is introduced as the container orchestration platform that handles container deployment. We will go through each of these tools and other supportive tools and build a microservices system with Ballerina's **code-to-cloud** feature, which helps developers to deploy Ballerina applications on Docker and Kubernetes environments.

The following are the topics that will be discussed in this chapter:

- Ballerina cloud native syntaxes and features
- Containerizing applications with **Docker**
- Building a Docker image with Ballerina
- Container orchestration with **Kubernetes**

At the end of this chapter, you will understand how Ballerina helps in building a cloud native application with its pre-built libraries and components.

Technical requirements

The example used in this chapter uses Docker as the container platform and Kubernetes as the container orchestration platform. You can download Docker from the official Docker website: `https://docs.docker.com/desktop/`.

The **Kubernetes CLI** tool can be installed on your computer from the Kubernetes website: `https://kubernetes.io/docs/tasks/tools/install-kubectl/`. You can use a cloud platform such as **Amazon EKS**, **Google Kubernetes Engine** (**GKE**), or **Azure Kubernetes Service** (**AKS**) to build a Kubernetes cluster. If you need to test a Kubernetes deployment on your local computer, you can install **Minikube**. This program will create a local Kubernetes cluster so that you can test the provided examples locally. **Kustomize** is used as a Kubernetes artifact management tool that you can download and install from the `https://kubectl.docs.kubernetes.io/installation/kustomize/` website.

You can find the code files for this chapter at `https://github.com/PacktPublishing/Cloud-Native-Applications-with-Ballerina/tree/master/Chapter03`.

The Code in Action video for the chapter can be found here: `https://bit.ly/2UzIER1`

Ballerina cloud native syntaxes and features

Ballerina includes a rich collection of syntaxes and features that are specially designed for cloud native applications. Ballerina has its own way of defining services, which is the basic building unit of microservices applications. You can handle data coming in and going out with **JSON** and **XML** data types that are natively supported by the Ballerina language. In the upcoming sections, we will discuss these features in more detail with code samples. First, we'll look at the role of Ballerina in cloud native applications.

The role of Ballerina in cloud native architecture

Ballerina is an open source, general-purpose programming language built specifically to program distributed systems running as services. Ballerina offers a range of libraries and tools to build cloud native applications easily. The Ballerina language is compatible with all the factors of the **twelve-factor application**, which is essential in building cloud native applications. In the previous chapter, you learned about basic syntaxes in the Ballerina language. From this chapter onward, we will be discussing building cloud native applications with the Ballerina language.

Here are some facts behind why Ballerina can be used to build cloud native applications:

- **Open source**: Ballerina is a free, open source programming language released under the **Apache 2.0** license. Therefore, there is no vendor locking, expensive license fees, or sudden price increase. Developers have transparency over the language and there are no unforeseen limitations or unwelcome surprises. The global community is always watching and improving its capabilities.

- **Built-in container support**: You can write business logic or integration logic with Ballerina and containerize easily with built-in Docker support. Instead of writing a separate **Dockerfile**, you can write containerization instructions on the Ballerina configurations. Kubernetes artifacts can also be automatically generated with the Ballerina program. When Ballerina code is built, the related Docker and Kubernetes artifacts are generated automatically so that you can simply deploy a Ballerina program on Kubernetes.

- **Syntax**: The purpose of programming languages is not just to provide instructions for the computer. It also makes it easier for other programmers to understand the instructions given in the code. The Ballerina syntax style helps you to easily understand the instructions given. Since being cloud native heavily depends on network communication, Ballerina focuses on providing much more expressive syntax styles for network distributed systems. In the *Ballerina remote methods* section, we will discuss this syntax.

- **Messaging protocol support**: Cloud native applications are highly dependent on network communication protocols because of their distributed nature. The programming language that we are going to use to build cloud native applications should be able to support different messaging protocols and formats. Ballerina natively supports various messaging formats, such as **XML** and **JSON**. It also supports messaging protocols such as HTTP and WebSocket. You can easily manipulate XML and JSON message formats. It is easy to manipulate messages and send them through different protocols and various inbound and outbound connectors.

- **Built-in libraries**: Ballerina comes with lots of libraries that you can use to build cloud native applications. You can use Ballerina connectors to integrate applications with various cloud API systems. Connectors are available to consume different types of services, such as ERP systems, social media, storage services, and many more services besides. Community-built libraries can also be found in Ballerina Central, which you can reuse in your program.

- **DevOps support**: Ballerina provides a testing framework, build tool, and packaging system to support automated deployment. Ballerina has its own testing framework to write test cases for Ballerina programs. Developers can run automated tests before every deployment as well as build deployment artifacts with the Ballerina CLI tool. It has its own building system that generates deployable artifacts. In the automated deployment process, deployment tools can check out the code from **GitHub** and build deployable artifacts with Ballerina build tools.

Ballerina versioning follows **Semantic versioning (SemVer)** for package versioning. It has a strict versioning system for dependency management. Strict rules of dependency management generate fewer package versioning conflicts. Unlike other programming languages, you need to be specific with the version when you use a library.

> **Note**
> Semantic versioning is a versioning specification that is commonly used in versioning software releases. The version number is generated in the format of <major>.<minor>.<patch>. For example, Ballerina 1.2.13 has a major version of one, a minor version of two, and a patch version of 13.

- **Security**: Ballerina supports authentication and authorization with **JWT** and **OAuth2** protocols. Developers can build distributed applications easily by using Ballerina's security features. We will discuss security further in *Chapter 7, Securing the Ballerina Cloud Platform.*

- **Serverless Support**: Ballerina supports serverless computing on **Function as a Service (FaaS)** platforms. Maintaining larger infrastructures could be expensive for small businesses. Hence, Ballerina supports serverless functions with multiple cloud vendors such as **AWS Lambda** and **Azure Functions**.

We will go through each of the aforementioned factors throughout this book with samples. To begin developing a cloud native program with Ballerina, first, you need to create a service that is the building block of microservices applications. To start developing a cloud native application with Ballerina, we need to understand how to create simple HTTP services and how to send data over it and get a response. In the next section, we will learn how to build a simple HTTP service with the Ballerina language.

Building a Ballerina HTTP service

Cloud native applications depend heavily on network communications and HTTP is a widely used protocol for communicating between servers and clients. HTTP is the most common protocol used to implement web servers through various types of communication protocols. The HTTP protocol is simple by design, and it is the main protocol used in a web browser.

Network communication is a first-class concept in the Ballerina language. Ballerina has built-in HTTP libraries to support building HTTP services. Here, we discuss different ways of building an HTTP server that responds to the client with a payload.

Ballerina uses the concept called *service*, which is a collection of resources. Each resource has its own implementation that can be uniquely identified by a **Universal Resource Locator (URL)** address. A service exposes that collection of resources on a specified port. This will enable another service or client application to access these services and get the job done.

Check out the following example that receives an HTTP request with the service name `hello`, and the resource name `greeting`:

```
import ballerina/http;
service /hello on new http:Listener(9090) {
    resource function get greeting() returns error|string {
        return "Hello, World!";
    }
}
```

This code contains a single service and resource function. The `greeting` resource will simply send a `Hello, World!` response back to the caller. When you run this code with the Ballerina `run` command, it will start a server in port `9090` and expose the `hello` service. You can use the `curl` command as follows or navigate to the `http://localhost:9090/hello/greeting` page on the browser to test the service. It will print `Hello, World!` on the screen:

```
curl -X GET http://localhost:9090/hello/greeting
```

The following is the template of defining a resource in the Ballerina language. You can define multiple resources per service:

```
resource function <resource_HTTP_method> <resource_path> ()
returns<return_types> { }
```

The `resource` function always starts with the `resource` keyword. `resource_HTTP_method` is the HTTP method used to access this resource. This method can be `get`, `put`, `post`, and so on. `resource_path` is the context path used to access this resource. The `resource_path` content can be set to the root path by setting `resource_path` to `.` (a dot).

For example, the following resource can be accessed with the `http://localhost:9090/hello` URL, which does not have a `resource` context path:

```
resource function get .() returns error|string {
    return "The root context of hello service";
}
```

Also, you can have multiple contexts in the URL followed by the service context. For example, if you need to match the `http://localhost:9090/hello/customer/customer_details` URL pattern, then you can modify the `resource` definition as follows:

```
resource function get customer/customer_details(http:Caller
    caller, http:Request req) returns error? {
}
```

Ballerina matches the `resource` path with the incoming URL and invokes the corresponding `resource` function. You can invoke this endpoint with the following `curl` command:

```
curl -X GET http://localhost:9090/hello/customer/customer_details
```

For this example, we didn't send any data to the service. In the next section, we will discuss how to pass data to an HTTP resource with the `GET` request method.

Passing data to HTTP resources

Path parameters are usually used in transferring simple data to the server. Path parameters include data in the URL itself. The Ballerina HTTP service definition provides a way to access these path parameter values in the resource function. Check out the following example regarding the reading order ID from the incoming HTTP `GET` request:

```
resource function get 'order/[string orderId] () returns
    error|string {
      return "Order ID: " + orderId;
}
```

From the preceding code, the `resource function` name contains the URL parameters templated as the path parameter. In this example, the Ballerina service searches for the URL with the `order` string followed by another variable string. You can also see that the resource path contains a single quotation mark before the `order` string. Since Ballerina takes `order` as a keyword, you need to mark the keyword with a single quotation in order to escape the keyword. The context for this resource will be `order`, which is a resource path without the quotation mark. We can invoke this service with the following `curl` command with a `GET` request and it will respond with the order ID as a plain text response:

```
curl -X GET http://localhost:9090/hello/order/3ac327e9a8b9
```

You can have a combination of URL patterns to match resources as well. Check out the following example about reading two parameters from the URL pattern:

```
resource function get orderItem/[string orderItemId]/'order/
    [string orderId] () returns error|string {
      return "Order item ID: " + orderItemId + " in " + orderId;
}
```

The relevant `curl GET` command to invoke this service is as follows:

```
curl -X GET http://localhost:9090/hello/orderItem/3ac327e9a8b9/
order/34b3a342c3
```

This command will invoke the HTTP server and send the response back to the `orderItemId` and `orderId` parameters from the given URL. In the next section, we will learn how to use query parameters, which are also commonly used in passing data to a server.

Using query parameters on an HTTP server

Ballerina also supports the extraction of query parameters. **Query parameters** are passed along with the URL itself by appending key-value pairs at the end of the URL. Query parameters are commonly used in filtering a result by different parameters in table views on **user interfaces** (**UIs**). Check out the following example, which reads a set of query parameters from the URL:

```
resource function get orderQueryParam (http:Request req)
  returns http:InternalServerError|string {
    string? orderId = req.getQueryParamValue("orderId");
    string? customerId = req.getQueryParamValue("customerId");
    if orderId is string && customerId is string{
        return "Order ID: " + orderId + " customer ID: " +
          customerId;
    } else {
        http:InternalServerError internalError = {};
        return internalError;
    }
}
```

The resource function, `orderQueryParam`, reads the URL that has the `orderId` query parameter. The `getQueryParamValue` function is used to read the query parameter by giving the query parameter key as the argument. This function returns a value to the respective query keys. If no query value is found for the given key, it returns `nil`.

This resource can be invoked with the following `curl` command:

```
curl -X GET 'http://localhost:9090/hello/
orderQueryParam?orderId=324c324a2&customerId=433a23324'
```

Here, you can see that the end of the URL contains `orderId` as a key-value pair. You can add more query parameters to the URL by combining key-value pairs with the & symbol. Here, `customerId` is added to the end of the URL by using the & symbol. The same resource function can be implemented as follows to read query parameters as well:

```
resource function get orderQueryParam2 (string orderId, string
    customerId) returns error|string {
    return "Order ID: " + orderId + " customer ID: " +
        customerId;
}
```

Here, instead of the `getQueryParamValue` function, we can use the resource function definition to accept query parameters. `orderId` and `customerId` are taken from the resource function as input arguments.

In previous examples, we used URLs to pass data into the application. Rather than sending simple plain text data, we can use HTTP headers to pass structured data into an application. In the next section, let's discuss how to pass structured data into a Ballerina application.

Passing structured data to HTTP services

Compared to the GET request, HTTP methods such as POST and PUT are used to send various types of data over HTTP requests. POST and PUT headers can have text, **JSON** and **XML** content, images, and so on. Check out the following example, which reads JSON content from the request and converts it to the Ballerina JSON data type:

```
@http:ResourceConfig {
    consumes: ["application/json"]
}
resource function post createOrderJson(@http:Payload json
    message) returns error|json {
}
```

The `message` variable can be used in the `createOrder` function to access the input payload. You can use JSON manipulation to access the `Request` content.

You can also access XML content with the Ballerina XML type in the same way. Check out the following example, which reads an incoming XML payload:

```
@http:ResourceConfig {
    consumes: ["application/xml"]
}
resource function post createOrderXml(@http:Payload xml
    message) returns error|xml {
}
```

This resource function reads incoming XML messages and passes them to the function body with the XML message variable. Instead of using XML and JSON formats, you can also use Ballerina records to access a request payload. For example, the input payload of the incoming request is as follows:

```
{
    "customerId": "d929f59a-35a4-4153-cceb-0f432efa44c1",
    "shippingAddress": "New York"
}
```

We can create the following Ballerina record to hold this content as a record type:

```
type CreateOrder record {|
    string customerId;
    string shippingAddress;
|};
```

This record definition holds customerId and shippingAddress as string data types. Now we can use this record definition to access the incoming request content in a CreateOrder record type, given as follows:

```
@http:ResourceConfig {
    consumes: ["application/json"]
}
resource function post createOrderRecord(@http:Payload
    CreateOrder message) returns error|string {
}
```

Here, the `createOrderRecord` function reads the incoming message with the `CreateOrder` record format. You can access the `CreateOrder` record structure instead of accessing the XML and JSON formats.

So far, we have discussed building HTTP services with the Ballerina language. If you need to call an HTTP endpoint, then you need to have a Ballerina client endpoint to send a request to the target endpoint. In the next section, let's discuss how to use a Ballerina HTTP client to access an HTTP endpoint.

Invoking remote HTTP services

We have discussed how to create simple HTTP services with the Ballerina language. In some cases, the Ballerina program might need to call a remote endpoint to perform certain tasks. As microservices applications depend on other services, it is very common to access other services in order to get a job done.

This can be simply built with Ballerina by using a client endpoint. The HTTP client connector can be used to send requests to an HTTP server. Look at the following example code, which invokes a remote HTTP server with a GET request and prints the response:

```
http:Client clientEndpoint = check new
    (http://showcase.api.linx.twenty57.net);
json msg = check clientEndpoint->get("/UnixTime/
    tounix?date=now");
io:println("Response is: " + msg.toString());
```

Here, `clientEndpoint` is the endpoint definition that we are going to send the request with, via the preceding Ballerina code. The `get` method is used to invoke the service with the given resource URL. The response returned can be either data or an error message. Therefore, we need to validate the response irrespective of whether it is an error message. Here, we just print the message to the terminal output.

If you need to send a POST request instead of a GET request, you can use the `post` method. The payload can be provided as the second argument for the `post` function. For example, `clientEndpoint->post("/latest", {"base": "USD"})` can be used to send a JSON payload with POST methods. Here, `{"base": "USD"}` is the payload for the POST resource. Also, you can send custom REST methods with the `execute` function. For example, the `clientEndpoint->execute("POST", "/latest", {"POST", "base": "USD"});` statement can be used to call the same POST request to the backend server.

The JSON and XML formats are common message formats that are used to send and receive messages. As we discussed earlier, we can use both JSON and XML formats to send data to an external endpoint. Also, you can read the incoming message and transform it. In the next two sections, we will explore how to manipulate JSON and XML message formats.

Ballerina JSON support

JSON is a popular message format that is heavily used in the HTTP protocol. Since JSON is a human-readable message format, it is easy for developers to work with it. Ballerina natively supports using the JSON type with the built-in JSON data type. You can define, convert, and manipulate JSON easily in the Ballerina platform.

Look at the following example about creating a simple JSON object:

```
json customer = {
    name: "Tom",
    age: 26,
    shippingAddress: {
        street: "2307  Oak Street, Old Forge",
        city: "New York",
        country: "USA"
    }
};
```

JSON can be easily converted into a map. Once you create the JSON object, you can access its content with maps. Check out the following example that reads the name from the customer JSON object and prints it on the terminal:

```
map<json> customerMap = <map<json>> customer;
io:println("Customer name is " + customerMap["name"].
    toString());
```

The preceding example will cast the customer JSON variable to a map of JSON. Then, you can access each of its fields with indexes. Here, we can convert the incoming message to a Ballerina object since we defined the Ballerina record with the corresponding schema.

JSON fields can also be accessed by the dot operator. Check out the following example, which accesses the name field with the dot operator:

```
json customerName = check customer.name;
io:println("Customer name is " + customerName.toString());
```

You can use the dot operator to access fields within a field as well (**chained field access**). For example, you can access the country with the customer.shippingAddress. country operation. If Ballerina does not find the attribute that you are trying to access with the dot operator, this operation returns an error.

Two different JSON objects can also be merged with the mergeJson functionality provided by the JSON type. Check out the following example, which merges two different fields into a single JSON object:

```
json customerData1 = {name: "Tom"};
json customerData2 = {age: 26, shippingAddress: "1st Lane,
    New York"};
json customerData = checkpanic customerData1.
    mergeJson(customerData2);
```

The final result of customerData contains three fields defined in the JSON variables. This kind of JSON manipulation is important for building simple mediation scenarios, especially when you need to integrate two services that require a JSON value to be in a particular structure.

Other than using a JSON type to access JSON data, you can also parse the JSON variable into the Ballerina record data structure. Here, we create the following record to hold the address details:

```
type ShippingAddress record {|
    string street;
    string city;
    string country;
|};
```

Now we can use this record to create the `Customer` record, which holds the `Customer` data, given as follows:

```
type Customer record {|
    string name;
    int age;
    ShippingAddress shippingAddress;
|};
```

The `Customer` object contains `name`, `age`, and `shippingAddress` fields. JSON content can be converted to a record format with the `cloneWithType` function provided by the JSON type. Check the following code example for converting the JSON type into the record type:

```
Customer customerRecord = check customer.cloneWithType(Customer);
```

Here, we need to provide the output type similar to `Customer` in the `cloneWithType` function argument. This function clones the `customer` JSON object into the `Customer` record type. Now you can use the `customerRecord` variable to access the `customer` JSON content like how the record data type does.

Working with XML format in the Ballerina language

Just like how Ballerina supports the JSON format, Ballerina has built-in XML support as well. Messaging protocols, such as the SOAP protocol, use the XML message format to build web services. Therefore, the XML message format is common in building services that are cross communicating with web services. Check out the following definition of order with the Ballerina XML type:

```
xml orderDetail = xml '
<order orderId="342345">
<customerName customerId = "34254234">Tom</customerName>
  <shippingAddress>2307  Oak Street, Old Forge, New York,
    USA</shippingAddress>
</order>';
```

The Ballerina XML data type that you will be defining should be in accordance with the XML format. The XML variable that you are going to define should not be a half-defined XML text segment. The Ballerina XML value should be properly formatted with the correct XML syntax.

Accessing an XML value is a little different to accessing a JSON value. An XML element can have both its root elements and a child element. You can use / to access the child element on the XML data type. For example, you can use the `orderDetail/<customerName>` XML navigation expression to access the XML element with the customer's name. This will access the child XML element with the `customerName` tag.

You can use the dot operator to access attributes. For example, to access `customerId` on the `CustomerName` tag, you can use the `orderDetail/<customerName>.customerId` expression.

With Ballerina, you can use the plus (+) operator to add two XML values. Check out the following example, which defines two order items that need to be added to the order value that we have created:

```
xml orderItem1 = xml '<orderItem itemId="3244543">
    <quantity>12</quantity>
    <price>100</price>
</orderItem>';
xml orderItem2 = xml '<orderItem itemId="988443">
    <quantity>10</quantity>
    <price>320</price>
</orderItem>';
```

Both the XML results can be combined with an `orderItem1 + orderItem2` operation. This operation generates an XML object that contains an array of XML values.

Ballerina supports both XML and JSON message formats as it is widely used in communication between services. You can easily interact with message payloads and transform those messages. It is a common use case in integrating two endpoints.

In the previous two sections, we discussed JSON and XML messaging formats. When it comes to building cloud native applications, we need to focus on network communication between services. When a service becomes small, this results in many network calls. We need to handle this with care in the source code. In the next section, we will discuss the special Ballerina syntax that is used to define network calls.

Ballerina remote methods

A remote method is another syntax approach introduced by Ballerina. Cloud applications are heavily dependent on network communication. In a microservice architecture, each service needs to call other services to produce the end output. The Ballerina syntax style enforces the network communications representation in the source code. Ballerina remote methods can be used to represent a method call to another remote endpoint.

The remote method calls are represented by an arrow (->) operator. This usage of this operator can already be found with the HTTP request invocation in the previous examples. We used the endpoint->get() function to send a GET request to a remote endpoint. The arrow in the function call indicates that it is a remote method call. This representation visualizes network communication for programmers and allows them to understand the code easily.

You can define a remote method by yourself when you are developing a client library that is used to call a remote endpoint. The following Ballerina code sample creates a client object that can be used to get a currency exchange rate for a given currency type:

```
public client class StadardTimeClient {
    remote function getCurrentTime() returns string|error{
        http:Client clientEndpoint = check new (http://
            showcase.api.linx.twenty57.net);
        string msg = check clientEndpoint->get("/UnixTime/
            tounix?date=now");

        return msg;
    }
}
```

From the preceding code, Client is an object type provided by Ballerina to implement libraries that can access different services. In this example, a class with a client type named StandardTimeClient is created to read the current UTC time. This class contains a method called getCurrentTime, which returns the current time from a remote server. This method sends a request to a public API to get the current time as a JSON message. The getCurrentTime method returns this response as a string.

In this method definition, the `remote` keyword was added at the beginning of the `getCurrentTime` method. This keyword specifies that this function is a remote method that sends a request across the network. When you need to access this method, you should use an arrow operator instead of using the regular dot operator. Check out the following example about calling this method to get the currency rate for USD:

```
StadardTimeClient standardTimeClient = new();
string result = check standardTimeClient->getCurrentTime();
```

The first line, from the preceding code, creates a new object with the `StandardTimeClient` class. If you need to access the `getCurrentTime` method in that class, you need to use the arrow operator. The result of this method call is assigned to the `result` variable.

You can use remote methods, in particular when you need to create libraries to connect with remote APIs. All the methods with network API calls can be marked as remote methods, such that the code reader can easily read the code and identify network calls.

In this section, we discussed the features and syntaxes that the Ballerina language uniquely provides for developers who develop cloud native applications. We have discussed building HTTP services and passing data to services in different ways, including path parameters, query parameters, request headers, and so on. XML and JSON formats are commonly used in network communication. We discussed Ballerina XML and JSON native support as this plays a crucial part in building any web application. In the next section, we will discuss how to deploy Ballerina programs in the cloud with Ballerina's built-in cloud deployment features. First, we will discuss how to containerize a Ballerina application with Docker and then we will discuss how to orchestrate containers with Kubernetes.

Containerizing applications with Ballerina

Dependency resolution is a headache for developers in software development and deployment. You must be aware that a microservices application is not a single piece of software. The whole system is made up of hundreds of libraries and dependencies.

All these libraries and dependencies should be correctly linked together for a proper function program. Containers solve this problem by providing an isolated environment for applications. In this section, we will discuss how to use containers to develop a Ballerina application with the most famous container platform – Docker. Let's begin by introducing containers.

Introduction to containers

The first time you learned about containers in cloud native environments, I'm pretty certain the first thing that came to your mind was, *Does this have something to do with a shipping container?*. The idea of containers in a microservice architecture is pretty much the same as shipping containers.

Nowadays, there are several languages and libraries available to developers to build software applications. Each of these applications has its own dependencies that need to be resolved in order to run the application correctly. The common problem that developers end up facing when there are lots of dependencies is that the application works on their environment, but when it's deployed to another platform, the system breaks up and crashes. Well then, *why did you just send the code to the target platform? What if you send the environment and other build artifacts to the target platform along with the code?*

Similar to shipping container items, developers can collect all the code, runtime configurations, system tools, system libraries, and settings and package them into a container. Containers are nothing more than an encapsulated environment to run the programs. Containers provide an isolated environment to run a process on its hosted machine within its own filesystem image. This image is used to create a running instance of the application. This image contains all the code and libraries that are required to execute the image.

Another advantage of using containers is the ability to manage the resources allocated to containers. Since containers act much like VMs, you can set a limit for CPU and memory usage. This makes it much easier for maintainers to allocate resources for optimizing the overall performance of the system.

You might be familiar with a VM that runs on a host OS. In a microservice architecture, we mostly consider using containers rather than VMs since containers are lightweight. In the next section, we compare containers and VMs.

Containers versus VMs

Both containers and **virtual machines** (**VMs**) provide isolated environments for applications running on them. However, the biggest downside of using a VM is that it takes a long time to start, and it drains the host machine's performance.

VMs provide **operating system** (**OS**) virtualization by using a **hypervisor**, which separates the resources of the host machine. However, containers use **container engines** to provide an OS-level abstraction to the container runtime. Containers use OS functionalities instead of providing virtual hardware in VMs. Check out the following diagram, which compares the architectures of containers and VMs:

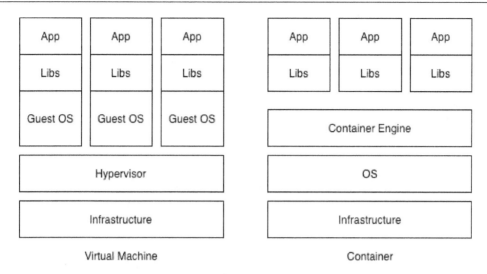

Figure 3.1 – VM versus container architecture

Infrastructure is the physical computer that contains the hardware. VMs contain a hypervisor to virtualize infrastructure for the VM's OS. On top of the VM guest OS, you can run apps with the required library dependencies. On the other hand, containers are running on top of the host OS with the container engine. It uses a host OS rather than a hypervisor to access resources.

Containers are much smaller in size than VMs and depend on the OS rather than providing separate OS functionality. Sharing OS resources makes containers smaller in size. Therefore, containers use a smaller amount of space to keep and run the images. They also have faster start up times compared to VMs since they consume OS resources. This is an important factor in microservice architecture since services can die at any time and start at any time. The following table compares different aspects of containers and VMs. You can decide in which scenarios you would use one or the other:

Container	VM
It can start in a matter of seconds.	It takes a long time to start.
It shares the host's OS.	It has its own OS.
It has a higher performance.	It has limited performance.
It has a reduced level of security due to only process-level isolation.	It has a higher level of security since it is fully isolated from the host machine.

Table 3.1 – Container versus VM comparison

There are multiple container technologies available, including **Docker**, **CoreOS RKT**, **Mesos**, and **LXC Linux**. However, the most popular choice within the community is Docker. As its name suggests, a docker is a person employed in a port who loads and unloads ship containers. In a container environment, Docker does the same by creating containers that allow it to run services in an isolated environment.

Containerizing applications with Docker

In the same way as VMs, container images start to build with a base image. For example, the Ballerina Docker base image is **Alpine Linux**. This is a lightweight Linux-based OS that was developed especially to be used in a container environment. You can copy all your application dependencies to the container and build a new container that contains all the dependent libraries and start up instructions.

This newly created image can be versioned in order to simplify the deployment process so that anyone can easily switch between the different versions of the image. This works similar to the Git versioning system, where it keeps a snapshot of the current work along with a tag that can be later used to pull a specific version.

Docker Registry provides a way to keep all these containers in a remote repository, the same as the **Version Controlling** system. After a new Docker image is created, it can be pushed to a remote repository to persist as an image. When you need to deploy it, you can pull it to any deployment platform. Docker itself includes a Docker registry called the **Docker Hub**, a place where you can keep all the Docker containers and pull them to deployment. This makes the deployment process so much easier in a container environment since it pretty much works the same as Git.

Docker provides **Dockerfile**, which contains all the instructions that are required to build the image. Starting with the base image, you can specify the running instructions, copy application dependencies to the Docker image, and build it. Ballerina makes this process easy with built-in Docker support to create a Docker image.

To create a Ballerina Docker image, you can use Ballerina's built-in Docker image that already contains the Ballerina runtime. You can simply use this built-in image and include the Ballerina executable in it. You can do this without using any code or command. The default Ballerina Docker artifact can be generated by Ballerina itself. In the next section, we will discuss how to create a Docker image with Ballerina artifacts and how to customize it.

Building a Docker image with Ballerina

Building a Docker image with the Ballerina platform is easy with the **code-to-cloud** feature. This feature lets you create a Docker image from the Ballerina code by inferring some details from the source code as well. Check out the following example of a hello world service that responds with plain text content:

```
import ballerina/http;
service /hello on new http:Listener(9090) {
    resource function get greeting() returns error|string {
        return "Hello, World!";
    }
}
```

To generate Docker artifacts, we need to define a container image, tags, and a repository. This can be configured by using the `Cloud.toml` file in the Ballerina project's home directory. Go to the project home page and create a new toml file named `Cloud.toml`. Add the following content to that file to define the `repository`, `name`, and `tag`:

```
[container.image]
repository="dhanushka"
name="code_to_cloud"
tag="v0.1.0"
```

You can build this project with the `bal build --cloud=docker` command. This command generates a Docker image inside your local Docker repository. Once you execute the `build` command and check the list of Docker images with the `docker images` command, you can ascertain that the Ballerina `build` command generates a Docker image with the package name shown as follows:

REPOSITORY	TAG	IMAGE ID	CREATED	SIZE
dhanushka/code_to_cloud	v0.1.0	285cb053b5db	4 seconds ago	278MB

Ballerina generates this Docker image by creating a Dockerfile, which contains all the instructions that are needed to generate the Docker image. This Dockerfile is generated in the `<project_home>/target/docker/<package_name>/Dockerfile` directory when you run the Ballerina `build` command. The sample Dockerfile contains the following instructions regarding Docker deployment that are used to generate the Ballerina Docker image (please note, the list of library copying instructions has been removed from the following content to make this code segment more readable):

```
FROM ballerina/jre11:v1
LABEL maintainer="dev@ballerina.io"


COPY snakeyaml-1.26.0.wso2v1.jar /home/ballerina/jars/
COPY netty-transport-4.1.63.Final.jar /home/ballerina/jars/
RUN addgroup troupe \
    && adduser -S -s /bin/bash -g 'ballerina' -G troupe -D
      ballerina \
    && apk add --update --no-cache bash \
    && chown -R ballerina:troupe /usr/bin/java \
    && rm -rf /var/cache/apk/*
WORKDIR /home/ballerina
COPY code_to_cloud.jar /home/ballerina
EXPOSE   9090
USER ballerina
CMD java -Xdiag -cp "code_to_cloud.jar:jars/*" 'dhanushka/code_
   to_cloud/0_1_0/$_init'
```

As you can see, the first line of the Dockerfile starts with the FROM keyword, which specifies the base Docker image that we are going to use. Here, it is specified as the `ballerina/jre11:v1` Docker image. This Docker image contains the Ballerina runtime that is based on **JRE 11**.

The next line gives a `dev@ballerina.io` label with the LABEL keyword to the Docker image as the maintainer. The next few lines of instruction contain the COPY commands that copy Ballerina-dependent `.jar` files to the Ballerina dependency directory. All the required dependencies for your project will be automatically copied to the Ballerina dependency directory.

After that, the RUN instructions create the environment for Ballerina to execute. This creates a new group with a new user. Ballerina uses **Docker Alpine Linux** as the base image to build the Ballerina runtime. Since the Alpine image does not contain Bash, here, it sets Bash by means of the apk add and rm -rf commands. Finally, it executes the program with the CMD instructions with **Java** by setting up the classpaths.

We can start this Docker image with the following Docker command:

```
docker run -d -p 9090:9090 -it <docker_image_name>:<docker_tag>
```

For this example, the command would be docker run -d -p 9090:9090 -it dhanushka/code_to_cloud:v0.1.0. This command starts a Docker image with port mapping 9090. You can access the Ballerina program running on a Docker container with the localhost:9090 URL. In this example, you can execute the following GET request to invoke the greeting resource function with the curl command-line tool:

```
curl -X GET http://localhost:9090/hello/greeting
```

You can check all the running Docker containers with the docker ps command. There, you can find the Docker container you just started. This command will show you a list of Docker images that are currently running on your computer. This list contains the details of port mapping, status, and image details.

Other than these given Docker configurations, we can copy files into the Docker container as well. In some scenarios, we might need to include configuration files, libraries, and static content files in the Docker container. In those cases, we need to include those files in the Docker container while building the Docker image. Ballerina's file copying functionality provides a way to add files to the Docker file. Let's understand this by building a simple Ballerina program that includes a static HTML file in the Docker image and responds to it when invoked. The content of the Ballerina code to serve the HTML file is as follows:

```
service /hello on new http:Listener(9090){
   resource function get page(http:Caller caller,http:Request
      request) returns error? {
      http:Response response = new;
      response.setFileAsPayload("/home/ballerina/index.html",
         contentType = "text/html");
      check caller->respond(response);
   }
}
```

This Ballerina service reads a file from `/home/ballerina/index.html` and serves it as a text/HTML content type. The `setFileAsPayload` function lets us serve file content to the client directly. Since it should serve as a web page, we need to mark `contentType` as `text/html`.

Now we need to define the copy instruction to copy the `index.html` file to the Docker image. We can define these instructions in the `Cloud.toml` file as follows:

```
[[container.copy.files]]
sourceFile="index.html"
target="/home/ballerina/index.html"
```

Here, `sourceFile` is defined as the `index.html` file location. Here we have given the location relative to the program execution directory. `target` is the target path where the file should be copied into the Docker container. You can build the project to generate Docker artifacts with the `bal build --cloud=docker` command.

If you check the Dockerfile in the `target` directory, you can find the copy file instructions as `COPY index.html /home/ballerina/index.html`. All the changes that we have done in relation to generating a Docker image on the Ballerina project are reflected in this Dockerfile. You can run this Docker image and invoke this endpoint from a web browser. When you invoke it, you will get an HTML page that contains the `index.html` file's content.

Ballerina runtime containers can be used to create Ballerina Docker images that contain all the required Ballerina runtimes. Ballerina provides a Ballerina compiler Docker image that can be used to compile Ballerina source code. The Ballerina development Docker image contains a Ballerina compiler and other necessary CLI commands that can be used to develop the Ballerina program. In the next section, let's discuss how to use the Ballerina development Docker image.

Using Ballerina's development Docker image

Other than the Ballerina runtime, the Ballerina team also provides you with a Dockerized Ballerina image that you can use to build and test a Ballerina application. This development Docker image contains the Ballerina CLI tool with the compiler that you can use to build a Ballerina program. Ballerina's development Docker image is highly important when it comes to building a Ballerina deployment pipeline. This Docker image can be used with deployment pipelines such as Jenkins and Wercker to build and test Ballerina programs. Since the Docker image contains all the dependencies, you can use this image directly to build the Ballerina program and perform tests.

This Ballerina development Docker image can be pulled into your development environment with the following Docker command:

```
docker pull ballerina/ballerina
```

This command pulls the latest Ballerina development Docker image to your local computer. With the `docker images` command, you can make sure that the Docker image is listed locally. To start development with Ballerina in the Docker image, first, you need to create a directory to keep the Ballerina project. Then, create a directory on your host computer. Once you have created the project directory, execute the following command to start the Ballerina shell on the Docker container:

```
docker run -v <path_to_host_directory>:/home/ballerina -it
ballerina/ballerina:latest
```

You need to replace `path_to_host_directory` with the directory path you have just created. This command will start a Ballerina container on your terminal where you can use the Ballerina CLI commands.

For example, if you need to create a new Ballerina project, first you need to change the current directory to the mounted path, which is `/home/ballerina`. For this example, it would be `cd /home/ballerina`. You can now create a new Ballerina project by running the `bal new <project_name>` command. You can also use the `build` and `run` commands in the same way. However, if you are building an HTTP server, then remember to expose the relevant ports when you are running the program from the Docker image.

Instead of executing a command on the Docker image using the terminal, you can issue commands directly to run Ballerina CLI commands on the Docker containers. For example, the following command creates a new project in the `project` directory that you have just created:

```
docker run -v <path_to_host_directory>:/home/ballerina -it
ballerina/ballerina:latest bal build <project_name>/
```

Replace `path_to_host_directory` with the new directory you have created and `project_name` with the Ballerina project name. This command builds the Ballerina project. This command can be used to build deployment pipelines. A simple `build` command will build the source code and make sure that it has no syntax issues and test failures. With the same structure, you can issue test commands that allow you to run test cases. The only change you need to make to this command is to replace the `build` argument with `test`. You can do this for all Ballerina CLI commands as well.

Containerizing applications is an important step in building a microservices system. This will reduce the complexity of the deployment as it hides complex dependency management from developers. Ballerina provides a code-to-cloud feature that enables you to easily convert a Ballerina service into a Docker image. In this section, we discussed how to create a Ballerina Docker image and publish it on the Docker Hub, which you can pull whenever you need it.

A single service can be represented by a single container that executes the service. Handling multiple containers is the next challenge that we need to address. Microservices applications can be formed with thousands of containers. We need a proper container orchestration framework to handle all of these containers. Let's discuss Kubernetes, which is the most popular container orchestration system, in the next section.

Container orchestration with Kubernetes

Containers provide a lightweight solution to deploy applications that bundle all their required resources on them. There can be hundreds of containers that are running on a system. Now, we have a set of problems that we need to resolve, such as maintaining the number of containers running for each service, building connections between different containers, and maintaining security. This is where the container orchestration platforms come in.

Container orchestration platforms are capable of handling and maintaining containers according to the rules and policies specified by the system engineers. There are various tools available for container orchestration. Kubernetes is the most popular container orchestration platform provided by **Google** to build cloud native applications. In this section, we will learn how Kubernetes works by generating Kubernetes artifacts from the Ballerina compiler, perform a health check on the Kubernetes cluster, and generate custom Kubernetes artifacts.

Introduction to Kubernetes

Kubernetes is widely used as a platform for container orchestration. Kubernetes provides a rich set of features for handling containers with simple CLI commands. **Kubectl** commands provide an elegant way of handling Kubernetes operations, such as creating, updating, and maintaining Kubernetes clusters through the Kubernetes API. The name Kubernetes derives from Greek and means *helmsman*. Like a ship's helmsman, Kubernetes drives containers on the correct path.

A Kubernetes cluster can be run on multiple nodes connected to a network. Each of these nodes contains a program called a **kubelet**, which handles all running containers. A kubelet is responsible for handling the number of containers, which images should be running, and so on. The containers are running in a component called a **Pod**.

Pods are the atomic units, or the smallest *unit of work*, in the Kubernetes cluster. There can be one or more containers running on a single Pod. Kubernetes is designed to run a single container in a single Pod. In some situations, such as sidecar proxies, bridges, and adaptors, you can have multiple containers running on a single Pod. The overall Kubernetes architecture can be seen in the following diagram:

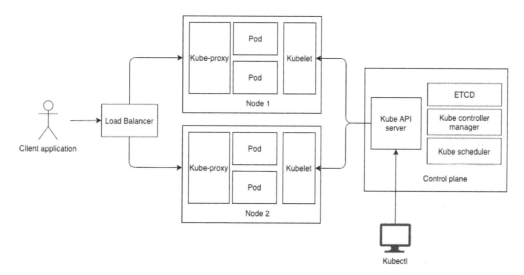

Figure 3.2 – Kubernetes architecture diagram

As you can see, the control planes of Kubernetes controlling all the nodes are connected to the cluster with a program called the **Kubelet**. The control plane contains the **Kube API server**, which exposes Kubernetes functionalities to the developers as an API endpoint. The **Kubectl** CLI client is a client program that uses the Kube API endpoint to configure the Kubernetes cluster. You can use both the CLI tool and the API endpoint to configure a Kubernetes cluster.

A Kubernetes cluster also has a distributed storage built with ETCD. The name **ETCD** is formed from the `etc` directory on the Linux OS to hold configurations. The **D** in ETCD stands for **distributed**. Therefore, ETCD is distributed configuration filesystem storage. This storage service is used to store key-value pairs with high data consistency. The Control Manager in the Kubernetes Control Plane is responsible for maintaining the state of the system. The Control Manager includes a replication controller, Pod controller, service controller, and endpoint controllers. The scheduler, on the other hand, handles resource management in the Kubernetes cluster. It handles resource requirements, data locality, and hardware/software constraints.

The **Kube-proxy** service runs on all nodes to handle request routing on the cluster. It handles both interservice communication and outside traffic. A Kubernetes Pod has its own IP address to connect with. Once the application needs to communicate with another service, it can use this IP address. However, Kubernetes Pods are not designed to persist for a long time and can die at any time. Therefore, accessing a Pod with an IP address is not a good solution since these addresses are changing rapidly.

Kubernetes introduced the concept of service to solve this problem. A service represents a set of Pods that has the same program running. Instead of accessing individual Pods, services give a way to access the same service running on different Pods with a service IP address or DNS name.

Kubernetes artifacts can be defined in the `.yaml` file format. It is easy to use, understand, and read. The artifacts can be run with the `kubectl apply -f <artifact_file_location>` command with the Kubectl CLI. To run this command, first, you need to set up a Kubernetes cluster and configure Kubectl to access that cluster. You can use **minikube**, which is a single-node cluster design, for testing purposes.

Another important concept that you should be aware of in Kubernetes is namespaces. A namespace provides a way to partition an existing Kubernetes cluster into multiple environments. For example, the same Kubernetes cluster can be used to deploy development and QA environments. Development and QA teams can have two separate testing environments that have the same Kubernetes cluster separated by two different namespaces. Another advantage of having a namespace is that you can allocate resources for each of the namespaces. While moving on to Kubernetes with Service Mesh, you will learn that Service Mesh implementations, such as **Istio** and **Consul**, use separate namespaces to attach sidecar proxies to each of the services.

Kubernetes is widely used in building cloud native applications in microservice architecture. Kubernetes stateless Pod architecture is strongly compatible with microservice architecture and provides an easy CLI and `.yaml` file configuration model to interact with the cluster. In the next section, we will discuss how to use the Kubernetes cluster with the Ballerina program with the help of examples.

Generating Kubernetes artifacts with Ballerina

Ballerina's **C2C** feature lets you work easily with Kubernetes deployment. As you learned in the previous section, we can use the C2C feature to automatically create a Docker image by using a Ballerina program. C2C can identify the exposed port and create a Docker image with the package name.

Similar to creating Docker artifacts, Ballerina's C2C feature is capable of generating Kubernetes artifacts as well. These artifacts can be found in the `<project_home>/target/kubernetes/<package_name>/<package_name>.yaml` file. In this file, you can find the default `.yaml` file, which can be used directly to perform a Ballerina deployment. This file contains Kubernetes instructions to deploy a Ballerina program with given default configurations.

If you need to override these values, you can create a `Cloud.toml` file in the project's home directory and change the default parameters. In the same way as the previous Docker example, we need to define the Docker image details in the `Cloud.toml` file. We will go through each of these default parameters and use the `Cloud.toml` file to override the default values. This file is not available by default, and you need to create this file in the project's home directory. This file allows you to configure image details, tags, and base image details as follows:

```
[container.image]
name = "code_to_cloud"
repository = "dhanushka"
tag = "v0.1.0"
base = "ballerina/jre11:v1"
```

The `name` and `repository` fields are used to identify the image name and the repository where you keep the image. If you didn't specify these values, then Ballerina adds the package name as the image name and the repository is left empty. The `tag` field specifies the version of the Docker image. Here, we set the version as `v0.1.0`. The `base` field can be configured to have a custom base image. The Ballerina runtime uses `ballerina/jre11:v1` as the base image. In case you need to have a custom Ballerina runtime, then you can specify it here.

You can build this project with the `bal build --cloud=k8s` command. Once you create the image, Ballerina generates a Docker image with these configurations. If you check the list of Docker images, you can find the Docker image created with the given name, and the repository with the tag ID. Also, the corresponding Kubernetes YAML file deployment details change according to these values.

If you check the Kubernetes YAML file in the `target/kubernetes` directory, you will find the `spec.template.spec.containers.images` attribute containing the image name along with the repository name and the tag as `dhanushka/code_to_cloud:v0.1.0`.

You can execute this Kubernetes file with the `kubectl apply` command to apply these changes to the Kubernetes environment. For this example, the following command can be used to execute Kubernetes YAML files:

```
kubectl apply -f target/kubernetes/code_to_cloud
```

If you check the number of Pods with the `kubectl get pods` command, it lists the Pod that we have just created, shown as follows:

NAME	READY	STATUS	RESTARTS	AGE
code-to-cloud-deployment-64cmn	1/1	Running	0	1h

If the status of the Pod is shown as `Running`, this means the service is up and running. You can also list the service with the `kubectl get services` command. The output of this command is shown as follows:

NAME	TYPE	CLUSTER-IP	EXTERNAL-IP	PORT(S)	AGE
code-to-cloud-s	ClusterIP	10.104.199.29	<none>	9090/TCP	1h
kubernetes	ClusterIP	10.96.0.1	<none>	443/TCP	1h

This output represents the services that we are running and we can find the services that we have just created on the list.

The `Hello world` service can be accessed through your web browser by entering the IP address of the Kubernetes cluster's NodePort IP address. If you are running these samples with minikube, you need to link the host machine with minikube to access the Ballerina service. We'll need to execute the `minikube service --url` command with the service name to generate the URL for the Ballerina service.

> **Important note**
> If you are trying these samples with minikube, make sure to attach the Docker container image to the minikube's VM. Simply, you can run the `eval $(minikube docker-env)` command to point the local Docker environment to minikube. Then, you can execute the Ballerina `build` command to push the Ballerina Docker image to the minikube Docker repository.

For this example, the command would be as follows:

```
minikube service --url code-to-cloud-s
```

This will give you a URL with a port and you can access the Ballerina service by using it. Then, you can send a GET request similar to running the Ballerina program locally. We will discuss how to expose the Kubernetes port outside the cluster with NodePort in the *Using Kustomize to modify Ballerina Kubernetes configurations* section.

Once we have deployed the Ballerina service on the Kubernetes cluster, then we need to manage the cluster. Scaling the deployment is the main feature provided by the Kubernetes cluster with which you can scale the system up and down. In the next section, let's discuss how to utilize the Kubernetes cluster for Ballerina applications.

Utilizing Kubernetes resources

You can apply resource limitations to container environments with Kubernetes. By using these configurations, you can specify memory and CPU resource allocation for each Pod. You can have maximum and minimum resource allocation for memory and CPU resources for each Kubernetes deployment. If you select a minimum resource limitation on a Pod, then Kubernetes makes sure this deployment can only be configured in nodes that can provide minimum resource allocation as requested by the configuration. On the other hand, Kubernetes can apply maximum resource allocation on the Pod as well.

Resource utilization can be achieved by limiting CPU resources and memory allocation. CPU resources are measured by a unit called **millicores**. A single CPU core is equal to 1,000 millicores, and two CPU cores can be represented as 2,000 millicores. In the same way, you can specify half of a CPU core by 500 millicores.

> Tip
> For ease of reading, millicores is written as m.

The memory limit is measured by the number of bytes allocated to a container. You can use **Kilobytes (K)**, **Megabytes (M)**, **Gigabytes (G)**, **Terabytes (T)**, **Petabytes (P)**, and **Exabytes (E)** to represent the number of bytes allocated. You can also represent these values with bibytes, such as **Kibibytes (Ki)**, **Mibibytes (Mi)**, **Gibibytes (Gi)**, **Tebibytes (Ti)**, **Pebibytes (Pi)**, and **Exbibytes (Ei)**.

With Kubernetes, you can perform resource allocation at the container and namespace levels. The Ballerina configuration lets you define a container-level resource allocation for each Ballerina program deployment. Check out the following example of a `Cloud.toml` configuration file that configures container resource allocation:

```
[cloud.deployment]
min_memory="100Mi"
min_cpu="1000m"
max_memory="256Mi"
max_cpu="1500m"
```

The `min_memory` and `min_cpu` fields are used to define the minimum resource requirements to schedule a Pod in a given node. If the Pod does not find sufficient `min_memory` and `min_cpu` resources in a particular node, then Kubernetes does not allocate this deployment to that node.

On other hand, the `max_memory` and `min_cpu` parameters are used to define the maximum resource usage for a given Pod. The Pod should not exceed the `max_memory` amount of memory and `max_cpu` amount of CPU power. Altogether, these parameters act as a maximum and minimum resource limit for a given deployment.

Once you build the Ballerina program, you will get the following **YAML** file content for the `spec.template.spec.containers.resources` YAML section:

```
resources:
  requests:
    memory: "100Mi"
    cpu: "1000m"
  limits:
    memory: "256Mi"
    cpu: "1500m"
```

The **TOML** configurations that we have defined in the `Cloud.toml` file are mapped into the Ballerina compiler-generated Kubernetes YAML artifact file as shown in the aforementioned code. Here, `resources.requests.memory` is mapped to the `min_memory` field, and `resources.requests.cpu` is mapped to the `min_cpu` field. Similarly, the `max_memory` and `max_cpu` fields are mapped with the `resources.limits.memory` and `resources.limits.cpu` fields, respectively.

By applying these configurations, you can set a resource limitation for a given deployment over a Kubernetes cluster.

Ballerina also provides autoscaling support to handle the number of replicas for each deployment. This feature is known as **Horizontal Pod Autoscaler (HPA)**. HPA is capable of scaling the number of Pods based on computation requirements.

The following `Cloud.toml` configuration example specifies that the minimum number of Pods to replicate is 2, while the maximum replicated Pods is 6. This configuration makes sure that we have between two and six running Pods. You can define the resource threshold as well to generate a new Pod. The following TOML configuration example specifies that we have a CPU threshold of 60% and a memory threshold of 80%:

```
[cloud.deployment.autoscaling]
min_replicas=2
max_replicas=6
cpu=60
memory=80
```

If the Pod exceeds the CPU threshold of 60% or the memory threshold of 80%, it spawns a new Pod. However, it does not create a Pod if the total number of Pods is six. These configurations can also be found in the target YAML file as follows:

```
spec:
  maxReplicas: 6
  metrics:
  - resource:
      name: "cpu"
      target:
        averageUtilization: 60
        type: "Utilization"
    type: "Resource"
  minReplicas: 2
  scaleTargetRef:
    apiVersion: "apps/v1"
    kind: "Deployment"
    name: " resource-utiliz-deployment"
```

In this YAML file, you can see that the minimum number of replicas is set to 2 with the spec.minReplicas parameter, and that the maximum number of replicas is set to 6 with the spec.maxReplicas parameter. You can check the number of replicas deployed by means of the kubectl get deployments command. This will print the two Pods that are deployed for this Ballerina service under the READY column as follows. 2/2 means that two Pods are available and that they are both up and running:

NAME	READY	UP-TO-DATE	AVAILABLE	AGE
resource-utiliz-deployment	2/2	2	2	65s

Now we know how to build a Kubernetes cluster and manage it to scale based on system requirements. In some scenarios, you may need to provide dynamic configurations to applications that are running on a Kubernetes cluster. In the next section, let's discuss how to pass configurations into the Kubernetes cluster by using config maps.

Using config maps in the Kubernetes cluster

Kubernetes uses config maps to store non-confidential data, such as key-value pairs. Config maps are not designed to hold larger data chunks. The maximum amount of data you can store on config maps is 1 MiB. A key benefit of using config maps is that you can separate configuration data from the source code and read it when necessary. For example, if your application needs to connect to another service with an HTTP call that might be its URL, then you can have a config map to store the connection data and use that URL when you need to connect to it.

There are two ways in which you can use config map data from a Pod. One way is to use the volume mount, and the other way is to use the environment variables. To start using config maps, first, you need to create a file that contains all of your configurations. For this example, we will create the following file containing some key-value pairs in the <project_home>/resources/Config.toml file:

```
key1="value1"
key2="value2"
```

Now, we can instruct Ballerina to generate Kubernetes config maps artifacts with this configuration file. This configuration is defined in the [[cloud.config.files]] section in the Cloud.toml file, which should be created in the project's home directory. The TOML configurations for this are as follows:

```
[[cloud.config.files]]
file = "resources/Config.toml"
```

Since this is an array of files, you can define multiple config files as well. Here, we have only defined a single TOML file that contains all the required configurations. Optionally, you can define a mount path to where these configurations should be saved in the Docker container. This can be configured using mount_path.

By building this project, you can generate the Kubernetes artifact to deploy the service. The Kubernetes configuration of the YAML file in the project's target directory will be updated as follows to set the config maps in the Kubernetes cluster:

```
apiVersion: "v1"
kind: "ConfigMap"
metadata:
  name: "config-manageme-ballerina-conf-config-map"
data:
  Config.toml: "key1=\"value1\" \nkey2=\"value2\" "
```

This YAML definition contains the config map configurations. In the data section, you can see all the configurations defined in the Config.toml file, which is also available here as a single text value. As described earlier, these config map configurations are given to the Pod by the volume mount. In the specs.template.spec.containers[0].volumes section, you can find the config map used as a volume:

```
volumes:
    - configMap:
      name: "config-manageme-ballerina-conf-config-map"
      name: "config-manageme-ballerina-conf-config-map-volume"
```

Then, it also mounts this volume in the specs.template.spec.containers[0].volumeMounts YAML section with the mount path where Cloud.toml file by default stored into

```
volumeMounts:
    - mountPath: "/home/ballerina/conf/"
      name: "config-manageme-ballerina-conf-config-map-volume"
      readOnly: false
```

Next, we can focus on creating the Ballerina service to read this configuration. In the Ballerina program, you can directly refer to the configurations that we have defined in the `Config.toml` file with a configurable string `key1 = "default"` statement. Here, we are reading the `key1` value directly from the `Config.toml` file that was copied into the container. Now we can use this `key1` variable to access the `Config.toml` content. If the Ballerina program does not find the configuration, it uses the assigned value of default as the `default` value. You can use `bal build --cloud=k8s` to build this project and deploy it with `kubectl` commands.

Configuring a Kubernetes health check for Ballerina services

While a Ballerina service is running, it may go to an unexpected status where the program itself is unable to recover. In this situation, the program might keep on running without performing productive tasks. Therefore, we need to identify those instances and stop those services so that Kubernetes can generate another Pod to replace the erroneous Pod. To achieve this, Kubernetes provides a **liveness probe** to check the health of the Pod.

The Kubernetes liveness check can be easily implemented with Ballerina by exposing a Ballerina health check endpoint. Kubernetes regularly calls this endpoint to check the status of the Ballerina service. If the service does not send back a response, then Kubernetes will kill this particular Pod and create another one. The following `Cloud.toml` configuration can be used to define the liveness health check logic:

```
[cloud.deployment.probes.liveness]
port = 9090
path = "/hello/liveness"
```

Here, we specify the port as `9090` and the path to the service as `hello/liveness`. In your `hello` service, you need to add a new resource with the `liveness` path to reply to some messages for indicating that this service is working fine. For this example, we consider the service name as `hello`, which has a context URL, `/hello`. If Kubernetes is able to read the response from this resource, it will mark this Pod as a healthy Pod. Check out the following example resource function that responds to a payload for a liveness check inside the `hello` service:

```
resource function get liveness() returns error|string {
    return "pong";
}
```

This resource function goes inside your service definition that responds to a payload if the service is up and running. If the service does not respond, Kubernetes will stop this Pod and generate a new one based on the Kubernetes scaling policies.

There may be certain situations where you need to keep your application alive, but make sure that it doesn't receive any traffic until certain conditions have been met. Therefore, readiness defines when to expose the service to the end user. Check out the following example, where a Pod is waiting until a backend service is available:

```
[cloud.deployment.probes.readiness]
port = 9090
path = "/hello/readiness"
```

Here, we set the port to `9090` and the resource path to `//hello/readiness`. The resource we specified is used to check the availability of another backend endpoint by sending a `GET` request. If this service is unable to find the address in the given URL, it returns an error status code. Check out the following example `resource` definition, which sends a health check of another service:

```
resource function get readiness() returns error|string {
    http:Client clientEndpoint = check new ("http://
        localhost:9091");
    string response = check clientEndpoint->get("/backend/
        health_check");
    return "pong";
}
```

This service can be used to check the availability of another service, database, or anything that is required to start before starting a given service. If Kubernetes does not return a 200 status code response, then Kubernetes marks this Pod as not ready and does not accept any traffic.

Readiness and liveness can be implemented by running the `curl` command on the given Pod as well. In this way, instead of sending a request to another server, you can execute a shell script on the Pod to check the readiness and liveness of that particular Pod. Ballerina only supports the use of an HTTP request to check the readiness and liveness. However, you can modify default Kubernetes artifacts to execute a shell script to do the readiness and liveness check. In context, readiness is used to define when your service is ready to handle traffic and liveness is used to define when to terminate the Pod in the event of Pod failure.

So far, we have discussed the Kubernetes features provided by the Ballerina language. But what if you have a custom modification that you need to apply to autogenerated Kubernetes artifacts. In the next section, we will discuss how to combine your custom Kubernetes artifacts with Ballerina-generated YAML artifacts.

Using Kustomize to modify Ballerina Kubernetes configurations

Ballerina natively supports a limited number of Kubernetes functionalities that are essential configurations to be deployed in Kubernetes. Kubernetes contains different configurations, features, and policies that you can set up in the Kubernetes cluster. Since you already created the deployment artifacts with Ballerina, we need to find a way to combine Ballerina autogenerated configurations with your custom configurations. You can use **Kustomize**, which is a tool provided by Kubernetes, to merge two configurations into a single deployment file.

To understand how Kustomize works, let's add `NodePort` for a simple Ballerina `hello world` service. We will use `NodePort` to expose the Ballerina service on the outside. For this example, you can create a simple `hello world` service that returns a text message to the caller. We will name this project `kustomize_sample`.

You can add the `Cloud.toml` file to the project's home folder if you have any Kubernetes configurations such as config maps and replication to add. Then, you can build the project to generate the Docker image and the Kubernetes artifact files.

Now, we need to update the generated Kubernetes artifact file to have `NodePort` for the service we created. You can use Kustomize to merge these generated artifacts with the artifacts that we need to configure. We can create a file in the `<project_home>/resources/deployment.yaml` file to hold the `NodePort` YAML artifacts. The content of this file is as follows:

```yaml
apiVersion: "v1"
kind: "Service"
metadata:
  labels:
    app: "kustomize_sample"
    name: "kustomize-sampl"
spec:
  type: NodePort
  ports:
    - port: 9090
      nodePort: 32269
```

In this configuration, we defined a service with the label information derived from the Ballerina-generated Kubernetes artifacts. In the `spec` section in the YAML file, we set `type` as `NodePort`. The `port` field represents the internal port number for the Kubernetes cluster, which is `9090`, and `nodePort` is the port number that is exposed to the other endpoint outside the Kubernetes cluster. This instructs Kubernetes to expose `nodePort` outside the cluster with port `32269`.

The next step is to combine the `deployment.yaml` file's content with the Ballerina-generated artifacts. To do this, you need to create another file in the project's home directory as `kustomization.yaml`. This file contains information, such as what are the Kubernetes YAML files you need to merge and how they need to be merged. Check the following `kustomization.yaml` file:

```
resources:
- target/kubernetes/kustomize_sample/kustomize_sample.yaml
patchesStrategicMerge:
- resources/deployment.yaml
```

In this configuration file, `resources` is the Ballerina-generated file that contains the default Kubernetes artifacts. `patchesStrategicMerge` is the file location of the file that contains changes that we need to apply to the generated Ballerina Kubernetes artifacts. Here, we can point to multiple files for both `resources` and `patchesStrategicMerge` sections.

Now, we can generate new artifacts by running the `kustomize build` command on the project's home directory. This will print the modified Kubernetes `.yaml` file in the terminal window, which contains `NodePort` configurations. Instead of generating a result on the terminal, you can deploy those changes directly to the Kubernetes cluster by running the following command:

```
kustomize build | kubectl apply -f -
```

This command generates new artifacts that contain a combination of autogenerated Ballerina Kubernetes artifacts and your custom Kubernetes artifacts. The `kubectl apply` command is also executed as a pipe and it applies new Kubernetes artifacts to the Kubernetes cluster.

With Kustomize, you can combine Ballerina-generated artifacts with custom Kubernetes artifacts. You can easily add additional configurations to the Kubernetes cluster using this tool. We will use this tool in the next chapter to modify Ballerina-generated Kubernetes artifacts.

Summary

In this chapter, we discussed the features provided by the Ballerina programming language to simplify the cloud application development process. Ballerina's syntax style and built-in libraries provide an easy way of building Agile cloud applications. We discussed how Ballerina provides a service definition for HTTP services with its own syntax style. Also, we discussed the use of remote functions and message formats such as JSON and XML. We discussed how to create and manipulate JSON and XML data formats with multiple examples.

To simplify the deployment process, we discussed Docker, which is the most popular container platform. In this chapter, we discussed what a container is and how it differs from a VM. Docker Hub provides a central location for storing Docker images that we can use to deploy in production when needed. Ballerina provides a `Cloud.toml` file that we can use to build custom Docker images and Kubernetes deployment artifacts.

We also introduced Kubernetes, the most famous container orchestration platform, to handle containers in the desired state. Also, we discussed the fact that Ballerina's built-in code-to-cloud feature can be used to generate Kubernetes artifacts automatically. These artifacts can be used to maintain your container cluster in the desired state with a set of policies. We also discussed Kustomize as regards adding more configurations to the Kubernetes artifact files that cannot be generated by Ballerina.

In this chapter, we focused on creating and deploying Ballerina applications on microservice architecture. In the next chapter, we will focus on building a communication network of services in a microservice architecture. There, we will discuss different types of communication protocols and architectures that can be used to communicate with services.

Questions

1. What is the status of a Docker container?
2. What is the difference between Docker Swarm and Kubernetes?
3. What is the list of objects in Kubernetes?
4. What is the difference between Kustomize and Helm?

Further reading

- *Mastering Kubernetes: Level Up Your Container Orchestration Skills with Kubernetes to Build, Run, Secure, and Observe Large-Scale Distributed Apps*, by Gigy Sayfan, available at `https://www.packtpub.com/product/mastering-kubernetes-third-edition/9781839211256`

- *Learning DevOps: The Complete Guide to Accelerate Collaboration with Jenkins, Kubernetes, Terraform, and Azure DevOps*, by Mikael Krief, available at `https://www.packtpub.com/product/learning-devops/9781838642730`

- *Docker for Developers: Develop and Run Your Application with Docker Containers Using DevOps Tools for Continuous Delivery*, by Richard Bullington-McGuire, available at `https://www.packtpub.com/product/docker-for-developers/9781789536058`

Answers

1. Docker has a container life cycle much like threads. A Docker container can be in Running, Paused, Restarting, and Exited states. Running is the state where the Docker container is executing. While it is running, you can pause the container. The Restarting state comes when the Docker image is restarting. When the Docker container is stopped, it goes to the Exited state.

2. Both Docker Swarm and Kubernetes can be used for container orchestration. Docker Swarm comes with Docker itself, which has built-in features to perform container orchestration. Kubernetes, on the other hand, provides multiple features, including autoscaling, load balancing, and log monitoring. Deciding which platform to use depends on the architectural requirements of the system. Both products are good at container orchestration and are widely used in production systems.

3. Kubernetes has different types of objects for performing different tasks. The following are some of the objects used in Kubernetes. Some of these have already been discussed in this chapter:

    a) **Pod**: This is the smallest unit of deployment in a Kubernetes cluster that contains one or more containers.

    b) **ReplicaSet**: This object maintains the number of replicated Pods at a given time.

c) **Deployment**: This object updates Pods and ReplicaSets according to given deployment policies.

d) **StatefulSets**: This is the same as Deployment, but instead of stateless, these Deployments can keep their state. This type of Deployment is used in stateful services such as databases.

e) **DaemonSet**: This type of object is used to deploy Pods in all the nodes. For example, a log collector for a given node should be run on all nodes. Then, this service can be deployed as a DaemonSet.

4. Helm is a package manager for Kubernetes. Helm uses a template engine to generate Kubernetes artifacts. You can use templated Kubernetes YAML artifacts with Helm and replace those values with your own replacement values, whereas Kustomize does not use a template engine. Instead, it combines multiple YAML files to generate the Kubernetes artifacts.

4

Inter-Process Communication and Messaging

Microservice architecture is a trend that has emerged with the recent growth of cloud application technology. Microservice architecture is built by combining small service components. These services should be able to communicate with each other and external services to provide the final output. Communication, therefore, is a key component of cloud native technology that needs a lot of focus and attention.

In this chapter, we will discuss how to use different types of communication protocols with the Ballerina language. We will learn how to find service endpoint connection URLs with service discovery, using different types of synchronous communication methods, such as GraphQL, WebSocket, gRPC, and asynchronous communication protocols such as RabbitMQ and Kafka.

The following is the list of the main topics that we are going to discuss in this chapter:

- Communication between services in a microservice architecture
- Synchronous communication methods and protocols
- Asynchronous communication methods and protocols

By the end of this chapter, you should understand what the key factors are that you need to consider before building communication mechanism in a cloud native application. Those key factors include how to improve resiliency, service discovery, and the usage of different communication protocols. We will go through each scenario with a sample Ballerina implementation.

Technical requirements

This chapter contains multiple examples to demonstrate different types of communication methods and patterns. For service discovery examples, Docker is used as the container platform and Kubernetes is used as the container orchestration platform. Installation instructions for these were provided in the previous chapter.

We've used Consul as the service discovery tool and a **Helm Chart** to install it on the Kubernetes cluster. For service communication samples, we will use **Kafka** and **RabbitMQ** as the messaging services. Kafka can be download from the `https://kafka.apache.org/downloads` website, and RabbitMQ can be downloaded from the `https://www.rabbitmq.com/download.html` website. Other than these tools, you will need a web browser to run **HTML** and **JavaScript** code. The latest versions of common browsers, including **Chrome**, **Firefox**, and **Edge**, support previewing the samples provided in this chapter.

You can find the code files for this chapter at `https://github.com/PacktPublishing/Cloud-Native-Applications-with-Ballerina/tree/master/Chapter04`.

The Code in Action video for the chapter can be found here: `https://bit.ly/3k36E7y`

Communication between services in a microservice architecture

In the previous chapter, we discussed the development of simple services with the Ballerina language. We addressed the use of the HTTP server to manage HTTP requests. But, when it comes to production systems, we need to consider several factors in dealing with communication between services, such as how each service finds other services, how to manage failures, and so on. In this section, we will discuss how to solve these problems and build a reliable platform with microservice architecture.

Communication over services in the Kubernetes cluster

When two services need to communicate with each other, the client should know the address of the server to be connected with it. In the HTTP, we use a URL to represent this unique address. The URL consists of the hostname and the port number to connect to.

The hostname is a human-readable text that can be used to uniquely identify a node in a network. It acts as a label to identify the node and it is unique for a given network. On the other hand, an IP address can be used to identify a node with a unique combination of numbers separated by a dot. For example, `localhost` is the hostname of your loopback address, while `127.0.0.1` maps the corresponding IP address. When you are building a microservices application on a Kubernetes cluster, each service will run in separate pods that will have a different IP address. When one service wants to find another service to connect with, it needs to find a particular IP address to access the service.

For example, think of the order management system that we discussed in *Chapter 1, Introduction to Cloud Native*. When the order service needs to verify that an order can be fulfilled by the inventory service, the order service sends a request to the inventory service along with the quantity of the order items to confirm the availability of given order items. In this situation, in order to send the request, the order service should be able to find the address of the inventory service.

Kubernetes gives you a simple way to manage the service addressing problem with environment variables and DNS resolving. A deployment can have several replications of services on the Kubernetes cluster. Each of these services is exposed by a Kubernetes service. The Kubernetes service is responsible for the routing of traffic between service pods. Check the following diagram, which has three replicas of the **Backend Service Pod** providing the backend service. The **Caller Service Pod** is trying to access these backend pods through the **Backend Service**:

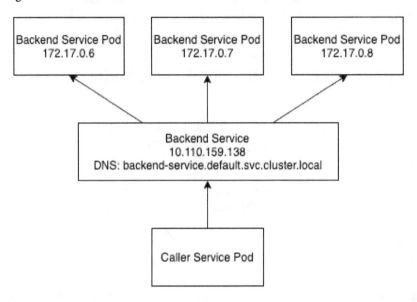

Figure 4.1 – Interservice communication

As you can see, the **Caller Service Pod** can directly call the **Backend Service Endpoint** instead of calling individual pods. This service has an IP address that can be used by the caller service. On the other hand, Kubernetes also provides DNS resolving to call a service by its service name.

To gain a better understanding of the system, we will create a sample to demonstrate this functionality. The following code creates a backend service that responds to a string when it gets invoked:

```
import ballerina/http;
service /backend on new http:Listener(9091) {
    resource function get greeting() returns error|string {
        return "Hello from backend server!";
    }
}
```

You can simply build the preceding Ballerina source code to generate Docker and Kubernetes artifacts. Then, you can use the following command to run this service on a Kubernetes cluster:

```
kubectl apply -f <Kubernetes_artifact_yaml_file
```

We can create another service to call this service and get the response back. Check the following Ballerina service sample code, which invokes a backend service and forwards the result back to the client:

```
service /caller on new http:Listener(9090) {
    resource function get sayHello() returns error|string {
        string payload = check clientEndpoint->get("/backend/
            greeting"); // clientEndpoint need to defined
        return "Response from backend server: " + payload;
    }
}
```

This service implementation sends a GET request to the backend service and sends back a response by appending another text to the backend response. Now, we need to define the clientEndpoint variable with the proper client endpoint address to access the backend service.

In the Kubernetes environment, there are multiple ways that you can use to find the address for the clientEndpoint variable. As mentioned previously, we can define the clientEndpoint variable with either environment variables or with Kubernetes DNS resolving. In the next section, we will learn how to use environment variables to find the backend endpoint address.

Using environment variables for service discovery

Each pod running on the Kubernetes cluster has its own IP address to communicate with. You can list down these endpoints with the kubectl describe service <service_name> command. For the backend service example, the command would be as follows:

```
kubectl describe service backend-service
```

You can see from the following command output that each of the endpoints contains an IP address with the port that we have exposed from the `kubectl describe` command output:

```
Name:              backend-service
Namespace:         default
Labels:            app=backend_service
Annotations:       <none>
Selector:          app=backend_service
Type:              ClusterIP
IP:                10.108.4.114
Port:              port-1-backend   9091/TCP
TargetPort:        9091/TCP
Endpoints:         172.17.0.14:9091
Session Affinity:  None
Events:            <none>
```

This is the IP address of the pod that contains a backend service. If you scale up the backend service deployment with the `kubectl scale deploy backend-service-deployment --replicas=2` command, then you can see that there are two endpoints listed under the `endpoints` section on the terminal command output of the `kubectl describe` command.

For each of these pods, Kubernetes adds an environment variable that can be used to identify each service. You can easily check all available environment variables by executing the `env` command on an already running pod.

To list all environment variables defined in a pod, you can use the `kubectl exec <pod_name> env` command. You'll need to replace `pod_name` with the backend service pod name by listing all of the pods with the `kubectl get pods` command. The list of environment variables for a given pod is as follows:

```
PATH=/usr/local/sbin:/usr/local/bin:/usr/sbin:/usr/bin:/sbin:/bin
HOSTNAME=backend-service-deployment-7759db7854-7hvnp
BACKEND_SERVICE_PORT_9091_TCP_ADDR=10.110.159.138
BACKEND_SERVICE_PORT_9091_TCP_PROTO=tcp
BACKEND_SERVICE_PORT_9091_TCP_PORT=9091
BACKEND_SERVICE_SERVICE_PORT=9091
BACKEND_SERVICE_PORT=tcp://10.110.159.138:9091
```

```
BACKEND_SERVICE_PORT_9091_TCP=tcp://10.110.159.138:9091
BACKEND_SERVICE_SERVICE_HOST=10.110.159.138
BACKEND_SERVICE_SERVICE_PORT_ANONSERVICE_1_S=9091
HOME=/home/ballerina
```

This environment variable list contains the IP address of the backend service with the `BACKEND_SERVICE_SERVICE_HOST` key. Kubernetes has the environment variable naming convention, where `_SERVICE_HOST` is appended to the end of the service name in uppercase letters separated by an underscore. In the same way, you can see that the port is also specified in the `BACKEND_SERVICE_SERVICE_PORT` environment variable.

Now, we can define the `clientEndpoint` variable to have its endpoint value from the environment variables. The `clientEndpoint` variable definition should be as follows, which takes the endpoint IP address and the port from the environment variable on the pod:

```
http:Client clientEndpoint = new ("http://" +
os:getEnv("BACKEND_SERVICE_SERVICE_HOST") + ":" +
os:getEnv("BACKEND_SERVICE_SERVICE_PORT"));
```

The `os:getEnv` function in Ballerina can read the environment variable value with the given key. The client endpoint address is generated by reading the environment variables for the backend service host and the service port. This environment variable is attached to a pod by Kubernetes while deploying it on the Kubernetes cluster. You can read this environment variable to find a particular service.

The problem with using environmental variables for service discovery is that the backend service should be deployed earlier than deploying the caller service. Otherwise, the caller service would not have the backend service address as it was not deployed at that time. We can use the Kubernetes DNS solution to solve this problem. In the next section, we will look at how to use the Kubernetes DNS to call the backend service.

Using the Kubernetes DNS resolver for service discovery

Kubernetes generates a DNS record for each given service. Instead of directly calling the pod using the IP address, the caller service can now call a pod by a DNS name. The DNS name is inherited from the service name given from the `Cloud.toml` file or the package name if there is no `Cloud.toml` file. Kubernetes generates the DNS name using the following template:

```
<service_name>.<namespace>.svc.<cluster_domain>
```

service_name is the name of the service that needs to be accessed. For namespace, you need to provide the namespace where the service is deployed. cluster_domain is the domain given for the Kubernetes cluster. The default cluster domain is cluster. local. The Fully Qualified Domain Name (**FQDN**) for the backend service is backend-service.default.svc.cluster.local. Instead of using the IP address to access the backend service, we can use this FQDN to access the backend service.

Kubernetes provides an internal DNS service to resolve DNS names. You can check this DNS service with the kubectl get service -n kube-system command. You can verify whether this DNS is correctly resolved by the Kubernetes cluster by executing the nslookup command on a pod in a Kubernetes cluster. For example, you can execute the following command to check the DNS resolve logs on the backend service pod:

```
kubectl exec <backend_pod_name> -- nslookup backend-service.
default.svc.cluster.local
```

This command gives you the resolved IP address of the given service. A sample output of this command is as follows:

```
Server:    10.96.0.10
Address:   10.96.0.10:53


Name:      backend-service.default.svc.cluster.local
Address:   10.108.4.114
```

Now, we can modify the Ballerina caller service, clientEndpoint, as follows to use the DNS name to resolve the backend service:

```
http:Client clientEndpoint = new (http://backend-service.
    default.svc.cluster.local:9091);
```

Once you execute the caller service on Kubernetes and invoke the caller service, then you will get a response as a combination of the caller service and backend service output.

Other than using IP address and environment variables to resolve services on Kubernetes, you can use third-party services to find available services. A **service mesh** is one such solution that handles all the complexities of inter-service communication between microservices in a microservice architecture. A service mesh can be used to implement service discovery while you can add policies to interservice communication. In the next section, we will discuss what a service mesh is and what its architecture looks like.

Using a service mesh to simplify inter-service communication

Container orchestration platforms such as Kubernetes provide an elegant way of handling multiple containers with a collection of policies. In a microservice architecture, services need to communicate with other services. There are multiple challenges associated with inter-service communication that are common to all services. The following is a list of challenges in inter-service communication on microservice architecture and how they are handled in a service mesh architecture:

- **Handling encryption between services**: Each service needs to encrypt data to prevent possible vulnerabilities. An SSL certificate is commonly used for validating endpoints. The service mesh handles encryption between services connected to the service mesh.

- **Retries to resolve failures**: Resiliency patterns such as timeouts, retries, and circuit breakers can be attached to the service mesh itself such that you don't need to handle it in your source code for each service.

- **Smart load balancing**: Rather than having round-robin and random load balancing, you can have least connection-based load balancing that requests directly to the service with the least number of connections. A service mesh is capable of doing smart load balancing rather than redirecting traffic dumbly.

- **Collecting metrics, logs, and traces**: The service mesh can collect metrics, logs, and traces from the mesh to debug the request flow.

- **Access controls**: In the service mesh, you can define policies on how each service is connected to other services. For example, you can restrict access to certain services called by another service. You can restrict access to a service if a health check has failed and so on.

In a basic Kubernetes pod, there will be just the service containers running on it. The service mesh is built with a sidecar proxy, which is injected into the service pod. This proxy handles all incoming and outgoing requests from and to the service. The Ballerina application performs all the communication between services within the cluster through this sidecar proxy. Therefore, the sidecar proxy handles application routing logic, access control, and TLS termination to simplify the system development process. In service mesh architecture, a service will never communicate directly with another service. Check the following architecture diagram to see how the service mesh communicates with each service:

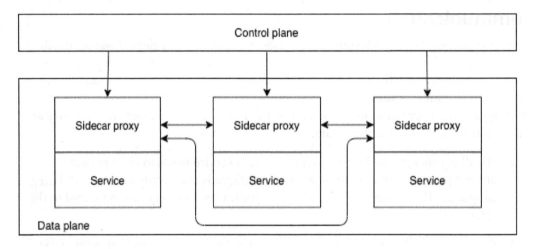

Figure 4.2 – Service mesh architecture

As you can see, the data plane has deployment-related components such as services and sidecar proxies. Sidecar proxies handle all incoming and outgoing traffic from each service. Sidecar proxies communicate with other services to create the service mesh. The control plane is used to control sidecar communication by adding policies and security features on sidecar proxies. You can also have logs, traces, and metrics collection on sidecar proxies to make the service mesh more observable. In the next section, we will explore how to implement a service mesh with **Consul** and perform service discovery for the sample order management Ballerina application that we discussed in *Chapter 1, Introduction to Cloud Native*.

Service discovery in a service mesh by using Consul

Consul is a service mesh tool, provided by *HashiCorp*, that can be used to build a service mesh on a Kubernetes cluster. A Consul sidecar proxy is built with the **Envoy** proxy, which is widely used to implement a service mesh. Consul provides multiple features to build a microservices application on Kubernetes. These features include the following:

- **Service discovery**: Services can register on the Consul server as services and other nodes in the cluster are able to access these services by querying a service list from Consul. Consul has DNS and REST API support that you can use to discover services:

- **Health checking**: A cluster's health status can be monitored by performing health checks on each node. Consul clients can be set up to collect health metrics from the node and publish them on the Consul server. This information can be used to check the node health and route traffic accordingly.

- **KV store**: Consul provides distributed key-value stores to store data on a distributed system. You can easily access the KV store with the REST API provided by the Consul server.

- **Secure service communication**: Interservice communication can be made more secure with mutual certificates. Consul is capable of handling all the certificate validation among services and preventing a **man-in-the-middle** attack.

Consul has the capability of building service discovery along with health check monitoring. You can use the **User Interface** (**UI**) to check the current status of the service and nodes. Further, you can add route restrictions to limit access to some services. You can easily integrate Consul with a Ballerina program to build a communication network on a Kubernetes cluster. We can use the following steps to do this:

1. First, you need to install Consul on the Kubernetes cluster to start the development with Consul. The easiest way to install Consul on a Kubernetes cluster is by using a Helm Chart. A **Helm Chart** is a tool that is widely used as the package manager for Kubernetes. Helm can be used to install different Kubernetes components on a Kubernetes cluster. You can follow the Helm installation documentation at `https://helm.sh/docs/intro/install/` to install Helm on your computer.

   After you have installed Helm, you can use the Helm CLI tool to install different packages on the Kubernetes cluster. To install Consul on the Kubernetes cluster, first, you need to add the **HashiCorp Helm repository** with the following `helm` command:

   ```
   helm repo add hashicorp https://helm.releases.hashicorp.com
   ```

2. Next, we can install Consul with the following `helm install` command:

   ```
   helm install consul hashicorp/consul --set global.
   name=consul
   ```

3. Once the installation is completed, we can check the running pods on the Kubernetes cluster by using the `kubectl get pods` command. In the output for this command, we can identify that there are four new pods created by Consul on the Kubernetes cluster as follows:

```
pod/hashicorp-consul-connect-injector-webhook-deployment-7ddcfdh9k6    1/1    Running    1    1h
pod/hashicorp-consul-controller-6569dd7fcf-ft6cv                       1/1    Running    1    1h
pod/hashicorp-consul-server-0                                          1/1    Running    1    1h
pod/hashicorp-consul-vxjss                                             1/1    Running    1    1h
pod/hashicorp-consul-webhook-cert-manager-965bd556d-jtskz             1/1    Running    1    1h
```

Figure 4.3 – List of Consul pods running on the Kubernetes cluster

We will use the same caller and backend implementation to demonstrate the Consul service registry with small modifications. Like the previous example, the caller service calls the backend service and sends the message back to the client by concatenating the two responses.

For this example, we created a new project named `consul_backend`. For the backend service, we need to inject the Consul sidecar proxy into the `.yaml` configuration file. We can use **Kustomize** to merge two different configurations for generating a single `.yaml` config that can be directly deployed into the Kubernetes cluster. So, create a new project and add the following YAML configurations to the `<project_home>/resources/deployment.yaml` file:

```yaml
apiVersion: apps/v1
kind: Deployment
metadata:
  labels:
    app: "consul_backend"
    name: "consul-backend-deployment"
spec:
  replicas: 1
  selector:
  matchLabels:
    app: "consul_backend"
  template:
  metadata:
    annotations:
      'consul.hashicorp.com/connect-inject': 'true'
```

Just like a regular Kubernetes file, we now have the YAML file that contains deployment information. Here, we need to provide the metadata similar to the metadata generated in the Ballerina-generated Kubernetes artifacts. For example, the `metadata.labels.app` property should be the same for both `deployment.yaml` and the Ballerina-generated YAML file. The YAML configuration with the `consul.hashicorp.com/connect-inject:true` parameter injects a sidecar proxy into the Ballerina service. This is the sidecar proxy that creates a service mesh and manages all incoming and outgoing traffic of the Ballerina service.

4. Now, we need to create a `Kustomization.yaml` file to locate the resource files that we need to merge. For that, add the following content to the `Kustomization.yaml` file:

```
resources:
- target/kubernetes/consul_backend/consul_backend.yaml
patchesStrategicMerge:
- resources/deployment.yaml
```

5. Now, we can build the project in the same way as the previous example to generate the Docker images. Once the Docker images get created, we need to run `Kustomize build` with the following command to generate and apply the new Kubernetes artifacts:

```
kustomize build | kubectl apply -f -
```

This command merges both `.yaml` files and applies them to the Kubernetes cluster. Once deployed, Kubernetes will create a pod for the backend service as in the previous example.

6. The next step is to deploy the caller service. Since we are using Consul DNS routing to access the backend service, we can modify the client endpoint program in the caller service as follows to access the backend services. For this example, we will create a new Ballerina project called `consul_caller_service` and copy the caller service code along with the following client endpoint definition:

```
http:Client clientEndpoint = new (http://consul-
    backend:9091);
```

7. We will register the caller service on Consul as a service such that Consul will be able to identify the caller service on the service mesh. We will need to create a `deployment.yaml` file in the `<project_home>/resources/` directory with the following code:

```yaml
apiVersion: "apps/v1"
kind: "Deployment"
metadata:
  labels:
    app: "consul_caller_service"
    name: "consul-caller-s-deployment"
spec:
  replicas: 1
  selector:
    matchLabels:
      app: "consul_caller_service"
  template:
    metadata:
      annotations:
        'consul.hashicorp.com/connect-inject': 'true'
        'consul.hashicorp.com/connect-service-upstreams':
          'consul-backend'
```

From the preceding code, other than injecting the sidecar proxy into the caller service, we will also specify the backend service to which the caller service is connected. This property is set with the `consul.hashicorp.com/connect-service-upstreams:consul-backend` parameter in the YAML file.

8. We will need to create the `kustomization.yaml` file in the project's home folder, given as follows, for this YAML definition:

```yaml
resources:
- target/kubernetes/consul_caller_service/consul_caller_
  service.yaml
patchesStrategicMerge:
- resources/deployment.yaml
```

This YAML file defines the locations of the YAML files to be merged. `consul_caller_service.yaml` is the YAML file that is automatically generated by the Ballerina compiler. `resources/deployment.yaml` is the file that contains additional YAML configurations that we need to be merged with the Ballerina-generated YAML file.

9. Now, we can build the Ballerina project to push the caller service Docker image. Next, we can use the following command to merge and deploy the custom Consul definition with Ballerina-generated Kubernetes artifacts:

```
kustomize build | kubectl apply -f -
```

This command will create another pod in the Kubernetes cluster as a caller service.

Once all these services are deployed, you can invoke the caller service with the same instructions as those given in previous examples. The caller service responds with a concatenated string from the backend, the same as in the previous examples.

Consul provides a UI so that we can visualize the services that we have registered on it. This service is exposed as a node port service where we can access this UI with a browser outside the Kubernetes cluster. If you are using **minikube**, you can access this UI by executing the `minikube service hashicorp-consul-ui` command. This command gives you the IP address and the port you need to connect to.

The following screenshot of the Consul dashboard is displayed once you open the IP address in the browser:

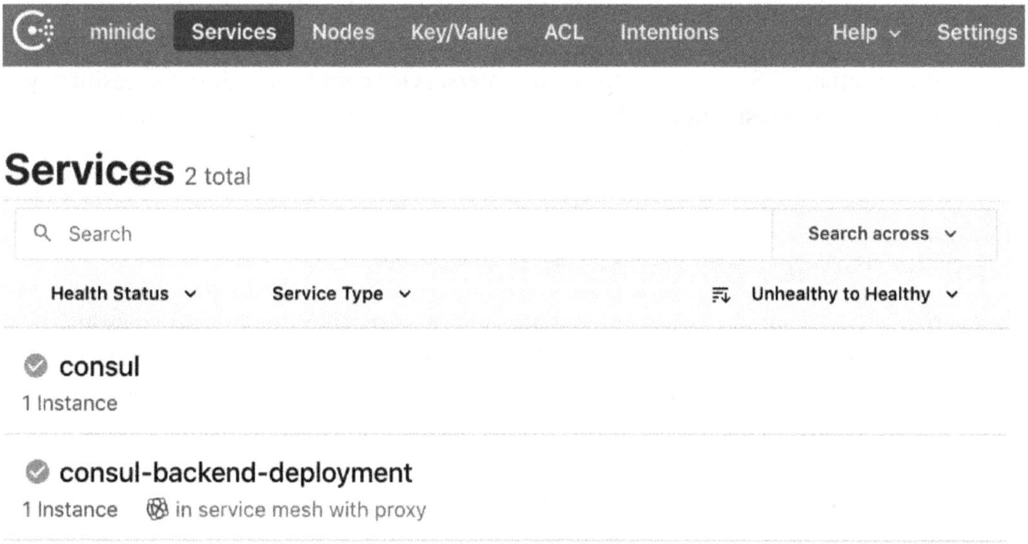

Figure 4.4 – Consul UI showing registered services

You can check the registered services and the health status of the services with this UI. Further, you can set route rules in the Consul UI with intentions. These routing rules can be used to restrict accessing particular services on a service mesh.

On the **Intentions** tab, you can either allow or block a service from calling another service, as shown in the following screenshot:

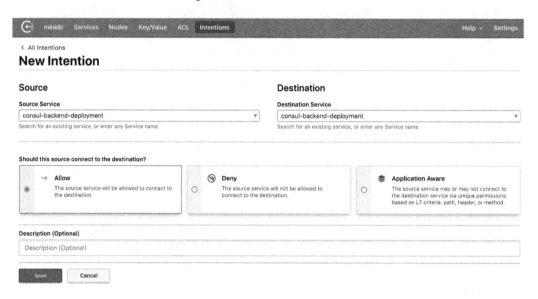

Figure 4.5 – Adding intentions in the Consul UI

Other than the service registry, you can also use Consul as a distributed key-value store. The Consul REST API provides a way to access the key-value store with HTTP calls. Also, you can add a mutual **TLS certificate** to secure interservice communication and resiliency patterns to improve the resiliency of the system with Consul. Other than providing resiliency through the service mesh, we can embed that information in Ballerina code itself. In the next section, we will discuss how to provide resiliency to a Ballerina program.

Using resiliency patterns in Ballerina

Failures are likely to occur in any kind of system. Failures can occur due to many reasons. Some of these failures can be due to issues with the program, while others can be due to infrastructural failures. **Availability** is a metric used to measure how well a system manages failures. The availability of an application can be calculated using the following formula:

$$Availability = MTTF/((MTTF + MTTR))$$

Mean Time to Failure (MTTF) is the total time used by the services that are up and running. **Mean Time to Recover (MTTR)** is the downtime of services. The obvious fact is that to improve the availability of an application, you need to increase the uptime and reduce the downtime.

But, in a production system, it is really hard to achieve an availability value of 1. What we can do is make it as close as possible to 1. Therefore, we need to accept the failure and recover the system as quickly as possible. This way, we can increase the availability of the system.

Timeout is the simplest way of handling failures in a program. We can specify a maximum time range that a program waits for an answer to be received. If a request is not completed within a given time, the next instruction will continue to run. This makes the downstream more responsive and prevents waiting in a loop. Ballerina provides a way to handle an endpoint timeout with a Ballerina client endpoint timeout. Check the following example, which sends a request to the backend endpoint and times out if the response does not come within the given time range:

```
http:Client backendClientEP =   check new ("http://showcase.api.
    linx.twenty57.net", {
        timeout: 2
});
```

In this program, we define the endpoint as a client endpoint and the timeout is specified in the `timeout` parameter in seconds. This parameter times out the program and generates a timeout error if the request does not complete within the given time. This can be used when you need to call an endpoint that should time out within the given duration.

On the other hand, you can retry the same endpoint several times when the service fails due to network failures. There could be network failures in the system that suddenly occur and get resolved in the next moment. This kind of failure can be easily managed with multiple retries. Ballerina provides built-in support to retry an endpoint in case of failure. Check the following example, which retries the given endpoint as given in endpoint configurations:

```
http:Client backendClientEP =   check new ("http://showcase.api.
    linx.twenty57.net", {
        retryConfig: {
            interval: 3,
            count: 3,
            backOffFactor: 2.0,
            maxWaitInterval: 20
        },
        timeout: 20
});
```

In this configuration, the `retryConfig` parameter contains the retry configuration details. The `interval` parameter describes the initial retry interval in seconds. The `count` parameter defines the maximum number of retries needed to be performed.

After the first failure of the program, it waits for an `interval` before the next try. After the next attempt to connect with the backend, it needs to wait until `interval` times the `backOffFactor` duration to retry the endpoint. This retry interval exponentially increases with the factor of `backOffFactor`. For the given example, retry timeouts would be 3,000, 6,000, 12,000, and so on. `maxWaitInterval` defines an upper limit for the retry timeout in seconds. Since `maxiWaitInteval` is 20 in this example, the maximum number of retries is 3. Here, we also set `timeout` to make sure that each of these requests does not wait longer than 20 seconds.

When you are running multiple services in a microservices application, there can be failures in services that take longer to recover. Unlike network connectivity problems, these failures might take some time to recover from failure. You may get errors when you try to access these services, and you may also slow down the recovery process if you keep on sending requests to such services. A **circuit breaker** pattern is a resiliency pattern that is used to connect endpoints that require some time to recover from failure. As its name suggests, the circuit breaker pattern is similar to what happens in your home electric circuit breaker system. The following figure shows a circuit breaker:

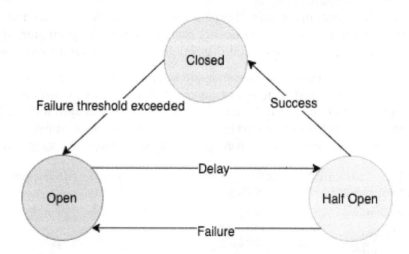

Figure 4.6 – Circuit breaker pattern state transition diagram

A circuit breaker can have the following three states:

- **Closed state**: This is the working state where the backend is working properly and can reach backend with no problems. The backend application is up and running, and the response from the backend is sent back to the caller correctly.

- **Open state**: When the backend server is not reachable, it comes to the open state. Just like the open state of a circuit that does not conduct current, the caller service cannot access the backend service and get a reply. When the system comes to an open state, it does not continue to send other requests to the endpoint and responds with an error message.

- **Half-open state**: After a given timeout, the application comes to another state called the half-open state. In this state, the program attempts to send a request to the backend application and check whether it is accessible. If it is accessible, then it goes back to the closed state and continues to send requests. If it is inaccessible, it will return to the open state.

Having a circuit breaker pattern for microservice applications is important in cases where they handle large amounts of traffic. The backend application might be in a recovery state after a failure. This pattern takes some time to get it to the operational state. When the program finds that there is a backend failure, it goes to the open state for some time. After a given timeout, the program attempts to call the backend again.

Ballerina provides built-in circuit breaker functionality to access a backend application. Endpoint configurations are integrated into the endpoint definition itself. Check the following example client endpoint definition, which is used to connect a currency exchange backend endpoint with the circuit breaker pattern:

```
http:Client backendClientEP =  check new ("https://showcase.
   api.linx.twenty57.net", {
    circuitBreaker: {
        rollingWindow: {
            timeWindow: 10,
            bucketSize: 2
        },
        failureThreshold: 0.2,
        resetTime: 10,
        statusCodes: [400, 404, 500]
    },
    timeout: 2
});
```

In this example, the backend client endpoint is set as the currency exchange endpoint. Other than the hostname of the endpoint, we also specified the connection status as a JSON parameter. The `circuitBreaker` parameter is used to instruct Ballerina to use a circuit breaker pattern for this particular endpoint call. The `failureThreshold` parameter in `circuitBreaker` is used to calculate when the circuit should be tripped. This is a ratio between the number of failures and the number of all the requests made. This ratio value is calculated based on the rolling window as defined in the `rollingWindow` field. If Ballerina finds that the endpoint exceeds the `failureThreshold` level, it comes to the open state.

A rolling window is used to calculate the failure threshold in a given window size. These window size configurations are placed under the `rollingWindow` configuration. It contains `timeWindow`, which specifies the time range used to measure the window size in seconds. This window again splits into smaller time ranges called **buckets**. The size of the bucket is defined in the `bucketSize` parameter. When sending a request to this endpoint, Ballerina calculates threshold values for each of these windows, which contain small buckets. This can be visualized in the following diagram.

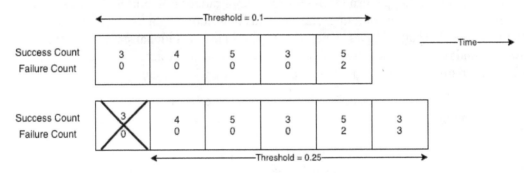

Figure 4.7 – Calculating the threshold for a circuit breaker window and buckets

As time flows, older buckets will be removed and a new bucket will be added to the window. If there are more request failures than the threshold level, the endpoint will go to the open state. The endpoint will be in the open state for `resetTime` seconds. After a `resetTime` timeout, it goes to the half-open state and tries to access the service again. `statusCodes` are a set of status codes that are considered as failures. Here, `404` (`Not found error`) and `500` (`Internal server error`) are defined as statuses that are identified as failures.

Another important aspect of increasing the resiliency of cloud applications is load balancing. In a microservice architecture, we can deploy multiple replicas of services and distribute traffic among those services. Ballerina client-side load balancing will help you to implement a load balancer. In the next section, we will discuss this more.

Ballerina client-side load balancing

Other than these resiliency patterns, you can also implement load balancing with Ballerina. Replicating services is a way of achieving high availability on a computer cluster. It also helps to distribute the load between multiple services. The component that is used to distribute the load between these services is known as a **load balancer**. The load balancer sits between the client and the backend implementation and distributes the traffic between services. The Kubernetes services and Consul programs that we described earlier have their own implementations of load balancing. When you call any of those given endpoints, they will automatically distribute traffic between available services. These implementations are called server-side load balancing where the client does not know the available endpoint and blindly calls the load balancer to handle the traffic.

Client-side load balancing, on the other hand, performs load balancing on the client service implementation itself. The client takes all the available services with the service discovery service and distributes traffic over the given endpoints.

Ballerina's built-in load balancing helps programmers to implement a load balancer in the Ballerina program itself. This load balancer uses the Round Robin algorithm to route requests. Requests will be fairly distributed over all the given endpoints. Here is a sample endpoint definition that routes traffic between multiple services:

```
http:Client backendClientEP = new ({
    targets: [
        {url: "http://192.168.43.1:9090"},
        {url: "http://192.168.43.2:9090"},
        {url: "http://192.168.43.3:9090"}
    ],
    timeout: 5
});
```

In this sample code, the backend endpoint has multiple targets defined with the same hostname. Each service is started in different services with different IP addresses.

> **Note**
>
> When you are implementing services in a Kubernetes cluster with client-side load balancing, do not hardcode IP addresses, since they are dynamic. You can use service discovery services such as Consul to obtain a list of endpoints for a given service and implement Ballerina load balancing.

In this section, we have focused on building communication mechanism in a microservice architecture. We have discussed finding services with service discovery and using resiliency patterns to improve the availability of the application. In the next section, we will discuss using different types of communication protocols. First, we will discuss synchronous communication methods and later we will discuss asynchronous communication methods.

Synchronous communication

Communication methods can be separated into two main categories, as synchronous and asynchronous communication. In **synchronous communication**, the client sends a request and waits for a reply. Examples of this type of communication method are HTTP, WebSocket, and GRPC protocols. **Asynchronous communication**, on the other hand, does not wait for a response. In this section, we will discuss different types of synchronous communication methods supported by the Ballerina language with sample code.

Handling HTML form data

Hypertext Transfer Protocol (**HTTP**) is the most common and simple synchronous method that is widely used to communicate with other services. We have already discussed how to create a simple HTTP server and how to invoke it in the previous chapter. In this section, we will discuss more practical aspects of HTTP.

When you are building web applications, HTML forms are a common way of collecting information from the end user. This information can be submitted to a backend server to perform specific operations on this data. HTTP can be used with different content types to transfer messages. Form data is submitted as the `application/x-www-form-urlencoded` content type. This content type gets converted from data key-value pairs to an encoded string and is submitted to the server. Check the following HTML form implementation, which reads multiple input values:

```html
<form action = "http://localhost:9090/form" method="POST">
<div>Name: <input name = "name" type = "text"/></div>
<div>Email: <input name = "email" type = "email"/></div>
<div>Birthday: <input name = "birthday" type = "date"/></div>
<div>Gender: <select name = "gender">
<option value = "male">Male</option>
<option value = "female">Female</option>
</select></div>
<input type = "submit" value="Submit"/>
</form>
```

This form submits a list of data as form URL-encoded data to the server running on the localhost 9090 port. This data can be read simply by a Ballerina server with a simple HTTP service as follows:

```
resource function post . (http:Request req) returns error|string
{
    map<string> data = check req.getFormParams();
    io:println("Name: " + data.get("name"));
    io:println("Email: " + data.get("email"));
    io:println("Birthday: " + data.get("birthday"));
    io:println("Gender: " + data.get("gender"));
    return "Hello " + data.get("name");
}
```

This code segment starts a new service on port 9090 and listens to the root context of /form. The data submitted from the form can be easily retrieved with the getFormParams function provided by the req variable. This returns a map of all available form data. You can get individual form data by calling the get method on the map by referring to the key.

There might be a requirement where you need to upload files to the backend server. In this case, you cannot use the x-www-form-urlendcoded content type to work with binary content. In this case, data should be sent to the server in the (MIME) multipart message format. The MIME messaging format uses encoding to encode binary content to a text representation and send it over an HTTP request. We will create a simple Ballerina server that can accept a file uploaded from a web browser as a multipart request. The following is the HTML form implementation to upload a file:

```
<form action = "http://localhost:9090/file/upload"
  method="post" enctype = "multipart/form-data">
    <div>Select image to upload: <input type = "file" name =
      "fileToUpload" id = "fileToUpload"></div>
    <br/>
    <div><input type = "submit" value = "Upload Image" name =
      "submit"></div>
</form>
```

This form submits the file to the Ballerina server endpoint with the action parameter. The encoding type, enctype, is set as multipart/form data that encodes the file content and is sent over the HTTP connection. The input box is provided to select the file and a submit button is provided to submit the form data to the Ballerina backend.

The Ballerina backend implementation contains a service that accepts form data with the context path of /file. The server listens to port 9090 for a POST request with the following resource definition:

```
service /file on new http:Listener(9090) {
    resource function post upload(http:Caller caller,
        http:Request request) returns error?{
        var bodyParts = request.getBodyParts();
        if (bodyParts is mime:Entity[]) {
            foreach var part in bodyParts {
                mime:ContentDisposition contentDisposition =
                    part.getContentDisposition();
                if (contentDisposition.name == "fileToUpload")
                {
                    byte[] fileContent = check part.
                        getByteArray();
                    check io:fileWriteBytes("/Users/user/
                        Desktop/" + contentDisposition.fileName,
                        fileContent);
                }
            }
        }
        check caller->respond("done");
    }
}
```

The request comes to the server containing the MIME media type. We need to go through each of the parts and select the correct ID, which is fileToUpload in this case. fileToUpload is the name that we have given to the file upload input box.

The next step is to get the content to a byte array. The fileContent variable is used to collect the byte array by calling the getByteArray function on the variable part. Now we can write this content to a file with the fileWriteByte function by giving the file location and the binary content as the function arguments. Here, we take the filename that we have uploaded from the contentDisposition variable and store it in a custom directory.

This sample is to demonstrate how you can use Ballerina to work with form and multipart form data. Here, we have given samples in simple HTML format to understand the concepts where you can use the latest JavaScript framework to create production-grade frontends.

In the next section, we will discuss another protocol that was built on HTTP, which is GraphQL. This will simplify communication with the backend service with GraphQL interfaces.

Building a Ballerina backend with GraphQL

We have discussed building applications that use HTTP. One thing you may notice about REST protocols is that there is a well-defined request and response format for each HTTP resource. An HTTP service always sends the payload in a predefined message format (JSON format in most cases). Even though some of these fields in messages are useless, you still get the whole message. This makes the message size large and the processing speed slower for the client application. What if the backend response can send only required data to be sent to the client application? This is where GraphQL comes in. With **GraphQL**, you can specify the output fields that should be in the response message. This reduces the size of the message and improves the performance of the frontend applications.

Another advantage is that you can reduce the network calls that need to be performed to grab details from the backend. For example, an order management system that we discussed in *Chapter 1, Introduction to Cloud Native*, contains the following record types:

- An `Inventory` record to hold a list of inventories that maintain products
- An `InventoryItems` record to hold each of the items available to sell in the inventory
- A `Product` record to hold details of a product

If you need to get both `Inventory` details and `InventoryItem` details, first you need to get the inventory. Then you need to call another request to get the relevant `InventoryItem` record and combine the result. With the GraphQL interface, you can do both of these things with a single call.

GraphQL is an HTTP-based protocol where you can call a GraphQL endpoint with HTTP requests. Nowadays, frontend applications are strongly supported by GraphQL endpoints since they make the development process much easier. Ballerina provides built-in features to implement GraphQL backend services. In context, GraphQL has the following properties:

- **Declarative** since the client is the party that specifies how the message should return.
- **Compositional** since data comes in hierarchical order in JSON format, which is easy to understand and to work with.
- **Strongly typed** GraphQL data has a predefined schema that defines the message structure. Therefore, the client application can handle well-defined, structured data, which makes the development process much easier.

To build the GraphQL backend, first, you need to create the schema and data structure of your application. The data structure for this sample is as follows:

```
public type Inventory record {
    string inventoryId;
    string name;
    InventoryItem[] inventoryItems;
};
public type InventoryItem record {
    *Product;
    string inventoryItemId;
    int quantity;
};
public type Product record {
    string productId;
    string name;
    float price;
};
```

We will initialize a sample value for the testing purpose as follows:

```
final readonly & Inventory[] inventory = [{
    name: "Tom's wears",
    inventoryId: "25234234",
    inventoryItems: [{
        productId: "5434323",
        name: "T-shirt",
        price: 20,
        inventoryItemId: "3423909340",
        quantity: 120
    }]
}, {
    name: "My Choice",
    inventoryId: "53542364",
    inventoryItems: [{
        productId: "6434432",
```

```
        name: "Shirt",
        price: 25,
        inventoryItemId: "3423654323",
        quantity: 50
    }]
}];
```

> **Note**
>
> You can change this implementation to read data from a database instead
> of hardcoding data. We will discuss how to access the database in the
> next chapter.

In the preceding code, two inventories are added to the `inventory` array that holds all
available inventories. Both inventories include a single inventory item that is available to
sell. Now, we will use this data to implement a GraphQL query endpoint as follows:

```
http:Listener httpListener = check new(9090);
service graphql:Service /graphql on new
    graphql:Listener(httpListener) {
    resource function get inventory(string inventoryId) returns
        Inventory|error {
        Inventory[] item = inventory.filter(function (Inventory
            value) returns boolean {
            if value.inventoryId == inventoryId{
                return true;
            } else {
                return false;
            }
        });
        if item.length() > 0{
            return item[0];
        } else {
            return error("No matching id found");
        }
    }
}
```

The service type is set to `graphql:Service` with the context of `/graphql` in the service definition. The listener server is defined to start on port `9090`. The resource function `inventory` takes `inventoryId` as an input argument to find the corresponding `Inventory` item. The `inventoryId` variable type is defined as a string and this resource function returns an `Inventory` object as the result. GraphQL functionality on Ballerina automatically converts this `Inventory` return object to the requested client schema. In this resource function, we simply filter the matching inventory ID with the request inventory ID. The filtered list contains the matching result. It returns the `Inventory` object if there are matched values found. If it does not find the item with the ID, it returns an error to the client.

This endpoint can be invoked with a GraphQL endpoint by specifying the required schema. Check the following sample schema, which only reads the inventory name and list of quantities of inventory items:

```
{
    inventory(inventoryId: "53542364") {
    name
    inventoryItems {
        quantity
    }
}
}
```

You can invoke a GraphQL endpoint with the `curl` command to read this schema as follows:

```
curl -XPOST -H "Content-type: application/json" -d '{
"query": " { inventory(inventoryId: \"53542364\") { name,
inventoryItems{quantity}} }" }' 'http://localhost:9090/graphql'
```

This `curl` command uses the same, previous schema to get the inventory with the given ID. You can query much more complex queries with GraphQL by using an alias, fragments, variables, and so on. These endpoints are important in building frontend applications that require access to backend services with a single backend data structure but with a different frontend schema. With the use of GraphQL, you can have a single endpoint that can serve multiple such frontend applications.

GraphQL helps developers to create a contract between frontend and backend services. In the same way, we can use the OpenAPI Specification to create a contract between backend and frontend services. In the next section, we will discuss what the OpenAPI Specification is and how it can be useful in API-first development.

Using the OpenAPI Specification with Ballerina

In *Chapter 1, Introduction to Cloud Native*, we discussed the use of the API-first approach to creating services, where an API definition is shared among software development teams and these teams implement the functionalities based on this API definition. In such an environment, since an API is finalized at an early stage of the software development process, developers can develop the system in parallel. The team that develops the API uses the API contract as the requirement document to generate the code. Both the API developer team and the API consumer team can do their development independently of each other. API consumers can also test the application by mocking the backend API since the API definition is already available.

OpenAPI is a specification used to define how an endpoint can communicate with a given endpoint using the REST protocol. **Swagger** is the implementation of the OpenAPI Specification. Swagger includes an editor to edit OpenAPI specs as a YAML file. You can use the Swagger UI to develop OpenAPI specs. You can download the Swagger UI from the `https://swagger.io/tools/swagger-ui/download/` website. Further, you can develop an API spec collaboratively at **SwaggerHub** as well.

Ballerina CLI tool converts OpenAPI definition to Ballerina source code and Ballerina code to OpenAPI definition. By using this tool, you can generate Ballerina code with the given OpenAPI specification. You can import an Open API spec with the Ballerina CLI command. Then, create a YAML file with Open API specs. A sample Open API YAML file is included in the GitHub repository at `https://github.com/PacktPublishing/Cloud-Native-Applications-with-Ballerina/blob/master/Chapter04/open_api/resources/open_api.yaml`. Then use the following command to generate Ballerina code from the given YAML spec:

```
bal openapi -i <openapi-contract-path> -o <output-path>
```

`openapi-contract-path` is the YAML file path containing the OpenAPI spec. `output-path` is the source code directory where your Ballerina code needs to be saved. Once you run the preceding command, it will generate all the services and resource definitions along with documents. Then you can continue to use that generated code and implement business logic.

You can also convert existing Ballerina code to the OpenAPI spec. For that, you can execute the following command by pointing to an already defined Ballerina service implementation:

```
bal openapi [-i <ballerina-file-path>] [-o <openapi-contract-
path>]
```

The Ballerina file path is the file that contains service definitions. The OpenAPI contract path is the location of the OpenAPI spec YAML file to be saved. This command generates a YAML file that represents the API in the OpenAPI spec.

So far, we have discussed building the Ballerina service with HTTP. HTTP is not scalable for building applications that require two-way communication. WebSocket is a TCP-based protocol that is specifically designed for two-way communication between the client and the server. In the next section, we will discuss what a WebSocket is and how to implement the WebSocket protocol with Ballerina.

Building a chat application with WebSocket

TCP socket is a protocol used for two-way communication. HTTP is used in one-way communication where the client sends the request to the server and the server sends the response to the client. In HTTP, the client can send a request to the server at any time. But the server is unable to send data to the client unless there is an HTTP connection available at that time. WebSocket, on the other hand, uses a TCP connection to connect the server and maintains the connection so that the server is also able to send messages to the client at any time. Two-way communication is important where the server needs to update the client application in real time.

The WebSocket protocol provides full-duplex communication over a single TCP connection. Most web browsers also support WebSocket communication being used in web pages. To understand how WebSocket communication can be used in the Ballerina language, let's create a simple chat application that can communicate over a WebSocket.

This program is capable of maintaining multiple users connected to the platform. If a connected user sends a message to the server, it broadcasts that message to all connected clients. The following are the steps to create a sample chat application with the Ballerina socket module:

1. To store the list of socket connections of each user, we will use a map of websocket:Caller defined globally as follows:

    ```
    map<websocket:Caller> clientConnectionMap = {};
    ```

We will define the following `clientName` variable in the global scope to keep the name of the connected clients. The `NAME` constant is used to store the attribute name of the caller object:

```
string clientName = "";
const string NAME = "NAME";
```

2. Next, we can define the WebSocket listener service that listens to the WebSocket on port `9091`. The context for this service is given as `/chat`:

```
service /chat on new websocket:Listener(9091) { }
```

3. The next step is to add a resource function to accept a new socket connection with a given URL. For this resource function, we will use the `GET HTTP` method that also accepts the `path` parameter. The `path` parameter here is assigned to the variable `name`. We will use this variable to identify each of the logged-in user's names. In each of the WebSocket connections, the URL contains the name appended at the end of the connection URL:

```
resource function get [string name](http:Request req)
    returns websocket:Service|websocket:Error { }
```

The source code of the Ballerina resource function is as follows and handles incoming HTTP requests and broadcast messages:

```
clientName = name;
return service object websocket:Service {
    remote function onOpen(websocket:Caller caller) {
        clientConnectionMap[caller.getConnectionId()] =
            caller;
        caller.setAttribute(NAME, clientName);
        broadcastMessage(caller.getAttribute(NAME).
            toString() + " join the conversation");
    }
    remote function onTextMessage(websocket:Caller
        caller, string text) {
        broadcastMessage(caller.getAttribute(NAME).
            toString() + ": " + text);
    }
    remote function onClose(websocket:Caller caller, int
        statusCode, string reason) {
```

```
          _ = clientConnectionMap.remove(caller.
   getConnectionId());_VR
         broadcastMessage(caller.getAttribute(NAME).
         toString() + " left the conversation");
      }
   };
```

This resource function generates a `websocket:Service` object that manages the WebSocket connection. This service definition contains three events, which are `onOpen`, `onTextMessage`, and `onClose`. The `onOpen` function gets called when a new user is connected to the system. The first step when a new user joins is to add that user's caller connection to the `clientConnectionMap`. To keep this on the map, we used the call connection ID that is automatically generated by the system as the map key. Additional details that are associated with each caller can also be attached to the caller object itself. Here, we set the name of the user to the caller object. As the final step, a message is broadcast to all users to notify them that a new user has joined the chat.

The next function is `onTextMessage`, which triggers when the server receives a message from a client. Once this function gets triggered, the name of the user that sends the message is taken from the caller variable as we assigned it on the `onOpen` function with the `caller.getAttribute()` function. Then, that message is broadcast to all connected clients.

The last function defined on the client service is the `onClose` function, which is triggered when the socket connection gets closed. This socket connection can be closed due to the client actively closing the connection or due to a network failure. When a user is disconnected from the system, it removes the caller object from the `clientConnectionMap`. This function also sends a message to the existing clients that the user disconnected from the chat.

4. The broadcast function that broadcasts the message to all nodes is as follows:

```
function broadcastMessage(string message) {
   foreach var con in clientConnectionMap {
      var err = con->writeTextMessage(message);
      if err is websocket:Error {
          io:println("Error while sending message:");
      }
   }
}
```

This function simply loops through all of the socket connections on clientConnectionMap and writes the given message to the socket. If any error occurred during the writing process, it prints the error message on the terminal.

The frontend implementation of the chat application contains two textboxes that collect the name of the user and the message to send. Two buttons are used to connect to the socket server and to send the message to the socket server. Check the following screenshot of the HTML web application running on a web browser:

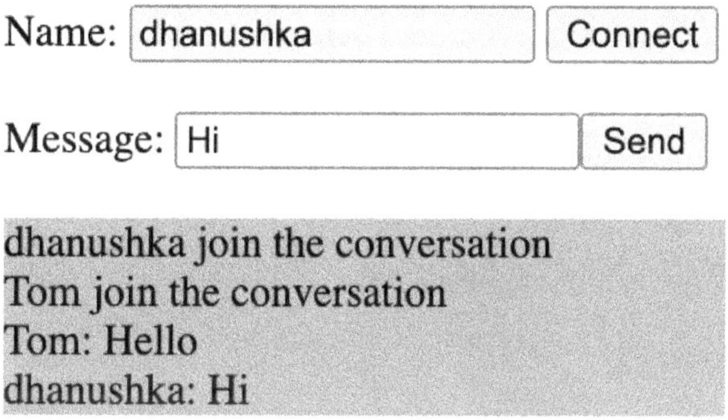

Name: dhanushka   Connect

Message: Hi   Send

dhanushka join the conversation
Tom join the conversation
Tom: Hello
dhanushka: Hi

Figure 4.8 – Chat application web page

5. When a user clicks on the **Connect** button, it calls the following JavaScript function:

```
function connectToServer() {
    ws = new WebSocket("ws://localhost:9091/chat/" +
        $("#name").val(), "xml", "my-protocol");
    ws.onmessage = function (frame)
        {$("#messageContent").html($("#messageContent").
        html() + "<div>" + frame.data + "</div>")};
}
```

This function creates a WebSocket connection with the given context URL. The name of the user is extracted from the input box with the jQuery library. In the next line of code, we assign a function to the WebSocket connection to specify what to do when it receives a message from the server. This function simply appends the content to the given HTML tag that represents the chat view. The message content is passed to this function as a variable called frame. We can access this variable to read the message content.

6. When you need to send a message, you can trigger the following function, which sends the message content to the server:

```
function sendMessage() {
    ws.send($("#message").val())
}
```

This sendMessage function reads the text in the textbox with the ID of the message and writes that text content to the WebSocket connection. Then the Ballerina service reads the message and broadcasts it to all connected users.

We can use a WebSocket connection to build a full-duplex communication requirement. The WebSocket protocol is simple by design and there are lots of libraries that you can use to work with the WebSocket endpoint. Other than WebSocket, you can also use long polling HTTP and HTTP/2 server push to archive full-duplex communication. In the next section, we will discuss the gRPC protocol, which is a lightweight protocol that is built on the HTTP/2 protocol.

Building Ballerina services with the gRPC protocol

We have already discussed HTTP, which is text-based. The gRPC (Remote Procedure Call) protocol is based on the HTTP/2 protocol. You can use the gRPC protocol with a web browser as well. You can find the JavaScript implementation of gRPC at https://github.com/grpc/grpc-web, which allows you to communicate with a remote gRPC server directly. In gRPC, standard REST protocol requests are handled by a server listening to an HTTP request. You can create different contexts to route the request and different resource functions to handle the request. The gRPC protocol can directly call remote functions along with arguments instead of parsing and redirecting HTTP request to different functions. gRPC stubs are used to invoke functions. Check the following diagram to see how a stub is used in a gRPC implementation:

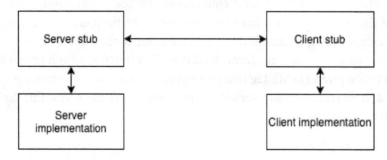

Figure 4.9 – Client and server communication via stub implementation

The gRPC server and the client can be implemented in four different ways. Since the gRPC contains streaming capability, rather than only sending a single message, the client and server can communicate with streams. gRPC is also a bidirectional protocol where both the server and the client have full-duplex communication. Here is an elaboration of the four different ways:

- **Unary blocking**: This implementation is very similar to the REST protocol where the client sends a request and waits for a reply. The server generates a response and responds to the client. The gRPC client sends a call to the remote procedure and receives the response to continue the program flow. This pattern can be used if you have a simple request-response use case.

- **Client streaming**: In the gRPC protocol, the client can send a stream of messages to the server. The server can listen to the stream and send a response after completely reading the stream. We can use this when the client needs to publish data to a server continuously.

- **Server streaming**: In this method, the client sends a request to the server and the server sends a data stream to the client application. Server streaming can be used where server need to send continuous data to the client application.

- **Bidirectional streaming**: In this method, both the client and the server communicate with each other using a stream of data. This is important when both the client and the server needs to send data as a stream.

Ballerina provides built-in support to implement gRPC communication. We will discuss building a sample Ballerina application that calculates a sum of items added to an order management system. To start building the application, follow these steps:

1. First, we need to create the protocol definition for the gRPC protocol. This can be implemented with **Protocol Buffers (protobuf)**, which was introduced by Google for lightweight data communication. Rather than using the JSON format, protobuf uses the binary format to transport data. The following is the protobuf definition that we are going to use to implement the sample:

```
syntax = "proto3";
import "google/protobuf/wrappers.proto";
service OrderCalculate {
    rpc calculateOneByOne (OrderItem) returns (google.
        protobuf.FloatValue);
    rpc calculateAsStream (stream OrderItem) returns
        (stream google.protobuf.FloatValue);
```

```
}
message OrderItem {
   string itemId = 1;
   int64 quantity = 2;
}
```

In the aforementioned code, in the first line of code, the syntax version is set to `"proto3"`. Next, we import `google/protobuf/wrappers.proto`, which contains predefined message structures. You can check all available message formats of this imported proto definition on the `https://github.com/protocolbuffers/protobuf/blob/master/src/google/protobuf/wrappers.proto` GitHub page. We create a service, `OrderCalculate`, as a protobuf service. Inside this service, we can identify a list of procedures that are listening to the gRPC client. Here, we have defined `calculateOneByOne` to implement unary blocking and `calculateAsStream` to implement bidirectional streaming procedures. The `calculateOneByOne` procedure reads `OrderItem` and returns a float value. The `calculateAsStream` procedure reads a stream of `OrderItem` and returns a stream of float values. Finally, we also need to define the `OrderItem` message structure to specify the procedure's input argument. This message structure contains `itemId` to hold the category of the item and quantity to hold the number of items.

The next step is to create the client and server code implementation. We will name the client project `grpc_client` and the server project name is `grpc_server`. After you create the project, use the following command to generate the stub files required for both the server and the client:

```
bal grpc --input <protobuf_file_location> --output
<outpu_file_location>
```

Make sure you copy the generated file to both projects' home directories.

2. Now we can start implementing the gRPC server. We will use the following `inventoryItem` data structure to hold the inventory item list in a table:

```
type InventoryItem record {|
    readonly string itemId;
    float price;
|};
type InventoryItemTable table<InventoryItem> key(itemId);
```

When the client sends a particular item, it checks for the item in this list and gets the price details. Then this price can be multiplied by the quantity to find the total price.

3. Now we can initialize this table with sample values as follows:

```
final readonly & InventoryItemTable itemList = table [
    {
        itemId: "item1",
        price: 120
    },{
        itemId: "item2",
        price: 20
    }
];
```

4. We can generate the Ballerina service implementation to implement procedure definitions. The service definition for the gRPC server, which is listening to port 9090, is as follows:

```
listener grpc:Listener grpcListener = new (9090);
@grpc:ServiceDescriptor {
    descriptor: ROOT_DESCRIPTOR,
    descMap: getDescriptorMap()
}
service "OrderCalculate" on grpcListener {
}
```

5. Now, we can add procedure definitions as resource functions to the OrderCalculate service. Check the following calculateOneByOne procedure implementation:

```
remote function calculateOneByOne(OrderItem item) returns
    float {
        return item.quantity * itemList[item.itemId].price;
}
```

This function simply reads OrderItem from the client and returns the price with the type of float. This resource function reads the quantity of the incoming message and multiplies it by the price taken from itemList. The response is sent back to the server stub.

6. Now we can build a client implementation to call this procedure with the gRPC protocol. To do this, we will first create a sample list of items that will be sent from the client:

```
OrderItem[] itemList = [
    {itemId: "item1", quantity: 10},
    {itemId: "item1", quantity: 5},
    {itemId: "item2", quantity: 12}
];
```

7. To call the gRPC service, we need to create the client connection and send each of these items one by one to the gRPC server as follows:

```
float total = 0;
OrderCalculateClient orderCalculateEP = check
  new("http://localhost:9090");
foreach OrderItem item in itemList {
    total += check  orderCalculateEP
      ->calculateOneByOne(item);
    io:println("Current total: ", total);
}
```

This program creates a new gRPC connection and assigns it to the `orderCalculateEP` variable with the `OrderCalculateClient` type. We can use `orderCalculateEP` to send data to the gRPC server. We can loop through all of the items with a `forEach` loop and send each item with the `orderCalculateEP ->calculateOnebyOne()` function call. This function calls the gRPC service and returns the total sum of each item. When we sum up all the returned values, we can calculate the total price of the item list.

gRPC communication can be implemented with bidirectional streaming as well.

To implement this, perform the following steps:

1. First, we need to add a new procedure, `calculateAsStream`, to the gRPC server with the following resource function:

```
remote function calculateAsStream(stream<OrderItem,
  error> itemStream) returns stream<float> {
    float[] responses = [];
    float total = 0;
    int i = 0;
```

```
error? e = itemStream.forEach(function(OrderItem value) {
    OrderItem = <OrderItem> value;
    InventoryItem = itemList[orderItem.itemId];
    total += orderItem.quantity * inventoryItem.price;
    responses[i] = total;
    i += 1;
  });
  return responses.toStream();
}
```

This function reads the data stream on the gRPC server side and loops through each of the items with the `forEach` loop. The total value is calculated by multiplying the quantity with the price corresponding to the given item ID taken from the `itemList` table. The result of each multiplication is sent back to the client as a stream by returning the responses float array as a stream.

2. The client implementation to send and read the stream of data is as follows:

```
OrderCalculateClient ep = check new("http://
    localhost:9090");
CalculateAsStreamStreamingClient streamingClient = check
    ep->calculateAsStream();
foreach OrderItem item in itemList {
    check streamingClient->sendOrderItem(item);
}
check streamingClient->complete();
float|grpc:Error result = streamingClient->receiveFloat();
while (result is float) {
    io:println("Current total: ", result);
    result = streamingClient->receiveFloat();
}
```

This program sends a stream of `OrderItem` record to the server and reads a stream of float values. The last stream float value is taken as the sum of the price of all the items sent.

The same as the unary blocking example, first we need to define the endpoint with the `OrderCalculateClient` type and get the response stream to the `CalculateAsStreamStreamingClient` type by calling the client stub with the `calculateAsStream` function. Then we can use the `streamingClient` variable to send the item list as a stream to the server. The end of the stream should be notified to the server with the `complete` function on the `streamingClient` variable. The result can be read with the `receiveFloat` function. This function can be called until the client reads all available streaming data.

Other than these two gRPC methods, you can also implement server streaming and client streaming in the same way. The gRPC protocol has become popular nowadays since it is fast and easy to use. You can replace the REST protocol with gRPC to create more responsive microservices applications. In the next section, we will discuss asynchronous communication methods, which are widely used in microservice architecture.

Asynchronous communication

In asynchronous communication, the caller does not wait for a response from the backend server. Instead, it sends the request to the backend server and continues to execute the program. In some cases, we cater an acknowledgement (ACK) to indicate that the message was delivered. The backend server handles those requests asynchronously and performs a task. Having a series of synchronous calls for microservices is usually not a good practice since if one of the services breaks, the entire series of transactions will fail and need to be rolled back. Usually, services in a microservice architecture should be stateless, responsive, and small in size as much as possible. Asynchronous communication methods help you to create stateless applications that are reliable and fast. In this section, we will discuss different types of asynchronous communication protocols and how the Ballerina language supports these.

Asynchronous communication over microservices

When you need to build a highly responsive service that produces the end output fast, you need to reduce the communication overhead. When applications become larger and larger, network communication also becomes more complex. Synchronous communication protocols such as HTTP create a chain of requests when they need to access multiple services. For example, the following diagram shows how data flows between multiple services with synchronous and asynchronous protocols:

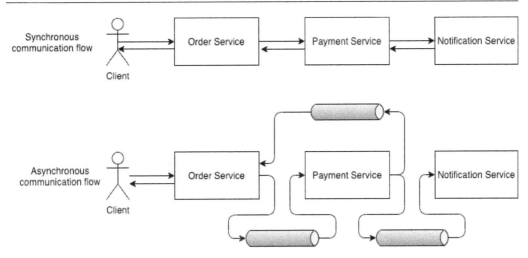

Figure 4.10 – Synchronous versus asynchronous communication flow

Message-based communication do not directly send messages to the target endpoint. Instead, they use a broker to publish messages. The client publishes messages to the broker and the target endpoint reads those messages from the broker. Message-based communication can be separated into two methods: message queues and publish-subscribe:

- **Message queues** are a point-to-point communication method. The client sends a message to the broker and one of the endpoints that are listening to the service reads the message. Message queues are important where a message is consumed by a single consumer service. Messages in a queue are kept until retrieved by a consumer.

- The **Publish-subscribe (Pub-sub)** method, on the other hand, publishes messages to a topic that subscribers are listening to. All the subscribers subscribed to the given topic read messages and perform an action based on them. The Pub-sub method is more a broadcast to all subscribers that are listening in rather than point-to-point communication like a message queue.

We will discuss both methods with Kafka and the RabbitMQ broker, which are widely used for messaging, in the coming sections. Ballerina supports both of these messaging protocols. We will discuss building a sample application with Kafka and RabbitMQ with Ballerina's built-in messaging interfaces.

Building the publisher and subscriber pattern with Apache Kafka

Apache Kafka is a distributed event streaming platform that is widely used in processing streams of data. Producers can publish data to a Kafka server and consumers can subscribe and read data. Kafka has high throughput, reliability, and availability. You can connect multiple Kafka nodes together to create a Kafka cluster by using ZooKeeper.

You need to download Kafka binaries to your machine first to start using Kafka. You can download Kafka from the official web page of Apache Kafka: `https://kafka.apache.org/downloads`. You can also use a Kafka Docker image from Docker Hub. After you download the distribution, extract files to a directory. Kafka uses Apache ZooKeeper to coordinate a Kafka cluster. A ZooKeeper node checks for the host machine's health to identify the Kafka server status.

You can start ZooKeeper by executing the command `./bin/zookeeper-server-start.sh config/zookeeper.properties` in the Kafka home directory. The `zookeeper.properties` file includes clustering configurations that you can customize. You can start the Kafka server with the command `./bin/kafka-server-start.sh config/server.properties`. Make sure you start the Zookeeper server first before starting the Kafka server as it fails while trying to connect to the ZooKeeper service.

The **Producer** is the component that publishes messages to the Kafka broker. Producers do not wait for an acknowledgment from the Kafka broker. Instead, they try to send messages as much as possible to offer higher throughput. Consumers, on the other hand, read messages from Kafka topics. Kafka is stateless since the messages read by the consumer are not kept there. Therefore, the consumer should provide the offset from where onward messages should be sent. Since Kafka reads data from the offset, we can rewind messages to take from any given offset. Kafka keeps messages for a given time as specified in the `server.properties` configuration file. You can have both time-based and message size-based message retention policies. In the time-based retention policy, you can set how long messages should be held in Kafka. In a message size-based policy, you can set the maximum size of messages that can be retained.

Kafka maintains messages in a topic where a consumer can pull messages from the required topic. The topic contains a set of partitions that store messages. A partition is an ordered, immutable sequence of messages that gets new messages appended continuously.

These partitions contain an offset. A consumer can set an offset value and read data onward from the offset. Kafka keeps its messages even after consumers have consumed those messages. You can change the offset to older messages such that you can read previously consumed messages again. Check the following diagram to see how Kafka producers and consumers connect with Kafka topics and partitions:

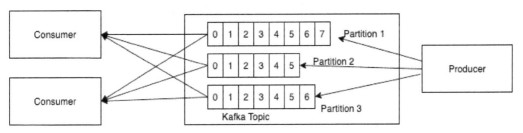

Figure 4.11 – Kafka producer-consumer architecture

Partitions are useful when multiple consumers need to consume messages parallelly. This increases the overall throughput of the system. Each message in Kafka can be uniquely identified by using the topic, partition, and offset value. If the producer hasn't specified a partition for the given topic, then it distributes messages in the partition in a round-robin manner. Consumers read messages from all the partitions. If multiple consumers are listening to the same topic, all of the consumers read all partitions in the Kafka topic.

Connecting Ballerina with Kafka

Ballerina's built-in Kafka client library lets you easily create Kafka producers and consumers. Now let's look at how to implement a simple Kafka message publisher with the Ballerina language. Check the following code, which publishes a JSON message to a Kafka topic:

```
kafka:ProducerConfiguration producerConfiguration = {
    acks: "all",
    retryCount: 3
}
kafka:Producer kafkaProducer = check new (kafka:DEFAULT_URL,
    producerConfiguration);
public function main() returns error? {
    json message = {"data": "Hello from Ballerina"};
    check kafkaProducer->send({topic: "TestTopic", value:
        message.toString().toBytes() });
    check kafkaProducer->'flush();
}
```

ProducerConfiguration allows you to define the connection details of the Kafka cluster. The number of acknowledgments acks field is set to all. This guarantees that the record will not be lost as long as at least one of the replicas is alive. Other options include 0, which does not wait for any acknowledgment. The 1 option allows only the leader to write a record to the local record and complete the transaction without waiting for other nodes to write it. retryCount is the number of times that a Ballerina producer needs to try sending a message if it gets failed.

For this example, a simple JSON object is generated and the JSON object is serialized into a byte array. Next, the byte array is sent to the `TestTopic` topic in the Kafka server. The `flushRecords` function in the `kafkaProducer` variable flushes the remaining messages to the Kafka server.

A Kafka consumer can also be implemented using the Ballerina language with Kafka listeners. The following is the endpoint definition used to create a Kafka listener:

```
kafka:ConsumerConfiguration consumerConfigs = {
    groupId: "consumer1",
    topics: ["TestTopic"],
    pollingInterval: 1
};
listener kafka:Listener kafkaListener = new (kafka:DEFAULT_URL,
    consumerConfigs);
```

Here, we connect the Kafka consumer to the Kafka server with the endpoint address `localhost:9092` by using the `bootstrapServers` property. `groupId` is a unique ID to identify the consumer group. The `topic` field lets you define topics that the client needs to listen to. Now we need to put the listener in place to listen and perform an action based on the received message:

```
service kafka:Service on kafkaListener {
    remote function onConsumerRecord(kafka:Caller caller,
        kafka:ConsumerRecord[] records) {
        foreach var kafkaRecord in records {
            string|error messageContent =
                string:fromBytes(kafkaRecord.value);
            if messageContent is string {
                json responseMessage = checkpanic
                    messageContent.fromJsonString();
                io:println("The message received: " +
                    responseMessage.toString());
            }
        }
        var commitResult = caller->commit();
        if (commitResult is error) {
            io:println("Error while commit");
```

```
            }
        }
    }
```

This service listens to the Kafka service and triggers the `onConsumeRecord` function when there are new messages. If there are new messages, we can loop through all messages and deserialize the message back to the JSON format. Here, we are just printing the decoded JSON message to the terminal output. You can perform business logic based on this message.

Consumer groups, on the other hand, have a parallel way of reading topics. Each consumer group in the Kafka cluster reads some specific partitions. If the number of consumers in a group is less than the number of partitions, then each consumer is allocated more than one partition to read. If the number of consumers is equal to the number of partitions, each consumer consumes messages from each partition. If there are more consumers than partitions, then there will be idle consumers. The following figure visualizes each of these three use cases where the number of consumers in a group is less than the number of partitions (a), the number of consumers in a group is more than the number of partitions (b), and both the number of topics and consumers are equal (c):

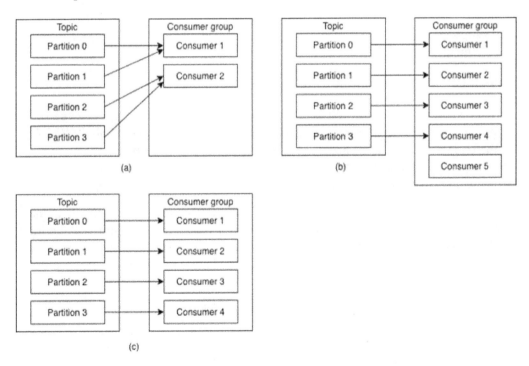

Figure 4.12 – Kafka consumer groups

In a Kafka listener definition, you can set multiple `groupId` values to mark that each listener belongs to a different consumer group. When you define multiple Ballerina programs with multiple `groupId` fields, the same `groupId` consumers behave as a single group of consumers.

In this section, we learned how to use Kafka as a message broker to implement the publish subscriber messaging pattern with Ballerina. In the next section, we will learn about another message broker, which is RabbitMQ, to implement a message queue.

Connecting Ballerina with RabbitMQ

RabbitMQ is an open source message broker that is widely used for asynchronous communication between different services. RabbitMQ is reliable when the message can be persisted, and the client can obtain an acknowledgment for each publishing message. RabbitMQ can be clustered together to create a highly available cluster such that if any of the RabbitMQ nodes die, another RabbitMQ node is able to handle the traffic. RabbitMQ supports AMQP, STOMP, and MQTT messaging protocols. You can also perform trace analysis with RabbitMQ by enabling Firehose traces.

You can download RabbitMQ binaries from the RabbitMQ download page: `https://www.rabbitmq.com/download.html`. Also, you can use the RabbitMQ Docker image to start a RabbitMQ server instance by pulling the RabbitMQ image from Docker Hub with the following Docker command:

```
docker pull rabbitmq
```

Once you download RabbitMQ, you can start it by running the `rabbitmq-server` script on the `<RabbitMQ_home>/sbin` directory. You can log in to the RabbitMQ UI using the URL `http://localhost:15672/` with the default username `guest` and the password `guest`. You will find an overview of the number of messages, message rates, and channel information on this web page.

We will now look at how Ballerina can be used in pub-sub architecture with the Ballerina language. The Ballerina service's listener provides an interface to create a simple Ballerina RabbitMQ message listener. This interface listens to messages from Ballerina and calls the function with the binary message content. Check the following example implementation of a Ballerina RabbitMQ message consumer:

```
listener rabbitmq:Listener channelListener =
new(rabbitmq:DEFAULT_HOST, rabbitmq:DEFAULT_PORT);
@rabbitmq:ServiceConfig {
    queueName: "TestQueue"
}
```

```
service rabbitmq:Service on channelListener {
    remote function onMessage(rabbitmq:Message message,
       rabbitmq:Caller caller) returns error? {
        string messageContent = check string:fromBytes(message.
           content);
        json responseMessage = checkpanic messageContent.
           fromJsonString();
        io:print("The message received: " + responseMessage.
           toString());
    }
}
```

Ballerina service annotations are used to define the queue name that the server is listening
to. This sample code listens to the queue with the name TestQueue. This service listens
to the RabbitMQ server hosted in localhost with the default port address 5672. When
Ballerina receives a message from the given queue, it triggers the onMessage function.
The message comes as a byte array that we need to convert to the required message
format. Since the message that we are going to publish is in JSON format, we need to
convert this message to JSON format. Here, we just print the message onto the StdOut.

Now we can create a producer that publishes messages to RabbitMQ. This code converts
the JSON message to a byte array and publishes it on the RabbitMQ server. Check the
following sample code, which publishes the JSON message to RabbitMQ:

```
rabbitmq:Client newClient = check new(rabbitmq:DEFAULT_
   HOST, rabbitmq:DEFAULT_PORT);check newClient-
      >queueDeclare("TestQueue");
json message = {"data": "Hello from Ballerina"};
check newClient->publishMessage({ content: message.toString().
   toBytes(), routingKey: "TestQueue" });
```

The RabbitMQ client creates a new connection with localhost by using port 5672. It
creates a new queue with the name of TestQueue. Then we can create the JSON message
that needs to be published. The JSON message is converted to a byte array and sent to the
queue using the publishMessage function.

Transactions can be used to handle message publishing and acknowledgments to roll
back changes in the event of failure. You can enclose the publishMessage function in
the producer and the basicAck acknowledge function in the consumer to mark them
as transactions. If any errors occurred in the transaction block, Ballerina will revert all
changes made to RabbitMQ.

Summary

In this chapter, we have discussed different methods, protocols, and patterns that are used to implement interservice communication. We have discussed how to find services on a Kubernetes cluster with multiple methods, including Kubernetes services and Consul. We addressed what a service mesh is and implemented a service discovery scenario with a Consul service mesh. Further, we have also discussed resiliency patterns, including retries, timeouts, failovers, and the circuit breaker pattern. We have also discussed client-side load balancing with Ballerina to improve the system's performance.

Then we discussed synchronous and asynchronous communication methods. Under synchronous communication methods, we explored more advanced concepts of the HTTP, including building form application, working with a multipart request, and the OpenAPI specification. We discussed the GraphQL interface to create a more sophisticated REST API to get only the required fields of the response message. The WebSocket interface was introduced as a full-duplex communication protocol and a simple chat application was created to demonstrate Ballerina socket server implementation. We also discussed the gRPC protocol, which is much faster and simpler than HTTP.

Under asynchronous communication methods, we discussed Kafka and RabbitMQ to implement publish subscriber patterns and message queues. Here, we sent a simple JSON message to and from a Ballerina application via Kafka and the RabbitMQ broker. By knowing about both asynchronous and synchronous method types, you can decide what type of communication methods to use for your system. The endpoint address for a given service can be found with service discovery tools. By combining service discovery with different types of communication protocols, you can build a sophisticated microservice architecture with the Ballerina language. More on asynchronous communication methods with sample code will be covered in the next chapter.

Questions

1. How would you implement an HTTP session with the Ballerina server? What are the tools that can be used to implement a service mesh other than Consul?

Further reading

- *Mastering Service Mesh: Enhance, Secure, and Observe Cloud Native Applications with Istio, Linkerd, and Consul*, by Anjali Khatri and Vikram Khatri, available at `https://www.packtpub.com/product/mastering-service-mesh/9781789615791`

- *Mastering Kubernetes: Level Up Your Container Orchestration Skills with Kubernetes to Build, Run, Secure, and Observe Large-Scale Distributed Apps*, Third Edition, by Gigi Sayfan, available at `https://www.packtpub.com/product/mastering-kubernetes-third-edition/9781839211256`

- *Learning Apache Kafka* – Second Edition, by Nishant Garg

Answers

1. An HTTP session is widely used in tracking a person's identity over multiple pages. Instead of sending authentication details again and again for each page request, the server sends a cookie to the browser and the browser sends it back, with all the requests that are made, to the server. At login time, you can set a session ID with the Ballerina cookie library. The `response.addCookie()` function can be used to add a new cookie to the response. You can keep the session ID on the server side and identify the user with request cookies from the next request onward. Other than Consul, you can use Istio and Linkerd to implement a service mesh. You can install both Istio and Linkerd by using a Helm Chart. You can use service mesh features such as security, log collection, tracing, and traffic management with these tools as well.

5
Accessing Data in Microservice Architecture

In this chapter, we will focus on accessing the database by using the Ballerina language with respect to monolithic 3-tier architecture-based applications. In the 3-tier architecture model, application logic is contained in the application layer. Single-database access for the entire system and database transaction queries are implemented in the application layer. Building an application that runs in a single database is simpler since we can join tables to query results. A microservice architecture, however, forces developers to have a database for each service. Therefore, we need to come up with a more scalable solution to handle problems related to data consistency in database per service architecture.

Here, we will explore various design patterns and how we can use them to implement the order management system that we discussed in the first chapter. Understanding these concepts and design patterns makes the system much simpler and more effective. The following are the topics that you will learn about in this chapter:

- Accessing data with Ballerina

- Managing transactions in Ballerina

- The role of **Domain-Driven Design** (**DDD**) in cloud native architecture

- Using event sourcing and *Command Query Responsibility Segregation* (*CQRS*) in a distributed system

At the end of this chapter, you should be able to understand the design concepts that we will use to implement a distributed application in a cloud environment with the Ballerina language.

Technical requirements

In this chapter, we are focusing on developing a distributed application with MySQL database as the database and RabbitMQ as the messaging platform. Therefore, you need to install the MySQL database and the RabbitMQ broker on your computer. In production environments, you can use containers, which can be directly run on container orchestration platforms such as Kubernetes.

You can find the code files for this chapter at `https://github.com/PacktPublishing/Cloud-Native-Applications-with-Ballerina/tree/master/Chapter05`.

The Code in Action video for the chapter can be found here: `https://bit.ly/2W2H9LZ`

Accessing data with Ballerina

Database access is a common requirement for almost all applications. Applications that need to maintain a state need to persist its status somewhere on the system. This status could be a permanent state or a temporary state. Databases are used for both permanent and in-memory data storage purposes since they provide a range of functionalities to read and write data. If data is to be permanent, we should keep data on permanent storage. If the data does not need to be permanently stored, this data can be persisted in an in-memory database.

Connecting the Ballerina application with MySQL

Data is a vital part of any software system. The database is the key component, which is used to store and query data. Programming languages have their own interfaces that can be used to communicate with databases. In short, these interactions can be divided into four parts known as CRUD operations. CRUD operations consist of the following:

- **Create**: Insert data into the database.
- **Read**: Read data from the database.
- **Update**: Update already existing data.
- **Delete**: Delete particular data from the database.

CRUD is simply a combination of all of these operations. CRUD is easily mapped with REST protocol operations such as PUT, GET, POST, and DELETE. Ballerina facilitates all of these operations with its own library collection. In the next section, we'll look at how to connect the Ballerina application to the popular MySQL database.

Ballerina built-in libraries help to operate with the MySQL database. MySQL is a commonly used relational open source database for data collection. MySQL offers a high level of security, scalability, reliability, and uptime. MySQL actively supports the building of a cloud native application with clustering capabilities. You can host several MySQL instances and link them to a cluster to provide high availability.

Ballerina uses the MySQL JDBC connector to connect to the MySQL database. You either need to place the MySQL database driver JAR in the Ballerina `lib` directory to import it as a platform library to run this project, or point the JDBC JAR file to the `Ballerina.toml` file. Ballerina's `lib` directory is located at `<Ballerina-Home> /bre/lib/distributions/jballerina-<VERSION>/bre/lib`. You can download MySQL drivers from the official MySQL Developer Zone. Placing the JAR file in the `lib` directory is important if you are running the Ballerina program as a single file.

Let's create a new Ballerina project for this example and import the JAR files instead of copying them to the `lib` directory. You need to add the artifact information to the `Ballerina.toml` file in order to include the drivers in the Ballerina runtime file. This approach is much more convenient than copying the drivers to the `lib` directory because you can point to different driver versions. On the other hand, this approach ensures that the Ballerina code mentions the version of the supported driver to be used. Therefore, always make sure to use this method to point to the library instead of directly copying the file to the `lib` directory.

First, you need to specify the libraries that are used as a TOML array under the `[[platform.java11.dependency]` section in the `Ballerina.toml` file. Since we are only using the MySQL library in this use case, there will only be a single item in the library array. It contains the following information about the library:

- `artifactId`: The artifact ID of the MySQL driver connector
- `version`: The driver version
- `path`: The path to the JAR file on your computer
- `groupId`: The group ID of the driver's JAR file

After setting all the previous properties, your complete `Ballerina.toml` file should look as follows:

```
[build-options]
observabilityIncluded = true


[[platform.java11.dependency]]
artafactId = "mysql-connector-java"
version = "8.0.x"
path = "/Users/libs/mysql-connector-java-8.0.x.jar"
groupId = "mysql"
```

You can then use the built-in Ballerina MySQL library to access MySQL functionality. First, we develop a simple application that can connect to the MySQL server and create a new database called `OMS_BALLERINA` as follows:

```
function initializeDB(mysql:Client mysqlClient) returns
    sql:Error? {
    sql:ExecutionResult result =
        check mysqlClient->execute("CREATE DATABASE IF NOT
            EXISTS OMS_BALLERINA");
}
public function main() returns error? {
    mysql:Client mysqlClient = check new (user = dbUser,
        password = dbPassword, host = "localhost", port = 3306);
```

```
sql:Error? dbError = initializeDB(mysqlClient);
check mysqlClient.close();
io:println("MySQL Client initialization for querying data
    successed!");
```

```
}
```

The `mysql:client` connector provides an object that you can use to interact with the database server. This initializes the connection with the given username and password. In the aforementioned example, the username and password are `root/root`. The host and port are optional parameters as it assigns the default host to localhost and the default port to `3306`, which is the default MySQL server port. Along with the optional parameters, the client initialization code for MySQL is as follows:

```
mysql:Client mysqlClient = check new (user = dbUser, password =
    dbPassword, host = "localhost", port = 3306);
```

Due to a database connection problem or due to authorization issues, this initialization can fail. If the database connection is not secure or broken, an error is returned as a communication failure. If an authorization failure occurs, it returns an error indicating that the access is denied. Here, we just return the error with the `check` keyword if any error does occur. If there are any errors, Ballerina will print them to the terminal.

You can now use the `mysqlClient` variable to perform some operations on the database. Here, we pass the `mysqlClient` variable to the `initializeDB` function, which initializes the database. To execute SQL queries, we can use the `execute` function given by the MySQL client. In the `execute` method, you can provide the query as a string. This function returns the result as either `ExecutionResult` or an error type. Since we have used the `check` keyword, if an error occurs while executing this command, it will return the error. It is possible to use `ExecutionResult` to verify the status of the execution query. The SQL database creation query is performed here to construct a new database called `OMS_BALLERINA`.

Once the query is executed, you can close the database connection with the `close` function provided by the Ballerina MySQL client library. Make sure you close the open connection once the database call is completed.

To create tables and insert the data, you can use the same `execute` function. After the database creation statement, you can execute `insert` statements as follows to create tables:

```
    result = check mysqlClient->execute("CREATE TABLE IF NOT
EXISTS OMS_BALLERINA.Customers(CustomerId INTEGER NOT NULL
AUTO_INCREMENT, FirstName  VARCHAR(300), LastName VARCHAR(300),
ShippingAddress VARCHAR(500), BillingAddress VARCHAR(500),
Email VARCHAR(300), Country  VARCHAR(300), PRIMARY KEY
(CustomerId))");
```

The following query inserts a sample value into the database. Inside the `execute` function, we added a query to insert a value into the `Customers` table:

```
    result = check mysqlClient->execute("INSERT INTO OMS_
BALLERINA.Customers(FirstName, LastName, ShippingAddress,
BillingAddress, Email, Country) VALUES('Sherlock', 'Holmes',
'221b, baker street', '221b, baker street', 'sherlock@mail.com',
'UK')");
```

Similarly, you can execute any SQL queries in the MySQL database.

In the previous example, we used a global level connection pool to connect with the database. With Ballerina, you can set up a local connection pool easily with connection pool parameters. This pool allows us to create a database connection based on the given connection pool parameters. The following parameters are used to define the connection pool:

- `maxOpenConnection`: Maximum number of open connections that the given pool can have. The default value is 15.

- `maxConnectionLifeTime`: This is the maximum lifetime that the connection pool can have in seconds. The default value is 1,800 seconds. You can set an unlimited time by setting this value to zero.

- `minIdleConnections`: This is the minimum number of active connections that can stay idle in the pool without creating additional connections.

Check the following example of setting custom pool parameters:

```
sql:ConnectionPool connPool = {
    maxOpenConnections: 5,
    maxConnectionLifeTime: 2000,
    minIdleConnections: 5
};
mysql:Client|sql:Error mysqlClient =
    check new (user = dbUser, password = dbPassword,
        connectionPool = connPool);
```

Here, we set the maximum open connection to 5, the maximum lifetime of the connection pool to 2000 seconds, and the minimum number of active connections to stay idle to 5.

Querying from MySQL

The Ballerina MySQL library provides the query function to query data from a database. This function returns a record stream that can be iterated over in order to fetch the output. Check the following example of reading query data into a record stream:

```
stream<record{}, error> resultStream = mysqlClient-
  >query("Select * from OMS_BALLERINA.CustomersTable");
error? e = resultStream.forEach(function(record {} result) {
    io:println("Customer detials: ", result);
    io:println("Customer first name: ", result["FirstName"]);
    io:println("Customer last name: ", result["LastName"]);
});
```

The foreach statement will iterate through the resultStream variable to retrieve each raw result. Each result can be accessed by the name of the record key. Here, we read the column value from the result and print it into the terminal. Once you read the values, you can close the connection if the connection is no longer required.

Instead of using an empty record to retrieve the result, we can define the record structure as follows to retrieve the table data:

```
type Customer record {|
    int customerId?;
    string firstName;
    string lastName;
    string shippingAddress;
    string billingAddress;
    string email;
    string country;
|};
```

Since this record structure does not change dynamically, we can add |. This record can be used directly to retrieve data from a database with the structure that follows:

```
function readData(mysql:Client mysqlClient) returns error? {
    stream<record{}, error> resultStream =
        mysqlClient->query("Select * from
          OMS_BALLERINA.Customers", Customer);
        stream<Customer, sql:Error> customerStream =
        <stream<Customer, sql:Error>>resultStream;
    error? e = customerStream.forEach(function(Customer
      customer) {
        io:println("Customer first name " + customer.firstName);
        io:println("Customer first name " + customer.lastName);
    });
}
```

Here, we read the query result as the `stream<record{}, error>` record stream. Then this stream is cast into the target record format `stream<Customer, sql:Error>` with the `<stream<Customer, sql:Error>>` cast operator. Then the `customerStream` variable can be used as a stream of records for further usage.

Accessing queries is hard with string values. Instead of using plain text, next, we will discuss how to use parameterized queries to execute a SQL expression.

Using parameterized queries

Parameterized queries are a handful when you are writing some complex SQL queries. Rather than using strings, the `ParameterizedQuery` type provides a nice way to build queries. Check the following example, which adds a new customer into the `Customers` table. Customer object initialization is as follows:

```
Customer = {
    firstName: "Tom",
    lastName: "Wilson",
    shippingAddress: "225, Rose St, New York",
    billingAddress: "13, 13 St New York",
    email: "tom@mail.com",
    country: "USA"
};
```

Next, we will look at how to insert this record into the MySQL table with the use of parameterized queries. Check the following example:

```
function insertCustomer(mysql:Client mysqlClient, Customer
    customer) returns error?{
    sql:ParameterizedQuery insertQuery = 'INSERT INTO
        OMS_BALLERINA.CustomersTable(FirstName, LastName,
            ShippingAddress,
        BillingAddress, Email, Country) VALUES(${customer.
            firstName},
        ${customer.lastName}, ${customer.shippingAddress},
        ${customer.billingAddress}, ${customer.email},
            ${customer.country})';
    sql:ExecutionResult result = check mysqlClient-
        >execute(insertQuery);
    io:println("Inserted Row count: ", result.affectedRowCount);
}
```

In this function, we used the `Customer` record to create the SQL expression. Customer properties are attached to the `customer` variable. This function generates relevant SQL expressions by using the given `customer` record. We can provide a SQL expression along with the template by adding variables with the `${ }` format. The SQL expression replaces `${ }` with the actual variable value and generates the final SQL expression.

This notation is a handful, so you can use it to write custom SQL templates to generate SQL expressions. You can serialize a given record with templates and generate SQL expressions. Instead of adding records one by one, in the next section, let's discuss using batch execution to execute SQL queries as a batch.

Using batch execution

The previous example is to insert a single raw entry into the database. Instead of running multiple SQL queries one by one, batch execution can run multiple queries as a single batch. This improves the efficiency of each query compared to running them one by one.

Think of adding several customers to the system for the previous example. You may call the `insertCustomer` function repeatedly to insert data into the database. Instead of multiple query executions, all statements can be executed once. Here, we will build an array of clients who need to be added to the database:

```
Customer[] customers = [
    {
        firstName: "Tom",
        lastName: "Wilson",
        shippingAddress: "New York",
        billingAddress: "New York",
        email: "tom@mail.com",
        country: "USA"
    }, {
        firstName: "Bob",
        lastName: "Ross",
        shippingAddress: "California",
        billingAddress: "California",
        email: "bob@mail.com",
        country: "USA"
    }
];
```

Instead of looping through this variable and sending it to the `insertCustomer` function, we create a separate function that accepts the customer detail array as follows:

```
function insertCustomer(mysql:Client mysqlClient, Customer[]
    customers) returns error?{
    sql:ParameterizedQuery[] insertQueries =
        from var customer in customers select 'INSERT INTO OMS_
            BALLERINA.CustomersTable(
        FirstName, LastName, ShippingAddress, BillingAddress,
            Email, Country)
        VALUES(${customer.firstName}, ${customer.lastName},
        ${customer.shippingAddress}, ${customer.
            billingAddress},
        ${customer.email}, ${customer.country})';
    sql:ExecutionResult[] result = check mysqlClient-
        >batchExecute(insertQueries);
    foreach var summary in result {
        io:println(summary);
    }
    io:println("Batch execution successful");
}
```

In this implementation, we will build a `ParameterizedQuery` array to hold customer insertion queries. The same as in the previous example, we use the parameterized string to insert variables into the SQL statement. The `BatchExecute` function can be used to create SQL expressions. This function inputs the `ParameterizedQuery` array and returns the `ExecutionResult` array or an error in the event of failure.

The `ExecutionResult` type contains the result of execution for each of the SQL queries provided. It has two properties: `affectedRowCount`, which returns the number of rows getting changed, and `lastInsertId`, which is the generated ID for each record. These attributes can be used optionally to check the status of execution.

MySQL data types and Ballerina data types

Applications need to deal with various data types, including numbers, strings, byte streams, and so on. MySQL has a rich data type structure that can be used to display any kind of data. So, first, we can divide the MySQL data types into the following categories:

- Numeric
- String
- Date time
- Binary data

We will discuss each of these data types in detail.

Numeric

The numeric category contains numbers that need to be represented. Numbers can be either decimals or integers. MySQL supports a different variety of number formats, including integers and decimals. Integers include TINYINT, SMALLINT, MEDIUMINT, INT, and BIGINT. The maximum value that a MySQL integer can have is 18,446,744,073,709,551,615 (the maximum value that can have 64 bits), which can be represented by BIGINT. Those values can be represented in Ballerina with the int data type. Since in Ballerina the int type has 64 bits, it is able to handle any of these integer datatypes.

MySQL contains FLOAT, DOUBLE, and DECIMAL to represent floating-point numbers. MySQL FLOAT is single-precision and DOUBLE is double-precision. You can use float or decimal in Ballerina to use these MySQL datatypes. The MySQL DECIMAL type is used to store exact numeric values with much more precision in the database. The DECIMAL (P, D) type can have precision (P) of 65 bits and a decimal scale (D) of 30 bits. In Ballerina, you can use decimals that can handle 128 bits to handle MySQL decimal values.

MySQL BOOLEAN is 1 bit and is used to store either a true or false value. Ballerina has the same mapping value as Boolean. You can use Boolean in Ballerina to access the MySQL BOOLEAN type directly.

String

You can use either CHAR (N) or VARCHAR (N) to represent a string in MySQL. CHAR is of fixed length, and it allocates a fixed size block (with a size of N) in the database memory. VARCHAR is of variable length and allocates characters until it reaches the defined maximum value (the value of N). You can use a string to manage the values of CHAR and VARCHAR in the Ballerina language.

Date time

The date time format is used to represent the date and time. DATE, TIME (N), DATETIME, and TIMESTAMP are the date time formats that are supported by the MySQL database:

- DATE represents the date in the *YYYY-MM-DD* structure.

- TIME (N) is represented in the *HH:MM:SS* format. N represents the fractional part, which can have a maximum of 6 digits.

- DATETIME is used to represent both date and time in the *YYYY-MM-DD HH:MM:SS* format.

- TIMESTAMP is also similar to DATETIME and keeps the time format the same as DATETIME. The difference between these two is the time range that can be stored. DATETIME has a range of *1000-01-01 00:00:00 to 9999-12-31 23:59:59* whereas TIMESTAMP has a range of *1970-01-01 00:00:01 to 2038-01-19 08:44:07*.

With Ballerina, you can access all of these MySQL types using Ballerina strings and time types. Create a table that contains time fields by executing the following SQL query:

```
CREATE TABLE OMS_BALLERINA.DatetimeTable (
    ID int(11) NOT NULL,
    Date DEFAULT NULL,
    Time time(6) DEFAULT NULL,
    Datetime DEFAULT NULL,
    Timestamp NULL DEFAULT NULL,
PRIMARY KEY (ID) ) ENGINE=InnoDB DEFAULT CHARSET=utf8;
```

This query contains all the different types of date format that MySQL supports. You can create the following record type to parse the data into the Ballerina variable:

```
type TimeTable record {|
    int id;
    time:Date date;
    time:TimeOfDay time;
    time:Utc datetime;
    time:Utc timestamp;
|};
```

Here, you may notice that some variables can use a string while some use the time type to retrieve MySQL data. If you choose the Ballerina time type to retrieve MySQL date time data, then you can use Ballerina built-in time functions to manipulate that data.

Binary data

MySQL uses BLOB as the data type to store binary data. The BLOB type includes TINYBLOB, BLOB, MEDIUMBLOB, and LONGBLOB. The only difference between each of these BLOB types is the amount of data they can store. TINYBLOB can hold 255 bytes, BLOB can hold 64 KB, MEDIUMBLOB can hold 16 MB, and LONGBLOB can hold up to 4 GB of storage.

You can use the Ballerina byte array to retrieve BLOB types. If you need to interact with images, files, or binary streams, you can use MySQL BLOB with Ballerina byte arrays. For example, the following SQL query creates a table with a BLOB field and inserts a sample value into it:

```
CREATE TABLE OMS_BALLERINA.BlobTable(
    ID INT NOT NULL,
    BlobValue BLOB NULL,
    PRIMARY KEY (id));

INSERT INTO OMS_BALLERINA.BlobTable(id, BlobValue) VALUES
    (1, b'0110100001100101011011000110110001101100');
```

You can use the following data structure to retrieve table output:

```
type BlobTable record {|
    int id;
    byte[] blobValue;
|};
```

Next, we will generalize accessing different types of databases by using the JDBC driver. We can connect any databases that provide a JDBC connector to connect with it. We will take the H2 database JDBC driver as an example in the next section.

Connecting databases with JDBC drivers

Ballerina provides support for multiple databases with unique connectors for each database. It also provides a common way of accessing relational databases with JDBC drivers. You can use JDBC drivers to connect to almost any kind of relational database that already has JDBC drivers.

There are different JDBC driver implementations for each type of database. The way it connects to a database is described by using a connection string. A connection string contains the host name, port number, database name, and other optional parameters. A relevant JDBC driver reads this connection string and connects to the database with the given parameters.

For example, we can connect a Ballerina program to the H2 database by using the JDBC driver. Since a JDBC driver is already included in Ballerina, you don't need to download and point to it as a dependency on the program.

H2 is a lightweight database written in Java. H2 supports both in-memory and disk-based persistence. If you need to use a lightweight database to access the H2 service, this would be an ideal solution. Check out the following example of creating a new H2 database connection:

```
jdbc:Client|sql:Error jdbcClient = new
   ("jdbc:h2:file:./target/customers", "root", "root");
```

Here, we need to provide a connection string that includes the location of the file. The file template will be as follows. Optionally, you can have the `file` prefix in the connection string:

```
jdbc:h2:[file:][<path>]<database_name>
```

If you need to use it as an in-memory database, the connection URL template will be as follows. In this configuration, data does not persist in permanent storage. You can use the database name to access the same database if you are running the Ballerina application in the same JVM:

```
jdbc:h2:mem:<database_name>
```

Optionally, you can also create an H2 database server and connect with it with the following connection URL template:

```
jdbc:h2:tcp://<server>[:<port>]/[<path>]<database_name>
```

You can also set the connection pool parameters in the same way as the MySQL example and give it as an option while creating the database.

You can use the `execute` function to execute SQL queries as well. The same as in the MySQL example, the `execute` function returns either an error or `ExecutionResult`. You can use this to check the status of the execution result.

By using the `query` function, we can extract data from the database in the same way we did with the MySQL database. If we were to use the same `Customer` record, the query code segment would be as follows:

```
function simpleQuery(jdbc:Client jdbcClient) {
    stream<record{}, error> resultStream =
        jdbcClient->query("Select * from Customers", Customer);
    stream<Customer, sql:Error> customerStream =
        <stream<Customer, sql:Error>>resultStream;
    error? e = customerStream.forEach(function(Customer
        customer) {
        io:println(customer);
    });
}
```

JDBC drivers can also be used to access data services provided by different cloud providers. Snowflake is a popular **Data-as-a-Service (DaaS)** provider. In the next section, we will learn how we can use the JDBC driver to access Snowflake's services.

Connecting Ballerina application with DaaS platforms

Snowflake is a popular DaaS provider. With Snowflake, you can create databases, tables, schemas, and views in the same way as other popular databases. Instead of hosting them on your own, you can use Snowflake's built-in servers to store databases. Optionally, you can use Snowflake with popular IaaS providers such as AWS, Azure, and GCP.

You can use the same JDBC connector to connect to Snowflake database services. You can create an account in Snowflake to start building a cloud-based data service. First, you need to create a new database called BALLERINA_TEST on the Snowflake management console. Once you create the database, you can create tables from the UI by providing a table schema.

For this example, we will create the same table that we used to implement the MySQL example. You can also use the JDBC interface to create databases and tables by running the SQL commands.

You need to download the Snowflake database drivers from their website or from the Maven repository. Here, I'm using the Snowflake 3.9.0 JDBC version. Once you download the JDBC driver, you need to define it as a dependency in the `Ballerina.toml` file. The `Ballerina.toml` configurations for the Snowflake dependency are as follows:

```
[[platform.java11.dependency]]
artafactId = "snowflake-jdbc"
version = "3.9.2"
path = "<snowflake_jar_path>/snowflake-jdbc-3.9.2.jar"
groupId = "net.snowflake"
```

Then you need to create a database connection string for the Snowflake database. The connection template for the Snowflake database is as follows:

```
jdbc:snowflake://<account_name>.snowflakecomputing.
com/?<connection_params>
```

The account name can be found from the management console URL that you used to create databases. You can submit databases and connection details along with the connection parameters as well.

The execute and query functionalities are the same as the JDBC connection. Check the following example of reading table entries with a `select` statement:

```
jdbc:Client jdbcClient = check new
    ("jdbc:snowflake://xxxx.snowflakecomputing.com/
     ?db=BALLERINA_TEST", user=dbUser, password = dbPassword);
stream<record{}, error> resultStream =
     jdbcClient->query("Select * from Customers");
error? e = resultStream.forEach(function(record {} result) {
    io:println("Full Customer details: ", result);
});
```

Here, we provide the database name as the connection parameter along with the username and password as input arguments for the database initialization function. The `query` function will execute the selected query and get the result as a record stream.

So far, we've only discussed how to create a simple Ballerina application that can interact with a database. When we implement a database access program in a distributed system or concurrently running applications, we also need to think about how each of these distributed services will simultaneously access the databases. In the next section, we will look at the details of handling transactions in the Ballerina language.

Managing transactions in Ballerina

It is a common requirement for applications to access databases concurrently. In this case, you need to focus seriously on the transaction management of the system to ensure that the system does not end up in inconsistent states. We will discuss more about transaction management systems with reference to the order verification use case in the order management system.

Building an order management system

To implement the order management system, first, we need to define the table structure that holds the data. The following diagram contains all the tables that we need to implement the order verification scenario.

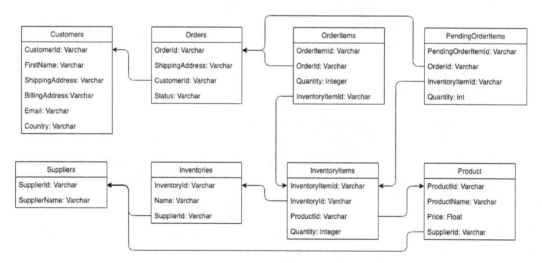

Figure 5.1 – Order management system database structure

The supplier is the entity that adds products to the system. The list of suppliers is saved in the `Suppliers` table. A supplier can have a supplier name and a supplier ID to uniquely identify it. Each supplier can have multiple inventories. Each inventory has a name and `SupplierId` to find who owns this inventory. These suppliers can add new products to the system. Each of these products is stored in the `Product` table. The product table keeps the name of the product and the price. `SupplierId` is used to track which supplier owns this product.

The `InventoryItems` table contains a list of items for each inventory. The `InventoryItems` table has `InventoryId`, which is used to track the relevant inventory, product ID, and quantity of each product ID. When the order is made by the customer, the number of items in inventory is decreased according to the order and a new entry is created in the `PendingOrderItems` table. This table holds the `inventoryItemId` and the quantity is reduced from the `InventoryItems` table.

The `Customers` table keeps all the data that is required to identify the customer. This table includes fields to store the customer's name, address, and email address. You can expand this table if you wish to store more information. The `Orders` table keeps the data for each order. The order contains data such as shipping addresses for the order and the order status. Each order row has a customer ID to track who owns the order. Orders can have one of the following statuses:

- **OrderCreated**: The order was successfully created.
- **OrderVerified**: Inventory has verified that the order can be fulfilled.
- **OrderVerificationFailed**: Inventory has rejected the order since it cannot fulfil the order.
- **Paid**: Payment was successful for the order.
- **PaymentFailed**: Payment did not succeed and the order has failed after the payment process.
- **OrderCanceled**: The customer canceled the order before proceeding to the payment or verification stage.
- **Shipping**: The delivery service has taken the order and sent the shipment.
- **Delivered**: The order has been delivered to the customer.
- **OrderDeliveryFailed**: The order has not been sent to the customer by the given date.

Those states can be visualized in the following state diagram:

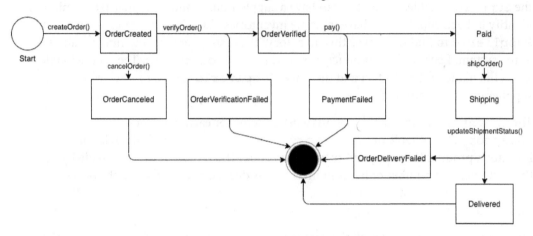

Figure 5.2 – Order management system state transition diagram

Each item in an order is stored in the `OrderItems` table. This table includes selected items from the inventory with their quantity. Each order item has an `orderId` field to track the order to which it belongs. Each order can have multiple order items to store each item in the order.

Building this type of order management is easy with the Ballerina programming language along with simple database calls. In the next section, we will learn how to use Ballerina with MySQL to implement this use case.

Initializing the MySQL database for the order management system

The order management system that we discussed can be implemented as a collection of database reads and writes. For that, we need to create database table schemas to store data in the table. For this example, we will use MySQL database, which is an RDBMS database. But you can also try with a NoSQL database as well. The steps are as follows:

1. First, we need to create a database named `OMS_BALLERINA` on the MySQL server using the following SQL query:

```
CREATE DATABASE IF NOT EXISTS OMS_BALLERINA
```

2. Next, we need to create tables to hold customers', orders', and suppliers'
    information. To do this, we can execute the following SQL queries to generate
    those tables:

```
CREATE TABLE 'Customers' (
    'CustomerId' varchar(50) NOT NULL,
    'FirstName' varchar(300) DEFAULT NULL,
    'LastName' varchar(300) DEFAULT NULL,
    'ShippingAddress' varchar(500) DEFAULT NULL,
    'BillingAddress' varchar(500) DEFAULT NULL,
    'Email' varchar(300) DEFAULT NULL,
    'Country' varchar(300) DEFAULT NULL,
    PRIMARY KEY ('CustomerId')
) ENGINE=InnoDB DEFAULT CHARSET=utf8;
```

3. Execute the following query to create the Orders table:

```
CREATE TABLE 'Orders' (
    'OrderId' varchar(50) NOT NULL,
    'ShippingAddress' varchar(500) DEFAULT NULL,
    'CustomerId' varchar(50) DEFAULT NULL,
    'Status' varchar(50) DEFAULT NULL,
    PRIMARY KEY ('OrderId')
) ENGINE=InnoDB DEFAULT CHARSET=utf8;
```

The following query generates the OrderItem table that stores each item of
the orders:

```
CREATE TABLE 'OrderItems' (
    'OrderItemId' varchar(50) NOT NULL,
    'OrderId' varchar(50) DEFAULT NULL,
    'Quantity' int DEFAULT NULL,
    'inventoryItemId' varchar(50) DEFAULT NULL,
    PRIMARY KEY ('OrderItemId'),
    KEY 'OrderId_idx' ('OrderId')
) ENGINE=InnoDB DEFAULT CHARSET=utf8;
```

4. The next query is to create the `Suppliers` table:

```
CREATE TABLE 'Suppliers' (
    'SupplierId' varchar(50) NOT NULL,
    'SupplierName' varchar(200) DEFAULT NULL,
    PRIMARY KEY ('SupplierId')
) ENGINE=InnoDB DEFAULT CHARSET=utf8;
```

Suppliers can add product types into the system by adding records to the `Products` table. The following SQL query generates a table to store a product list:

```
CREATE TABLE 'Products' (
    'ProductId' varchar(50) NOT NULL,
    'ProductName' varchar(200) DEFAULT NULL,
    'Price' float DEFAULT NULL,
    'SupplierId' varchar(50) DEFAULT NULL,
    PRIMARY KEY ('ProductId')
) ENGINE=InnoDB DEFAULT CHARSET=utf8;
```

5. Each supplier can have inventories. These inventories are saved in the following `Inventories` table:

```
CREATE TABLE 'Inventories' (
    'InventoryId' varchar(50) NOT NULL,
    'Name' varchar(100) DEFAULT NULL,
    'SupplierId' varchar(50) DEFAULT NULL,
    PRIMARY KEY ('InventoryId'),
    KEY 'SupplierId_idx' ('SupplierId')
) ENGINE=InnoDB DEFAULT CHARSET=utf8;
```

6. Items in inventories are kept in the `InventoryItems` table. That table is generated by the following SQL query:

```
CREATE TABLE 'InventoryItems' (
    'InventoryItemId' varchar(50) NOT NULL,
    'InventoryId' varchar(50) DEFAULT NULL,
    'ProductId' varchar(50) DEFAULT NULL,
    'Quantity' int DEFAULT NULL,
    PRIMARY KEY ('InventoryItemId'),
    KEY 'InventoryId_idx' ('InventoryId'),
    KEY 'ProductId_idx' ('ProductId')
) ENGINE=InnoDB DEFAULT CHARSET=utf8;
```

7. Pending orders that have not proceeded to payment are stored inside the PendingOrderItems table. This holds InventoryItemId along with the quantity:

```
CREATE TABLE 'PendingOrderItems' (
    'PendingOrderItemId' varchar(50) NOT NULL,
    'OrderId' varchar(50) DEFAULT NULL,
    'InventoryItemId' varchar(50) DEFAULT NULL,
    'Quantity' int DEFAULT NULL,
    PRIMARY KEY ('PendingOrderItemId')
) ENGINE=InnoDB DEFAULT CHARSET=utf8;
```

Now we need to define data structures to access these entries with a Ballerina record as it provides a more organized way of handling data. Here, we will be using the following data structures to access these tables:

```
public type Order record{
    string orderId;
    string customerId;
    string status;
    string shippingAddress;
};
```

The order record holds the mapping data for the Orders table:

```
public type OrderItem record {
    readonly string orderItemId;
    string orderId;
    int quantity;
    string inventoryItemId;
};
type OrderItemTable table<OrderItem> key(orderItemId);
```

The `OrderItem` record holds each row in the `OrderItems` table. We used a table here to hold a list of order items. Since each order has multiple products, the table structure keeps this list of products with an indexing key of its own ID. This makes it easy to work with multiple records, rather than using arrays, as it supports accessing data with indexes:

```
type InventoryItem record {
    readonly string inventoryItemId;
    string inventoryId;
    string productId;
    int quantity;
};
type InventoryItemTable table<InventoryItem>
  key(inventoryItemId);
```

The same as the `OrderItem` record, the `InventoryItem` record also contains a table type to hold a list of items. Each inventory can have a table of inventory items.

In the same way that you could create records for customers, suppliers, products, inventories, and pending order items, you can use these data structures directly to extract data from the database instead of accessing key indexes with streams.

Building an order management system with Ballerina

With monolithic architecture, we can use database queries to build an order management system. For that, first, we need to implement a server that accepts incoming requests. The following code creates an HTTP server that consumes `post` requests that come into the `addNewOrder` resource function:

```
service /OrderAPI  on new http:Listener(9090) {
    resource function post addNewOrder(@http:Payload json
        orderDetails)
    returns json|http:InternalServerError {
        // order adding logic goes here
    }
}
```

Then, we can implement the add order functionalities inside the addNewOrder function. This function takes order details as a JSON input. We can decompose that JSON to extract the data from the client request. Check the following example, which reads the content of the order details and sends a database call to add a new record to track the order:

```
do {
    json shippingAddress = check orderDetails.shippingAddress;
    json customerId = check orderDetails.customerId;
    string orderId = check createOrder(shippingAddress.
      toString(), customerId.toString());
    json response = {
        orderId: orderId
    };
    return response;
} on fail error e{
    log:printError("Error while adding order", e);
    http:InternalServerError internalError = {};
    return internalError;
}
```

Here, we read shippingAddress and customerId from the incoming request and send them to the createOrder function, which adds a new row entry to the Orders table. The createOrder function returns the order ID if it is successfully inserted into the row. Then we can do a type check to confirm the createOrder function has been successfully executed and send back the order ID. If successful, it creates a response with response code 200 along with the order ID that just got created and sends it to the client application.

The definition of the createOrder function is as follows and that adds a new entry in the Orders table. This function takes ShippingAddress and CustomerId as its input arguments. By executing an insert SQL statement, it adds a new entry into the Orders table. If any errors are detected, the createOrder function just returns back to the caller function:

```
function createOrder(string shippingAddress, string
  customerId) returns error|string{
    jdbc:Client jdbcClient = check new (jdbcUrl, dbUser,
      dbPassword);
    string orderId = uuid:createType1AsString();
```

```
    sql:ParameterizedQuery createOrder = 'INSERT INTO
      OMS_BALLERINA.Orders(
        OrderId, ShippingAddress, CustomerId, Status)
          VALUES(${orderId}, ${shippingAddress},
          ${customerId}, 'Created')';
    sql:ExecutionResult result =
      check jdbcClient->execute(createOrder);
    check jdbcClient.close();
    return orderId;
}
```

The `OrderId` field for the `Orders` table is generated by using the UUID generation function. You can use MySQL auto-increment instead of using the UUID generated for this example. But in upcoming implementations, you will realize that auto-increment generated IDs are not scalable with distributed systems.

You can use the same type of implementation to implement other functionalities such as adding new customers, adding products, updating customer information, updating product details, handling inventories' data, and so on. These are just a set of SQL operations that were executed against the database. For each operation, you can create a new resource that handles client requests.

Customers can place orders by selecting products from the suppliers. After the orders have been placed, the orders need to be verified with the suppliers, in terms of whether they are available or not, before the payment is made. Once the supplier verifies the items, the items are locked in such a way that another customer cannot add them to their list.

In order to implement this verification process, first we need to query all the inventory items that belong to the given order ID. We can implement this query with a SQL select statement on the `OrderItems` table to filter all order items with a given order ID:

```
public function getOrderItemTableByOrderId(string orderId)
  returns OrderItemTable|error {
    jdbc:Client jdbcClient = check new (jdbcUrl, dbUser,
      dbPassword);
    stream<record{}, error> resultStream = jdbcClient-
      >query('SELECT * FROM OrderItems WHERE
        OrderId=${orderId}', OrderItem);
    stream<OrderItem, sql:Error> orderItemStream =
      <stream<OrderItem, sql:Error>>resultStream;
```

```
        OrderItemTable = table [];
        error? e = orderItemStream.forEach(function(OrderItem
            orderItem) {
            orderItemTable.put(orderItem);
        });
        check jdbcClient.close();
        return orderItemTable;
    }
```

The result of the query `resultStream` is a stream of `OrderItems`. Since streams are not reusable, we can convert this stream into a table by iterating over it. All order items are added to a table with the type of `OrderItem`, which contains a key to the order item ID.

Now we have a table of order items that belong to an order. These items also have a reference ID to the inventory item that they have been taken from. Next, we can create another function to get the available quantity in the inventory by querying the `InventoryItems` table. This function takes `inventoryItemId` as its input argument and returns the available product quantity. This function filters the product list and returns the number of available products to the system. Let's name this function `getAvailableProductQuantity`:

```
function getAvailableProductQuantity(jdbc:Client
    jdbcClient, string inventoryItemId) returns int|error {
    stream<record{}, error> resultStream = jdbcClient-
        >query('SELECT Quantity FROM InventoryItems WHERE
            InventoryItemId = ${inventoryItemId}');
    record {|record {} value;|}? result = check
        resultStream.next();
    if (result is record {|record {} value;|}) {
        return <int>result.value["Quantity"];
    }
    return error("Inventory table is empty");
}
```

The next step is as simple as testing whether the number of items in the supplier inventory is larger than the number of items in the customer order. If so, the customer order can be fulfilled and continue to payment. Otherwise, the customer will have to reduce the number of items or cancel the order. This is a simple calculation and the logic can be written in the Ballerina language inside the `verifyOrder` function as follows:

```
OrderItemTable orderItems = check
    getOrderItemTableByOrderId(jdbcClient, orderId);// Line 2
boolean validOrder = true;
foreach OrderItem item in orderItems {      // Line 4
    int orderQuantity = item.quantity;
    int inventoryQuantity = check getAvailableProductQuantity
        (jdbcClient, item.inventoryItemId);
    if orderQuantity > inventoryQuantity {          // Line 9
        validOrder = false;
        break;
    }
}
if validOrder {
    check reserveOrderItems(jdbcClient, orderItems);
    check setOrderStatus(jdbcClient, orderId, "Verified");
} else {
    check setOrderStatus(jdbcClient, orderId,
        "VerificationFailed");
}
```

On line 1, take all the order items for a given order ID and assign them to a table. Then we can iterate over those order items as mentioned on line 4. Next, we can check the available quantity of the given item ID by calling the `getAvailableProductQuantity` function. On line 9, we check whether the inventory can fulfill the order. If it cannot be fulfilled, then it breaks. Otherwise, it keeps iterating over all of the order items and keeps the `validOrder` status as true.

If the order verification is successful, then the supplier should remove the items from the list with the `reserveOrderItems` function. Otherwise, it marks the order status as `VerificationFailed`:

```
public function reserveOrderItems(jdbc:Client jdbcClient,
    OrderItemTable orderItems) returns error?{
    foreach OrderItem item in orderItems {
        string inventoryItemId = item.inventoryItemId;
```

```
        int quantity = item.quantity;
    sql:ExecutionResult result = check jdbcClient-
        >execute('UPDATE InventoryItems SET Quantity =
            Quantity - ${quantity} WHERE
                InventoryItemId = ${inventoryItemId}');
    string pendingOrderId = uuid:createType1AsString();
    result = check jdbcClient->execute('INSERT INTO
        PendingOrderItems(PendingOrderItemId, OrderId,
            InventoryItemId, Quantity) VALUES
    (${pendingOrderId}, ${item.orderId},
        ${inventoryItemId}, ${quantity})');
    }
    return;
}
```

Also, the order state should be updated to the `Verified` state with the following `setOrderStatus` function:

```
function setOrderStatus(jdbc:Client jdbcClient, string
    orderId, string status) returns error? {
    sql:ParameterizedQuery updateOrder = 'UPDATE Orders
        SET Status = ${status} WHERE OrderId = ${orderId}';
    sql:ExecutionResult result = check jdbcClient-
        >execute(updateOrder);
}
```

This implementation works fine until this program runs as a single thread over a single server. If this application is run on multiple servers or multiple threads, then you need to consider the transaction management of the application.

Understanding ACID properties

ACID properties play a key role in the development of database-related applications. They are a set of properties to guarantee data validity, avoid data misshapes, and recover from power failure. The following are the properties that ACID applications should have:

- Atomicity
- Consistency
- Isolation
- Durability

We will go through each of these properties and how it applies to the system design.

Atomicity

Simply stated, atomicity implies that a transaction is meant to be atomic. Atomic transactions should be either committed or rolled back to the previous state if they fail. There can be no intermediate state in atomic transactions.

Remember the order management that we discussed earlier? The list of transactions in the order management system is represented in the following diagram:

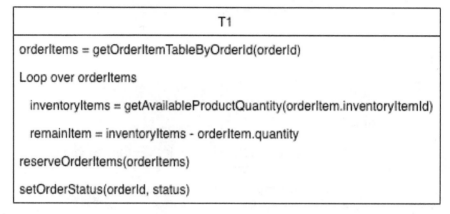

Figure 5.3 – Single transaction

Assume `reserveOrderItems` got executed but `setOrderStatus` failed due to a database connection failure. In this case, the database ends up in an inconsistent state where the supplier product list is updated with a reduced order quantity but the order state is still in the `Created` state. If another transaction is trying to read the `Orders` table, it shows that it has still not proceeded to the verified status, but the product items get reduced from the inventory.

If transactions are atomic, then either all the database transactions including `reserveOrderItems` and `setOrderStatus` get executed or the changes are rolled back if any of the transactions are unable to be performed. This makes sure that the database does not end up in an inconsistent state.

Consistency

Consistency makes sure that databases are always in a consistent state. For the aforementioned example, assume there are 100 products available in the inventory. Then the customer adds 30 products to the order. After the verification process, 30 products will be reduced from the inventory and added to the order of the customer with the `Verified` state. A total of 100 product items were in the inventory until verification. After the order verification process, 30 of them were transferred to the `order` table and the rest were kept in the inventory. The total number of products is still 100, which makes sure that the database is in a consistent state.

Isolation

So far, we have discussed running a single transaction in a single thread. What if multiple applications try to access the same database with the same transaction queries? The isolation property makes sure that the database status is consistent in concurrent transactions as well. Consider the previous example of running in a concurrent environment:

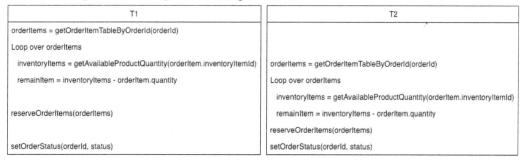

T1	T2
orderItems = getOrderItemTableByOrderId(orderId)	
Loop over orderItems	
inventoryItems = getAvailableProductQuantity(orderItem.inventoryItemId)	orderItems = getOrderItemTableByOrderId(orderId)
remainItem = inventoryItems - orderItem.quantity	Loop over orderItems
	inventoryItems = getAvailableProductQuantity(orderItem.inventoryItemId)
reserveOrderItems(orderItems)	remainItem = inventoryItems - orderItem.quantity
	reserveOrderItems(orderItems)
setOrderStatus(orderId, status)	setOrderStatus(orderId, status)

Figure 5.4 – Two concurrent transactions

In this example, there are two transactions doing the same order verification process. For example, assume there are only 100 products in the inventory. T1 transaction gets 100 as the remaining available inventory products. The T1 transaction order has 60 items to order. Since 100 > 60, the order can be fulfilled and it proceeds to reduce the quantity from the inventory and marks the order state as `Verified`. At the same time, the T2 transaction is also executed with the required order size of 70. At that point, the T2 transaction reads the number of products available in the inventory, which is still 100. Therefore, it also assumes that it can also order 70 items. Since T1 transaction updates the available inventory, this makes an inconsistent state where both orders get accepted.

To avoid these types of scenarios, the simplest solution is to complete the T1 transaction and then execute the T2 transaction. Running transactions sequentially enables databases to maintain a consistent state.

Durability

Durability is all about keeping data in a safe place. Data on the database should be protected even under power failure and physical damage. To prevent power failures, you can use databases that have a disk-based persistence mechanism. To make it more reliable, you can replicate databases using multiple data stores.

Ballerina transaction management

As described earlier, the database should be compatible with the ACID properties to keep data in a consistent state. You should make proper transaction management in the database framework to achieve atomicity, consistency, and isolation. Databases of cloud applications running on multiple instances might need to be accessed by multiple servers. These kinds of situations should be handled with care.

Ballerina provides a built-in transaction management syntax style to manage transactions. In Ballerina, you can specify a transaction that should be performed atomically. Ballerina provides the `transaction` keyword, which allows developers to create a transaction block. We can create a block of database queries together and create a single transaction with the Ballerina `transaction` keyword. We can modify the previous example as follows to execute database queries as a single transaction:

```
transaction {
    sql:Error? dbError = check initializeDB(mysqlClient);
    check commit;
}
```

With Ballerina transaction management support, we can secure transactions by using this transaction block. You can implement rollback functions and retry mechanisms as well with the Ballerina language. If there is any error or panic on the transaction block, it does not proceed to the `commit` instruction and rolls back to the initial state. With Ballerina, you can implement a retry as well by adding the `retry` keyword before the `transaction` keyword to retry transactions. You can specify the number of transaction retries as well. Check the following example template of using `retry` transactions:

```
retry(3) transaction {
    check doTask();
    check commit;
}
```

So far, we have discussed building a simple order management system on monolithic architecture. To extend this into cloud native concepts, we need to separate this application into multiple services and deploy it on separate servers. In the next section, we will learn how to separate this order management system and deploy it as multiple services.

The database-per-service design pattern

In microservices, single monolithic applications are separated into multiple services that can operate independently. Each of the microservices has its own databases rather than having a single database for the whole system. Having a database per service has multiple advantages in a microservice architecture.

Having services loosely coupled is the key benefit of the database-per-service pattern. Each service only operates with its own database. If another service needs to access that database, it cannot be directly connected to that database. Instead, it is supposed to go through the service that was supposed to deal with that service. The database-per-service design pattern helps to develop, deploy, and scale services independently of other services.

Another advantage is that the developer can select the most suitable database for the given service. For example, if the data contains a relational schema, then the developer can use a relational database to implement services. If you want to represent complex data in a database system, then you can use a graph database. You can select the database depending on the service requirement.

The sample order management system can be broken down as follows to support the database per service design pattern. Here, each service has its own database tables to work with:

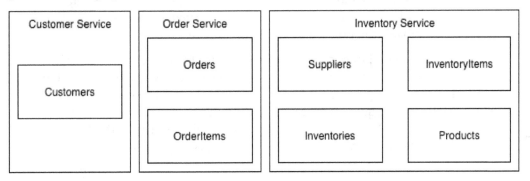

Figure 5.5 – Table separation for the database-per-service design pattern

Here, we have separated the `Customers` table into the **Customer Service**. The `Orders` and `OrderItems` tables are separated into the **Order Service**. The `Suppliers`, `InventoryItems`, `Inventories`, and `Products` tables are kept inside the **Inventory Service**. Now each table has a responsible service that it works with.

Having a database per service creates another problem with handling distributed transactions. Since databases are separated, a single transaction is not sufficient to perform all of the database changes. For example, to perform the verification process, we updated the `Orders`, `InventoryItems`, and `PendingItems` tables. But now, these tables are located on two different services. A single transaction is not sufficient to work with multiple services.

Now we need to find a solution to execute multiple queries for the same transactions on different databases. This can be achieved in two ways: either using synchronous lookups or asynchronous events. We will discuss more on these two methods in the next sections.

Synchronous lookup

Synchronous lookup calls each database separately and executes queries. Tables can be separated over different databases. But a single transaction is able to execute multiple queries over a distributed database. The most common protocol for implementing synchronous lookup is XA transactions.

XA is the abbreviation for the X/Open XA standard specification. The term **XA** is derived from **eXtended Architecture**. XA transactions give a standard for distributed transactions. The XA protocol is built with a two-phase commit protocol. As its name suggests, two-phase commits have two phases in the transaction. Transactions can have a coordinator that coordinates the transaction and participants that execute transactions. The two phases are as follows:

- **Phase 1**: The first phase is known as the voting phase. In this phase, the coordinator sends queries to participants. Participants evaluate the queries, lock the database, and send back whether that query is approved or rejected. If queries are approved, only the coordinator can proceed to the commit phase.

- **Phase 2**: In the commit phase, the coordinator sends the result to the participant on whether transactions should be committed or rolled back. When the transaction is completed, the participants will release the database lock. Participants maintain an undo log. When a query is processed in the first phase, the database keeps an entry in an undo log. If a second phase transaction is refused by the coordinator, a rollback transaction is carried out using undo logs.

Asynchronous events

Instead of sending synchronous messages, event-driven architecture gives a simple solution to execute distributed transactions using the publish-subscribe design pattern. When two services need to communicate with each other, the service publishes a message on the messaging channel. Another service that subscribes to this channel consumes messages and performs a database operation on its database.

The order management system that we discussed earlier contains some transactions that need to be executed on both the supplier and the order service. For example, when a customer places an order, the order service needs to be verified that the order can be fulfilled by the supplier service. The following diagram represents how two services communicate over brokers with asynchronous communication:

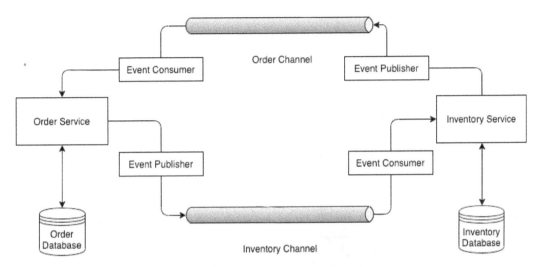

Figure 5.6 – Event-driven design for a distributed transaction

To verify the order, the order service sends a message to the inventory channel with the order quantities. The inventory service is listening to messages from the inventory channel. When the inventory service receives the message, it checks whether the order can be fulfilled. The result either confirms the order or rejects the order. The status of the order verification is published on the order channel. The order service listens to the order channel for any messages. When the order service receives the message, the order service can decide whether or not to proceed with the order.

As you may have noticed, the database-per-service design pattern is more complex than monolithic applications, which have a single database. Therefore, we need to have a proper design pattern that can be used to implement the database-per-service design pattern. In the next section, we will discuss more on the DDD pattern, which is widely used with the database-per-service design pattern.

The role of Domain-Driven Design in cloud native architecture

To overcome the problem of handling data between different services with their own databases, DDD comes to the rescue. The domain-driven development solution was discussed in the *Domain-Driven Design: Tackling Complexity in the Heart of Software* book.

DDD is a set of terminologies, requirements, and functionalities that can be used to design a system. The concept of DDD suits an event-driven system, where events handle the flow of the program. In this section, we will discuss the concepts of DDD and how to use DDD with event-driven design.

Object-oriented principles (OOP) make the system easy to understand and manageable. Designers can design the system by separating entities into different classes. This pattern is known as the Domain model, where business logic organized as an object model consists of a class having a state and a behavior. Let's look at the building blocks of the DDD architecture:

- **Entity**: Entities are objects that can have a persistent identity. For example, the `Customer` object in the order management system can be identified as an entity. Customers have their own identity and life cycle among other `customer` objects. On the other hand, entities are mutable objects that can be changed.

- **Value objects**: Unlike entities, value objects do not have a persistence identity. For example, think about the `Address` class. It can have multiple strings to keep address details. `Address` cannot exist on its own and depend on the `Customer` object. When the `Customer` object is created, the `Address` value object is also created.

- **Factory**: A factory creates objects with given specifications. If object creation is complex, you can use a factory to generate the relevant object instead of creating objects directly.

- **Repository**: A repository is an object that encapsulates persistence storage. Repositories provide a way to access the database and retrieve database entities.

- **Services**: A service is an object that implements business logic. This business logic does not belong to an entity or to value objects.

Entities and value objects are collectively treated as domain objects. This makes an important context of the DDD, which are aggregates. An aggregate is a collection of entities and value objects that can be treated as a single unit. Aggregates are loaded into the system as a whole and persisted in the storage as a whole.

The size of an aggregate should not be large since larger aggregates need to go through a long serialization process before being inserted into the database. Having a larger aggregate reduces the scalability of applications. When multiple aggregates are combined to generate a single aggregate, the system needs to handle the entire transaction at once.

If we can decompose larger aggregates into smaller aggregates, this ensures that the application can be separated and scaled independently. Multiple aggregates can work separately and perform aggregate related transactions as a set of atomic transactions. Therefore, reducing the size of an aggregate improves the concurrent execution of the application. In the case of providing atomic transactions over multiple aggregates, you can combine those aggregates together.

There are two major architectural patterns that you can use to build a DDD system. If you are building an application that has simple business cases, you can use the transactional script pattern. Rather than depending on OOPs, you can use functional language to build the transactional script pattern. In this pattern, transaction behaviors and states are separated. For example, `CustomerService` contains all the service implementations. It does not hold any entities. Instead, there is another object named `Customer` to hold the customer entity. The `Customer` entity does not have any functionality. It just keeps the customer's data. On the other hand, `Customer` aggregate holds the business functionality on its own, along with the customers' entity details. A **Data Access Object (DAO)** is used to access database entries. A DAO creates an entity from the database and provides it to the service.

Transactional script patterns are sufficient to implement basic services. When scenarios get complicated, we can shift to a much more object-oriented architectural pattern. Domain model patterns provide a way to implement DDD using an object-oriented pattern. In this type of model, objects can have both data and functionality. For example, the `Customer` class can have functions that manipulate the `Customer` object. Repository classes for each of these entities are used to interact with databases.

DDD contains four adapters that interact with outside services and endpoints. These adaptors' communication methods can be either synchronous or asynchronous:

- **API adaptors**: These are the REST server endpoints that read incoming requests. The Ballerina service interface can be used to create the API and receive the incoming payload.

- **Command handlers**: Command handlers are the inbound endpoints that listen to the incoming messages from different channels. For example, handler can read messages from a message queue and perform actions.

- **Database adaptors**: Database adaptors are outbound adaptors that are used to access databases. They can either read or write data in storage.

- **Event publishing adaptors**: These are outbound adaptors that are used to send messages as commands to other services. Another service's command handlers subscribe to these channels and perform actions.

Next, let's discuss how to create an aggregate with the Ballerina language and implement these communication adaptors.

Creating aggregates with Ballerina

Aggregates are the basic building units of DDD In this section, we are going to discuss building an aggregate with Ballerina classes. Ballerina class gives way to implement object-oriented programming. We can represent orders aggregate as a Ballerina class that can handle its own details. The basic definition of our `OrderAggregate` class is as follows:

```
public class OrderAggregate {
    private Order 'order;
    private OrderItemTable orderItems;
    private OrderRepository;
    function init(OrderRepository orderRepository,
        Order 'order, OrderItemTable orderItems) {
        self.orderRepository = orderRepository;
        self.'order = 'order;
        self.orderItems = orderItems;
    }
}
```

This class contains the `order` field to hold the order information and the `orderItems` field to hold a list of order items. We can also keep a reference to the `OrderRepository` implementation that contains database access logic. This class constructor takes `orderRepository`, `order`, and `orderItems` as constructor arguments in order to generate a new object. We can use this object to access each of the orders more easily.

There are two ways that this class can be instantiated. One method is to load existing order records from the database or create a new order. We will keep both implementations in the same Ballerina file where we keep the `OrderAggregate` but outside the class definition:

```
function getOrderAggregateById(OrderRepository
  orderRepository, string orderId) returns
    OrderAggregate|error{
    Order 'order =   check
      orderRepository.getOrderTableById(orderId); // Line 5
    OrderItemTable orderItems = check orderRepository.
      getOrderItemTableByOrderId(orderId);
    return new OrderAggregate(orderRepository, 'order,
      orderItems);
}
```

The `getOrderAggregateById` function reads the database for the order ID and generates the order aggregate. First, this function reads the order details on line 5, and then it takes the order items' details on line 7. Then a new `OrderAggregate` object is created by calling the `OrderAggregate` constructor, which returns the `OrderAggregate` object:

```
function createOrderAggregate(OrderRepository
  orderRepository, string customerId,
    string shippingAddress) returns OrderAggregate|error{
    OrderItemTable orderItems = table[];
    string orderId = uuid:createType1AsString();
    Order 'order = {
    orderId: orderId,
    customerId: customerId,
    status: "Created",
    shippingAddress: shippingAddress
};
check orderRepository.createOrder('order);
return new OrderAggregate(orderRepository, 'order, orderItems);
}
```

The `createOrderAggregate` function creates a new database record with the `createOrder` function and creates an order aggregate with an empty `OrderItems` table.

Now we can add more functionalities into the aggregate by adding functions into the class. The following `addProductToOrder` function that is defined inside the `OrderAggregate` class lets us add a new item into the order aggregate. It takes the item ID that needs to be added and the quantity of the products as an input argument. Then it creates a database entry to add new items and update the order aggregate:

```
function addProductToOrder(string inventoryItemId, int
    quantity) returns error? {
    string? orderId = self.'order?.orderId;
    if orderId is string {
        string orderItemId = check
            self.orderRepository.addOrderItem(orderId,
            quantity, inventoryItemId);
        OrderItem = {
            orderItemId: orderItemId,
            orderId: orderId,
            quantity: quantity,
            inventoryItemId: inventoryItemId
        };
        self.orderItems.put(orderItem);
        return;
    } else {
        return error("Error while finding the order id");
    }
}
```

Since we marked all class field values as private, you should provide getters and setters to access those fields. You can add more functionalities such as removing products, updating products, changing shipping addresses, and so on.

The next problem that we need to solve is how we can access another aggregate that is separated into another database. In the next section, we will discuss saga transactions and how to use them to build Ballerina applications that can interact with multiple aggregates in different services.

Distributed transactions with saga

Saga transactions are atomic. Either the transaction is successful and complete, or it fails and rolls back to the previous state. It guarantees that the database will be consistent at any point in time. A series of transactions can be connected with a sequence of events that pass through the messaging channels. If any of those transactions fail, compensation transactions will be generated to roll back changes or to change the status to a failed state.

ACID transactions are straightforward in terms of transaction handling. If any transaction fails, the whole transaction can be rolled back to the previous state. But with saga transactions, these transactions happen on multiple independent servers. Therefore, the design should be carefully designed to handle the atomicity and consistency of the system.

Saga transaction coordination can be built in two ways:

- Choreography
- Orchestration

Let's have a look at each one in detail.

The choreography pattern

As the name suggests, choreography is the same as coordinating each individual dancer in a dancing team. Each person in the group has to coordinate with each other to perform a dance. Choreography patterns also let each service in a microservice system interact with other services to perform a task. Let's take a look at how to build the order management system with a Saga choreography pattern:

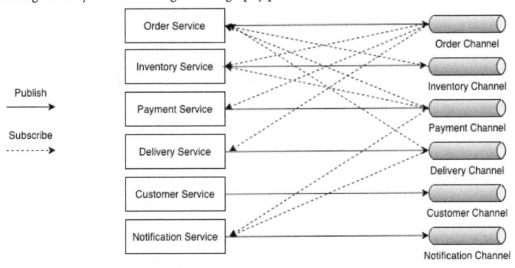

Figure 5.7 – Saga transaction model for the order management system

> **Important note**
> The publish-subscribe notation given in the diagram represents how messages flow into the system. The direction of the arrow represents the message flow direction. Solid lines represent message publishing and dotted lines represent message subscription.

In choreography-based systems, each service has a messaging channel for publishing events. These messaging channels can be a topic that has been subscribed to by multiple services. The services that depend on the order service can subscribe to the order channel and read events. For example, when the order needs to be verified with the inventory service, the order service publishes a message to the order channel. Here, the inventory service subscribes to the order channel and listens to the messages.

When the inventory receives the message, it can verify the order and publish the message with the status on the inventory channel. The order service is also subscribed to the inventory service. When an inventory sends the message to the inventory channel, the order service can read the message and mark the status of the order as either verified or verification failed. The same implementation can be extended to implement other scenarios for the remaining services as well. Each service that is dependent on other services can subscribe to the corresponding channel.

Compensation transactions can also be implemented in the same way. For example, when a payment is rejected by the payment gateway, the payment service can publish a message on the payment channel mentioning the payment is rejected. Both the inventory service and the order service listen to the payment channel. When the payment is rejected and the payment service publishes an event, the inventory service can remove pending order items from its database and the order service can update its status to the order rejected state. You may notice that there is no central component in this design that handles all transactions. Services are working together to perform transactions.

Choreography-based saga is simple by design and provides loose coupling between services. However, it can get complicated as the system grows. Understanding a choreography-based system might get complicated in a larger system due to its complex routing logic. Alternatively, we can use the orchestrator pattern to implement saga to manage transactions. In the next section, we discuss the orchestrator pattern in depth.

The orchestrator pattern

The orchestrator pattern is another saga pattern that handles transactions using a central unit. The same as music orchestration has a conductor who conducts the performance, the orchestrator pattern has an orchestrator that handles all the transactions.
The following diagram represents how messages are communicated over the services using the orchestrator design pattern:

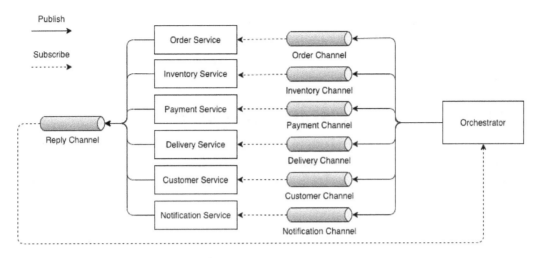

Figure 5.8 – Saga orchestration model for the order management system

Each service is subscribed to its own messaging channel. Only the orchestrator publishes messages to each channel. These channels can be introduced as message queues. These channels act as command channels that send commands to the services. The orchestrator produces commands and publishes them on the respective channel. When the service completes executing commands, it generates an event and publishes it on the reply channel. The orchestrator listens to the reply channel and generates the next command. The orchestrator is a state machine that generates commands based on incoming events. The state diagram that we discussed earlier is implemented on the orchestrator.

For example, when an order needs to be verified with the inventory service, the order service places a message on the reply channel. The orchestrator listens to the reply channel and receives a message from the order service. The order service checks the status of the order. Since the current status is OrderCreated and the action that needs to be performed is order verification, it publishes the message on the inventory channel. The inventory service listens to the inventory channel and reads the order verification command. Then the inventory service verifies the order details and publishes the order status to the reply channel. The state after order verification might be order verified or order verification failed. Then the orchestrator sends an update order status command to the order service to update the order status.

Compensation transactions can also be easily managed in the orchestration model. For the same example of a payment failure, the payment service can publish the failure message to the reply channel. The orchestrator can execute compensation transactions by sending commands to the inventory channel to remove pending items and to the order channel to update the order status as failed.

This kind of system can be easily implemented with Ballerina along with messaging services such as RabbitMQ or Kafka. In the next section, we will look into the implementation of saga orchestration with the Ballerina language along with the RabbitMQ messaging service.

Building a saga orchestrator with Ballerina

We use RabbitMQ as a messaging platform for this example. Each of the channels is implemented as a RabbitMQ message queue. For this example, we will build the order verification process where the order service interacts with the inventory service in order to verify orders. The following services are implemented in the Ballerina project to maintain order, inventory, and orchestrator services:

- **OrderAPI**: This service reads incoming HTTP requests that are related to the orders from the client. In this example, we will implement order creation, adding products to an order, and order verification functionalities.

- **OrderHandler**: This service is to read messages from the orchestrator. This service contains logic to update the order status after the order verification.

- **InventoryAPI**: This service handles incoming HTTP requests from the client that are specific to inventories. InventoryAPI includes functionalities to add items and products to the inventories.

- **InventoryHandler**: This service listens to commands from the orchestrator. Order verification logic is implemented on this handler.

- **Orchestrator**: This is the orchestrator service that listens to reply queues and runs the state machine. The orchestrator publishes messages by evaluating service responses.

The following `sendMessage` function is used all over the implementation to publish messages to RabbitMQ. You can use the `bal pull ballerinax/rabbitmq` command to pull the RabbitMQ library from Ballerina Central:

```
function sendMessage(string queueName, ChannelMessage
  channelMessage) returns error?{
    rabbitmq:Client newClient = check
      new(rabbitmq:DEFAULT_HOST, rabbitmq:DEFAULT_PORT);
    check newClient->queueDeclare(queueName);
    check newClient->publishMessage({ content: channelMessage
      .toString().toBytes(), routingKey: queueName });
}
```

This function takes the arguments `queueName` and `channelMessage` and publishes messages to the RabbitMQ server. Here, we used the following record to send messages over the queue:

```
public type ChannelMessage record {
    string serviceType;
    string action;
    json message;
};
```

The `serviceType` attribute is used to identify the target service. For example, when the order service sends a message to the orchestrator, the service type will be `Orchestrator`. The `action` field determines what action needs to be performed. The `message` field holds additional details that are required to perform the action.

The first step of the order verification is to send a message to the inventory service to verify the list of product items. The `OrderAPI` service contains the following function, which publishes messages to the reply queue with the list of order items:

```
function verifyOrder(OrderRepository, string orderId) returns
    error? {
    OrderItemTable = check orderRepository.
        getOrderItemTableByOrderId (orderId);
    OrderItem[] orderItems = orderItemTable.toArray();
    json data = orderItems.toJson();
    ChannelMessage = {
        serviceType: "Orchestrator",
        action: "OrderCreated",
        message: data
    };
    check sendMessage("ReplyQueue", channelMessage);
}
```

Here, we used the previous `getOrderItemTableByOrderId` function to get the order item list from the database using the order ID. The order ID is sent from the client to validate the order request. This list of order items includes `inventoryItemId` and the quantity of each item. Then we can submit this message to the inventory through the orchestrator with the `OrderCreated` action. The `sendMessage` function publishes the message to the RabbitMQ server.

Here, we use repository objects to hold the database information and use the same object to access databases. We used the following class definition to define all of the database transactions:

```
public class OrderRepository {
    jdbc:Client jdbcClient;
    public function init() returns error?{
        io:println("Initializing Order Repository");
        jdbc:Client tempClient = check new (jdbcUrl,
            dbUser, dbPassword);
        self.jdbcClient = tempClient;
    }
    public function getOrderItemTableByOrderId(string
        orderId) returns OrderItemTable|error {
        // content goes here
    }
    //... other database access functions
}
```

The messages are published by the order service to the reply queue. The Orchestrator service listens to these messages. The Orchestrator service has a listener service that listens to RabbitMQ queues:

```
listener rabbitmq:Listener channelListenerOrchestrator =
    new(rabbitmq:DEFAULT_HOST, rabbitmq:DEFAULT_PORT);
@rabbitmq:ServiceConfig {
    queueName: "ReplyQueue"
}
service rabbitmq:Service on channelListenerOrchestrator {
    remote function onMessage(rabbitmq:Message message,
        rabbitmq:Caller caller) returns error? {
        string messageContent = check
            string:fromBytes(message.content);
        json j2 = check value:fromJsonString(messageContent);
        ChannelMessage = check j2.cloneWithType(ChannelMessage);
        // State machine logic
    }
}
```

When the message is published in the reply queue, the orchestrator listener consumes the message and triggers the onMessage function. We can read the message's content as a ChannelMessage record after deserializing the message content. The next step is to read the message and forward it to the inventory service. We can use match statements to match the channel message actions and forward the request:

```
if channelMessage.serviceType == "Orchestrator" {
    match channelMessage.action {
        "OrderCreated" => {
            channelMessage.serviceType = "InventoryService";
            channelMessage.action = "VerifyOrder";
            check sendMessage("InventoryQueue", channelMessage);
        }
    }
}
```

Action matches the OrderCreated event. If so, it will publish a message to the inventory queue. The same message was forwarded into the inventory queue for verification.
You may notice that when we publish a message to a service, we use a command similar to the naming convention for actions. In this case, it would be VerifyOrder. But when the order service publishes a message into the orchestrator, the naming convention is similar to an event. Saga transactions use commands to send instructions to services and events as the responses to the orchestrator.

InventoryHandler listens to InventoryQueue for incoming messages. This handler contains logic to validate orders and respond with messages to the orchestrator. The validation process is the same as we described in monolithic implementations. First, it takes a list of items of the order that was sent by the order service through the orchestrator service. Then it iterates over each of the order items and compares them with the available items in the inventory. If any of the items are unable to be fulfilled by the inventory, the order is rejected. The following Ballerina code checks available inventory items' quantity against the ordered item quantity and marks isInventoryAvailable true if it can fulfill the order:

```
boolean isInventoryAvailable = true;
foreach json item in orderItemList {
    int quantity = <int>item.quantity;
    string inventoryItemId = <string>item.inventoryItemId;
    orderId = <string>item.orderId;
    int totalItem = check inventoryRepository.
```

```
        getAvailableProductQuantity(inventoryItemId);
    if totalItem < quantity {
        isInventoryAvailable = false;
        break;
    }
}
```

Then we can have an `if` condition to check the order status and send the result back to the order service through the orchestrator:

```
if isInventoryAvailable {
    check inventoryRepository.reserveOrderItems (orderItems);
    ChannelMessage = {
        serviceType: "Orchestrator",
        action: "OrderVerified",
        message: orderId
    };
    check sendMessage("ReplyQueue", channelMessage);
} else {
    ChannelMessage = {
        serviceType: "Orchestrator",
        action: "OrderVerificationFailed",
        message: orderId
    };
    check sendMessage("ReplyQueue", channelMessage);
}
```

Here, if the order is approved by the inventory service, it publishes a message to the orchestrator as `OrderVerified`. Otherwise, it publishes a message as `OrderVerificationFailed`. Now the orchestrator picks up this message and sends a command to the order service to update the order status. The following code segment should be appended to the orchestrator to match the statement to handle these actions:

```
"OrderVerified" => {
    string orderId = <string> channelMessage.message;
    channelMessage.serviceType = "OrderService";
    channelMessage.action = "UpdateOrderVerified";
    channelMessage.message = orderId;
```

```
        check sendMessage("OrderQueue", channelMessage);
}
"OrderVerificationFailed" => {
    string orderId = <string> channelMessage.message;
    channelMessage.serviceType = "OrderService";
    channelMessage.action = "UpdateOrderVerificationFailed";
    channelMessage.message = orderId;
    check sendMessage("OrderQueue", channelMessage);
}
```

The final piece of the puzzle is to update the order status to either `Verified` or
`VerificationFailed`. The following code on the order handler reads the message
from the orchestrator and updates the order status in the database:

```
if response.serviceType == "OrderService" {
    string orderId = <string>response.message;
    match response.action {
        "UpdateOrderVerified" => {
            check orderRepository.setOrderStatus(orderId,
                "Verified");
        }
        "UpdateOrderVerificationFailed" => {
            check orderRepository.setOrderStatus(orderId,
                "VerificationFailed");
        }
    }
}
```

In this section, we discussed creating a simple order management system by using the saga
transaction pattern. In the next section, we will improve this design by introducing event
sourcing and CQRS patterns.

Using event sourcing and CQRS in a distributed system

The order management system that we described earlier is built with a series of commands and events. Events are a series of states that come in and are generated by the system. Events are immutable. This means that when an event is generated, the event will not be changed and will remain as it is. The immutable nature of events can be used to implement a consistent database. A series of events can also be used to analyze the system in the case of failure. Event sourcing can be used with the CQRS pattern, which allows systems to perform much faster. In this section, we will learn about building applications with event sourcing and the CQRS design pattern.

Using events to communicate among services

Events are immutable where the event does not get changed later. Events don't follow the CRUD principles. Instead, the event source system stores all the events in the event store. In simple words, event sourcing means keeping all the events that an application receives and replaying all of the events to synchronize the current state.

Events can be represented simply as JSON or XML where they provide language independence over various service implementations in different languages. However, developers can decide to use a language-specific object format as well.

Remember to provide meaningful names for events when naming an event. For example, `CreateOrderEvent` makes more sense as a command rather than an event. A proper name for that would be `OrderHasBeenCreated`. Avoid using general names for events such as `DoStuffEvent`. Give a specific name and specific logic to events based on what is happening inside the event, rather than naming them with general names.

Developing with event sourcing

Previously, we discussed aggregates and how they are stored in the database. Different aggregates are decomposed into multiple tables and stored in the database. In the event sourcing system, we store events rather than storing data. An event represents the state changes in an aggregate. The current state of the system is calculated by aggregating the events.

The most common example that uses event sourcing is bank transaction ledgers. Instead of keeping the final balance, the bank account maintains a ledger that contains all the transactions made by the customer. When the user spends money, the banking system keeps a record of the amount the user sends from their account to another account. When the user receives money, the system keeps a record of the amount that they receive. When the user needs to calculate the balance, the system starts from a balance of zero and applies the sequence of credited and debited amounts to the current balance. The final value will be the current balance of the account:

Date	Description	Debit	Credit	Balance
29-10-2020	Interest		12.5	222.5
10-11-2020	Withdrawal	100		122.5
11-11-2020	Ln Recovery	50		72.5

Table 5.1 – Sample bank ledger

Only the current state of the system remains in the conventional way of holding records. When the aggregates are changed, only the current state is preserved and the previous state is lost. On the other hand, we keep all the transformations that happen in the system rather than the current state. This approach is useful when the user needs to look back on the past to analyze or debug the system.

In the event store, we keep a list of events on persistent memory. Persistent memory might be a SQL or NoSQL database. There are a few databases available that are specifically written to implement event stores, such as **EventstoreDB** by *Greg Young* and **AxonDB** by *AxonIQ*. These event stores provide multiple features such as keeping snapshots, atomicity, validation, optimization for recent events, and so on. But you can still use SQL and NoSQL to store a sequence of events.

Kafka also has its own storage that holds a sequence of events. You can keep events in the memory for a defined time or indefinitely in Kafka. But using Kafka as an event store has a problem since it needs to iterate over the entire topic to read a single entity of the state. For example, if you keep different topics for Order, Customer and Payment, if you need to find a particular order, you will need to iterate over all orders in the Order topic.

The order service database contains a list of orders that customers added. If we model the same application with events, we can represent the orders' current states as a list of events. We can represent the list of events as a table in the relational database. Check the following table, which stores a list of order events:

EventId	EventType	EntityType	EntityId	Data	Timestamp
1000	OrderCreated	Order	123	{shippingAddress: ....}	...
1001	ProductAdded	Order	123	{orderItemId: "343 ...}	...
1002	OrderVerified	Order	123	{}	...
1003	OrderCreated	Order	124	{shippingAddress: ....}	...

Table 5.2 – Event table example

Whenever we need to take details about an order, we can filter events that correspond to the order by aggregating over the entity ID. This response is a list of events for a particular entity. Now we can replay all of these events in chronological order and generate the final aggregate.

We will also keep a table that holds entity details. When a new entity is created, a new record is added to the entity table. Here, we used the EntityVersion column to handle concurrent access with the optimistic locking technique. The entry version is modified every time an event is added to the table with a particular entity ID. When the entity gets changed, the system should increase the version by one. While the service loads the aggregate into the memory, it also reads the entity version. If it needs to add another event to the database, first it needs to check whether the version has been updated. If the version has been updated, it means that while it holds the data, another service updates the entity. Therefore, existing data held in the aggregate is not valid now and the system needs to replay new changes for the entity and try again to add a new event. The version of the entity is the way this system manages the transactions in a distributed event source system:

EntityId	EntityType	EntityVersion
123	Order	4
124	Order	2

Table 5.3 – Entity table example

In the next section, we will discuss updating the previous order aggregate to use with event sourcing.

Creating aggregates with event sourcing

Instead of using the table structure that we discussed while building the order management system, now we are going to use an event store as the database that keeps all the data. We need to build a way to generate aggregates from the events stored to perform queries on aggregate. We can modify the previous `OrderAggregate` class as follows to read data from the event store.

First, we will create tables to store events and entity data with the following SQL queries:

```
CREATE TABLE 'Events' (
  'EventId' varchar(50) NOT NULL,
  'EventType' varchar(45) DEFAULT NULL,
  'EntityType' varchar(45) DEFAULT NULL,
  'EntityId' varchar(50) DEFAULT NULL,
  'EventData' varchar(1000) DEFAULT NULL,
  'Timestamp' timestamp NULL DEFAULT NULL,
  PRIMARY KEY ('EventId')
) ENGINE=InnoDB DEFAULT CHARSET=utf8;
```

These queries generates an `Events` table that stores a list of events for a particular entity. To keep entities, we can use the following SQL query to generate an `Entities` table:

```
CREATE TABLE 'Entities' (
  'EntityId' varchar(50) NOT NULL,
  'EntityType' varchar(45) DEFAULT NULL,
  'EntityVersion' int DEFAULT NULL,
  PRIMARY KEY ('EntityId')
) ENGINE=InnoDB DEFAULT CHARSET=utf8;
```

To handle these tables, we can create a Ballerina record for each table. The following is the record that holds events and entity related data:

```
public type Event record {
    string eventId;
    string eventType;
    string entityType;
    string entityId;
    string eventData;
    time:Utc timestamp;
```

```
};
public type Entity record {
    string entityId;
    string entityType;
    int entityVersion;
};
```

Now we can modify the order aggregate to load data from the event store instead of the previous table structure. The following function in the order repository implementation creates a new entity, and writes events into the database. The entity version was initially set as one. Events are also persisted in this function with the insert query on the Events table:

```
public function createEntity(string entityId, string
    entityType, Event event) returns error? { transaction {
        sql:ParameterizedQuery createEntity = 'INSERT INTO
            Entities(EntityId, EntityType, EntityVersion)
            VALUES( ${entityId}, ${entityType}, 1)';
//Formatting the timestamp to remove the T and Z characters
    added by the UTC timestamp function of ballerina.
        string timeString = regex:replaceAll
            ((time:utcToString(event.timestamp)), "T"," ");
        timeString = regex:replaceAll(timeString,"Z","");

        _ = check self.jdbcClient->execute(createEntity);
        sql:ParameterizedQuery createEvent = 'INSERT INTO
            Events(EventId, EventType, EntityType, EntityId,
            EventData, Timestamp) VALUES(${event.eventId},
                ${event.eventType}, ${event.entityType},
                ${event.entityId}, ${event.eventData.
                toString()}, ${time:utcToString
                (event.timestamp)})';
        _ = check self.jdbcClient->execute(createEvent);
        check commit;
    }
    return;
}
```

Since we now also need to pass the entity ID when creating the order aggregate, we need to modify the constructor to include the entity ID. To create a new order along with the aggregate, we can define the createOrder function as follows:

```
public function createOrder(OrderRepository, string customerId,
    string shippingAddress, function (Event) returns error?
      eventHandler) returns OrderAggregate|error{
    string orderId = uuid:createType1AsString();
    OrderItemTable orderItems = table [];
    Order 'order = {
        orderId: orderId,
        customerId: customerId,
        status: "Created",
        shippingAddress: shippingAddress
    };
    string entityId = uuid:createType1AsString(); // Line 12
    string eventId = uuid:createType1AsString(); // Line 13
    OrderAggregate = new OrderAggregate(orderRepository,
      entityId, 'order, orderItems);
    Event orderCreateEvent = {    //Line 16
        eventId: eventId,
        eventType: "OrderCreated",
        entityType: "Order",
        entityId: entityId,
        eventData: 'order.toString(),   // Line 21
        timestamp: time:utcNow(2)
    };      // Line 23
    check orderRepository.createEntity(entityId, "Order",
      orderCreateEvent);
    check eventHandler(orderCreateEvent);
    return orderAggregate;
}
```

This function creates an order object the same as the previous one to hold the order information. Next, we need to create an event to store in the database. From line 16 to line 23, we create a variable called `orderCreateEvent` to hold event data. The event type being set as `OrderCreated` signifies that the order now has the `order created` status. The entity type is marked as `Order` to represent that this event belongs to the order aggregate. The event ID and entity ID are generated by using the UUID generation function from lines 12 to 13. Here, we are serializing the order to keep it in the database in line 21.

Next, we can store entities in the database by calling the `createEntity` function. This function creates a new entity row with version one and a new event row with the event data. Here, we used `eventHandler` as an input argument for the function that can execute a function over the event that has just been created. For example, when a new event is added to the database, we can use this `eventHandler` to publish messages into message queues. We will use this event handler with CQRS to create an event handler functionality.

Since this is order creation, we only need to initialize order aggregates with order details. This does not contain any product item information. This function returns the order aggregate object to the caller to work with the order aggregate.

We need to create the following `createEmptyOrderAggregate` function that generates the empty order aggregate. We will use this empty aggregate to apply events on it and generate the current states:

```
public function createEmptyOrderAggregate(OrderRepository
    orderRepository, string entityId) returns OrderAggregate{
    OrderItemTable orderItems = table [];
    Order 'order = {
        orderId: "",
        customerId: "",
        status: "Created",
        shippingAddress: ""
    };
    return new OrderAggregate(orderRepository, entityId,
        'order, orderItems);
}
```

This function is as simple as creating an empty `OrderItemTable` that holds order items and an empty `Order` record.

Now we will look at the implementation of the `getOrderAggregateByEntityId` function that returns the order aggregate. This function can read events from the `Events` table and generate the aggregate by playing events:

```
public function getOrderAggregateByEntityId(OrderRepository
    orderRepository, string entityId) returns
    OrderAggregate|error{
    OrderAggregate = createEmptyOrderAggregate
      (orderRepository, entityId);
    orderAggregate.setEntityVersion(check
      orderRepository.getEntityVerison(entityId));
    Event[] events = check orderRepository.
      readEvents(entityId); // Line 8
    foreach Event in events {
        check orderAggregate.apply(event);
    }
    return orderAggregate;
}
```

This function takes `entityId` as its input to retrieve the order aggregate. First, it initializes an empty table by calling the `createEmptyOrderAggregate` function. On line 6, we also take the entity version, which is the last version that is stored in the database relevant to this aggregate, and assign it to the aggregate. We will use this attribute when we need to store this aggregate in the database.

Line 8 reads the list of events filtered by event ID. These events can be played on the aggregate to generate the current aggregate by calling the `apply` function on the aggregate. The `foreach` loop loops through all of the events and updates the aggregate status.

The `apply` function is implemented on the aggregate and it updates the aggregated status by executing each event on it. A simple `apply` function to implement this scenario can be implemented as follows:

```
public function apply(Event event) returns error?{
    if event.entityType != "Order" {
        return;
    }
    match event.eventType {
        "OrderCreated" => {
            json orderData =
                checkpanic event.eventData.fromJsonString();
```

```
        self.'order.orderId = check orderData.orderId;
        self.'order.customerId = check
          orderData.customerId;
        self.'order.status = check orderData.status;
        self.'order.shippingAddress = check
          orderData.shippingAddress;
      }
    "ProductAdded" => {
        json orderItemJson = checkpanic
          event.eventData.fromJsonString();
        OrderItem = check
          orderItemJson.cloneWithType(OrderItem);
        if self.orderItems.hasKey(orderItem.orderItemId) {
            OrderItem item = self.orderItems.remove
              (orderItem.orderItemId);
        }
        self.orderItems.add(orderItem);
      }
    }
  }
```

This function has a `match` statement that checks the `eventType` of the event. It updates the current aggregate attributes for each mapping event. For example, the `OrderCreated` event updates the current order status by reading the events data. The `ProductAdded` event updates the added order item list by reading the list of order items from the event data.

The next step is to implement the product adding features into the aggregate. Before that, we need to improve the repository to add new events to the database. We can have the following function definition to insert a new event into the table:

```
public function addEvent(Event event, int initialVersion)
  returns error?{
    sql:ParameterizedCallQuery sqlQuery = 'CALL
      AddEvent(${event.entityId}, ${initialVersion},
        ${event.eventId}, ${event.eventType},
          ${event.entityType}, ${event.eventData})';
    _ = check self.jdbcClient->call(sqlQuery);
}
```

Here, we used a stored procedure to perform a series of queries in a single function. Optionally, you can still use multiple queries to implement this function. But using stored procedures has multiple advantages, such as less network communication overhead and fewer security concerns. But do not implement more complicated queries on the stored procedure as it has become an anti-pattern for microservice architecture owing to the implementation of business logic inside the database. Since this is a simple SQL transaction, we use stored procedures to execute multiple queries at once. The stored procedure definition for this is as follows:

```
DELIMITER #
CREATE PROCEDURE AddEvent
(
IN EntityId VARCHAR(50),
IN InitialVersion INT,
IN EventId VARCHAR(50),
IN EventType VARCHAR(45),
IN EntityType VARCHAR(45),
IN EventData VARCHAR(1000)
)
BEGIN
DECLARE ExistingVersion INT DEFAULT 0;
DECLARE RecordVersion INT DEFAULT 0;
SELECT EntityVersion INTO ExistingVersion FROM Entities WHERE
  EntityId = EntityId;
IF ExistingVersion = InitialVersion THEN
  UPDATE Entities SET EntityVersion = EntityVersion + 1
    WHERE EntityId = EntityId;

  INSERT INTO Events(EventId, EventType, EntityType,
    EntityId, EventData, Timestamp) VALUES(EventId,
      EventType, EntityType, EntityId, EventData, NOW());
ELSE
SIGNAL SQLSTATE '45000' SET MESSAGE_TEXT = 'Version not found';
END IF;
END# end
DELIMITER ;
```

This stored procedure takes event details that need to be inserted into the `Events` table along with the current version of the entity. In stored procedures, first, we retrieve the `EntityVersion` column from the `entity` table and assign it to the `ExistingVersion` variable. The `if` condition checks whether `ExistingVersion` is equal to `InitialVersion`. If these versions do not match, it means that the entity changed while it was loaded into the memory. Here, we simply throw an error if the values do not match.

If matching values are found, update the version by one and add the new event to the `Events` table. We will use this function to implement the new `addProductToOrder` function in the `OrderAggregate` class:

```
public function addProductToOrder(string inventoryItemId,
    int quantity, function (Event) returns error?
    eventHandler) returns error?{
    string orderItemId = uuid:createType1AsString();
    string? orderId = self.'order?.orderId;
    if orderId is string {
        OrderItem = {   // Line 7
            orderItemId: orderItemId,
            orderId: orderId,
            quantity: quantity,
            inventoryItemId: inventoryItemId
        };   // Line 12
        string eventId = uuid:createType1AsString();
        Event productAddEvent = {   // Line 14
            eventId: eventId,
            eventType: "ProductAdded",
            entityType: "Order",
            entityId: self.entityId,
            eventData: orderItem.toString(),
            timestamp: time:utcNow(2);
        };   // Line 21
        check
            self.orderRepository.addEvent(productAddEvent,
                self.entityVersion);
        self.entityVersion += 1;
```

```
        self.orderItems.add(orderItem);
        check eventHandler(productAddEvent);
    } else {
        return error("Error while reading order id");
    }
    return;
}
```

This function is very similar to the `createOrder` function. This function takes `inventoryItemId` that needs to be inserted into the item list and the quantity of the items. It also takes the handler to the function to call handlers. Here, we create a new variable to hold order item information and event data on line 7 to line 12 and line 14 to line 21. Then it is called from the `addEvent` function that we have just created. This function adds a new event record to the database and updates the entity version. If the transaction is successful, then it also increases the entity version and updates the order item information. Finally, it also calls the handler function.

The problem with event sourcing is that it needs to reconstruct the aggregate by applying a series of events to the initial aggregate. If the number of events stored for a given aggregate is increased, the aggregate generation is also time-consuming. In the next section, we will discuss using snapshots to store the status of the aggregate by aggregating a list of events.

Creating snapshots

Since calculating events from zero is time-consuming, we can use snapshots to keep the status of the aggregate up to some point in time. The snapshot contains a serialized aggregate as a single row up to the given time. The snapshot table that we are going to implement is as follows:

```
CREATE TABLE 'Snapshots' (
  'SnapshotsId' varchar(50) NOT NULL,
  'EntityType' varchar(45) DEFAULT NULL,
  'EntityId' varchar(45) DEFAULT NULL,
  'EntityVersion' varchar(45) DEFAULT NULL,
  'Message' varchar(1000) DEFAULT NULL,
  'Timestamp' timestamp NULL DEFAULT NULL,
  PRIMARY KEY ('SnapshotsId')
) ENGINE=InnoDB DEFAULT CHARSET=utf8;
```

This table contains the relevant entity ID that holds the aggregate data. `EntityType` holds the type of aggregate, which is the same as `EntityType` on the `Events` table. It also holds `EntityVersion`, which is the last version of aggregated events. The `Message` field holds aggregate data for each snapshot.

We used the following record definition to hold the message content of the order aggregate. This record holds the order information, and we will use this record to store the order aggregate. We also use this record to hold order information and the order item list:

```
public type OrderDetail record {
    *Order;
    OrderItem[] orderItems;
};
```

The snapshot generation process is a periodic task that can be run at a given interval. You can implement a service that generates snapshots. To build the snapshot, first, we need to list all entities by calling the `getAllEntities` function. This function simply returns all available entities in the `Entities` table. Next, we can iterate over the list of entities to generate a snapshot for each entity:

```
OrderRepository = check new();
    Entity[] entities= check
      orderRepository.getAllEntities();
    foreach Entity in entities {
        // Generate snapshot for each entities
    }
}
```

The next step is to get all the events that belong to each of the entities. The following code segment reads all of the events into the `events` variable:

```
time:Utc entityTo = check time:utcFromString("2029-03-
    28T10:23:42.120Z")
Event[] events = check orderRepository.readEventsTo
    (entity.entityId, entityTo);
OrderAggregate = check createEmptyOrderAggregate
    (orderRepository, entity.entityId);
```

```
if events.length() > 0 {
    foreach Event in events {
        check orderAggregate.apply(event);
    }
}
```

The list of events is filtered by filtering the result up to a given time. For the aforementioned example, the date is set to March 28, 2019. Now we can apply the list of events to the empty aggregate and generate the status up to the given time.

Next, we need to generate the snapshot message that we are going to store from the generated aggregate. The following code generates `OrderDetail`, which holds order aggregate data:

```
OrderDetail = {
    orderId: orderAggregate.getOrder().orderId,
    customerId: orderAggregate.getOrder().customerId,
    status: orderAggregate.getOrder().status,
    shippingAddress:
      orderAggregate.getOrder().shippingAddress,
    orderItems: orderAggregate.getOrderItems()
};
```

Then we can generate the snapshot table that holds the information on the database as follows:

```
string message = orderDetail.toString();
string snapshotId = uuid:createType1AsString();
Snapshot = {
        snapshotId: snapshotId,
        entityType: entity.entityType,
        entityId: entity.entityId,
        entityVersion: entity.entityVersion,
        message: message,
        timestamp: time:utcNow(2)
};
check orderRepository.addSnapshot(snapshot);
check orderRepository.removeEventTo(entity.entityId, entityTo);
```

This code generates a snapshot and we can use this to call the repository to update the database with the `addSnapshot` function. This is a simple function that inserts a snapshot into the database. Now we can remove events from the database since it already creates snapshots and we don't need them anymore. The `removeEventTo` function removes all events related to `entityId` up to the time defined by the `entityTo` argument.

You should arrange all of these database calls inside a transaction and make these transactions atomic. On the other hand, you also need to update the `getOrderAggregateById` function to take the initial data from the snapshot and then apply events from the `Events` table to create the latest aggregate.

Querying in microservice architecture

Querying data is an important part of building applications. For example, in the order management system, we need to implement a UI that previews data such as user data, order data, and so on. In monolithic applications, you can directly call the backend server, read the database, and send it to the frontend to be previewed. You can use a foreign key to combine multiple tables and query the result. For microservice applications, it is challenging since the data is distributed over multiple databases that are handled by multiple services.

There are two patterns available to simplify the query implementation in microservices. Those patterns are as follows:

- The API composition pattern
- The **CQRS** pattern

We will discuss each of these methods in detail in the following sections.

The API composition pattern

The API composition pattern is the simplest way to implement queries over multiple services. The API composition pattern is used by an API composer to access different services to get a result. An API composer splits queries into multiple API requests that can be executed by each of the services. Each service in the microservices system contains a component called a provider service that executes a segment of the query. The provider service can only read data from its own database. When the API composer sends a request to the provider service, it executes the query, gets the result, and sends it back to the API composer. The API composer aggregates each of the responses from the provider service and responds to the caller. The following diagram visualizes how the API composition pattern is used to compose different results from different services:

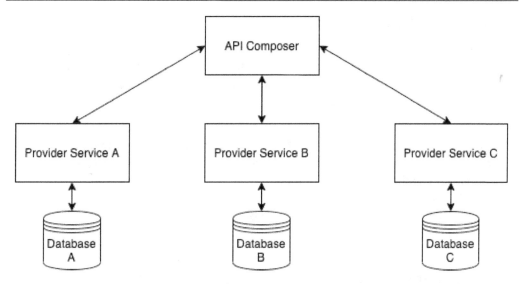

Figure 5.9 – The API composition pattern

This architecture is simple in nature, but there are issues such as the additional overhead of aggregating multiple results. Unlike monolithic applications, we cannot execute a single query and retrieve the result. This design needs to send multiple network calls to get the result and regular additional processing power to aggregate the result.

Also, this decreases the availability since if one service fails, it causes the whole query to fail. An API composer with a cache can be used to improve the availability by sending data over the cache. Sometimes it is OK to not have a result as well. For example, the user dashboard can view its data without order information to some extent. In these scenarios, API composers can ignore some of the results.

The key issue with the API composition pattern is handling the consistency of the system. The data in the services might be in an inconsistent state when the provider services read database entries. For example, customers might send order addition events to the system. These events should go through multiple services, such as the order service, inventory service, and so on to complete the request. If a query request comes in during this execution, the result might be in an inconsistent state such that the inventory amount gets reduced but still the order does not get verified.

Even though there are problems with the API composition pattern, this pattern can be used to build a simple querying scenario. However, there are some scenarios, such as performing a large table join in a dataset as an in-memory operation. For this type of scenario, the API composition pattern does not perform well. The API composer needs to join multiple aggregates to generate the result. This join operation requires a lot of computational power. The CQRS pattern is still a good candidate as an alternative to the API composition pattern.

Querying with the CQRS pattern

The core idea of the CQRS pattern is to separate the read and write operations of services. In the CQRS architecture, write operations are known as commands, and read operations are known as queries. CQRS separates read and write operations from database transactions. Compared to the CRUD model, the command part contains the CUD parts while the query part contains the R part.

CQRS enforces the system to have two different ways to read and write data but not both. Having two different services to read and write data means there are two services to scale independently. In applications, it is common to have a higher number of read than write operations. There can be two databases: one for the command side and one for the query side. The command-side database is for writing operations while the query-side database is for querying data.

We can use the same event store to improve the event sourcing system with the CQRS pattern. We will use the `eventHandler` handler function that we defined earlier to generate and send the events to a queue and update the view database. We can update the view database by using these events.

The `eventHandler` function simply sends a message to a queue named `OrderCommandQueue`:

```
function eventHandler(Event event) returns error?{
    OrderRepository orderRepository = check new();
    json eventData = check event.eventData.fromJsonString();
    ChannelCommandMessage channelCommandMessage = {
        serviceType: "OrderCommand",
        entityId: event.entityId,
        message: eventData
    };
    check sendMessage("OrderCommandQueue",
      channelCommandMessage);
}
```

This function creates a new `ChannelCommandMessage` object, converts it to JSON, and submits it to RabbitMQ. These events are handled by `OrderCommandHandler`. The `OrderCommandHandler` implementations are as follows to handle `OrderCommand`. This can be extended to handle other commands as well:

```
OrderRepository = check new OrderRepository();
match command.serviceType {
    "OrderCommand" => {
        OrderAggregate = check getOrderAggregateByEntityId
            (orderRepository, command.entityId);
        check orderRepository.updateOrder
            (orderAggregate.getOrder());
        OrderItem[] orderItems =
            orderAggregate.getOrderItems();
        if orderItems.length() > 0 {
            check orderRepository.updateOrderItems
                (orderAggregate.getOrderItems());
        }
    }
}
```

This handler reads events from RabbitMQ and updates the view database. `GetOrderAggregateByEntityId` gets the order aggregate by playing events on it. The `updateOrder` function update the `Orders` table and the `updateOrderItems` function updates the `OrderItems` table.

The CQRS and event sourcing design patterns are used together to create systems that have their own historical data. This event data is important for auditing purposes. You can get a complete idea of how a system enters a particular state since we have all transition information in the event store. Since event sourcing is a relatively new concept for cloud native design, it has fewer supported databases and frameworks that support event sourcing.

Summary

In this chapter, we focused on connecting a Ballerina application with databases and building a simple order management system that runs on a distributed system. We discussed how to connect different types of databases with Ballerina built-in libraries. We further discussed using MySQL as the backend database and how we can use different types in the Ballerina language to map to different MySQL data types.

The order management system that we discussed in the first chapter was implemented as an example by using the MySQL database. We also discussed transaction management, which is an essential part of building any application.

Further, we discussed writing applications in a distributed manner using different methodologies. To handle distributed transactions, we discussed the saga pattern. To hold the previous status of the system, we discussed using event stores. We also discussed CQRS patterns that separate read and write operations along with the event store. All of these design patterns are important aspects of building a distributed system that has distributed databases. These methods can be used in developing a cloud native application by combining these architectures.

In the next chapter, we will discuss how we can use serverless architecture to build applications with Ballerina. We will discuss using functions in different FaaS providers and build a sample FaaS application.

Questions

1. What is the difference between an entity and a value object?
2. How do you use DDD with test-driven development?

Further reading

- *Spring Microservices*, Rajesh, R.V., Packt Publishing, available at `https://www.packtpub.com/product/spring-microservices/9781786466686`

- *Cloud Native Architectures: Design high-availability and cost-effective applications for the cloud*, Laszewski, T., Arora, K., Farr, E. and Zonooz, P., Packt Publishing, available at `https://www.packtpub.com/product/cloud-native-architectures/9781787280540`

Answers

1. Value objects do not hold an identity, but an entity holds an identity. Entities exist on their own while value objects are used along with entities.

2. DDD is a design methodology that is used to design large-scale systems. Test-driven development is a practice that can be used to implement a system. You can start designing your system with DDD concepts and by writing tests before implementation. This makes development faster and fault-free.

Section 3: Moving on with Cloud Native

The last section discussed the enhancements that you can apply to your Ballerina cloud application. This section mostly focuses on improving the quality, security, deployability, and observability of Ballerina applications. We will further discuss using serverless architecture and the Choreo platform to develop a Ballerina application.

First, we will discuss serverless platforms as an alternative to microservice architecture to use with the Ballerina language. We will discuss building serverless Ballerina applications with AWS Lambda and Azure Functions.

Next, we will discuss securing the Ballerina platform with security features provided by the Ballerina language. We will look into using certificates, LDAP servers, JWT, and the OAuth 2 authentication/authorization mechanism with the Ballerina language.

Next, we will look into observing the Ballerina program with three pillars of observability. First, we will look at logs, then traces, and finally, metrics. We will discuss multiple tools that we can use to collect this information and how to visualize and monitor it.

Next, we will learn about integrating Ballerina applications. Here, we will learn about exposing the Ballerina service outside with an API gateway. We will use the micro-gateway that is built into the Ballerina language and you can customize it with the Ballerina language. Next, we will also discuss the Choreo iPaaS platform to develop integration scenarios on a cloud platform with the Ballerina language.

Finally, we will look into automating the building and deployment process of Ballerina applications. First, we'll look into creating automated testing for a Ballerina application by writing test cases. Next, we will discuss adding an automated pipeline to deploy a Ballerina application. We will also discuss Ballerina Central to simplify managing your source code.

This section has the following chapters:

6
Moving on to Serverless Architecture

Serverless architecture is a recent advancement in building systems that run on cloud service providers. Due to the ease of deployment, competitive pricing model, zero **Capital Expenditure (CAPEX)**, and scalability, developers use serverless architecture to build programs on serverless cloud platforms. In this chapter, we will learn about serverless architecture and how to build Ballerina applications with popular serverless platforms such as Azure Cloud and AWS. We will discuss the features provided by each of these platforms to build complex business applications that are running on the cloud.

The following are the topics that we are going to discuss in this chapter:

- Introducing serverless architecture
- Developing Ballerina applications with Azure Functions
- Developing Ballerina applications with AWS Lambda functions

By the end of this chapter, you should be able to create a serverless application by using the Ballerina language that can be deployed on either the Azure or AWS platforms.

Technical requirements

This chapter includes Ballerina program samples that will be deployed on AWS and Azure cloud platforms. Therefore, you need to have AWS and an Azure Cloud account to try out the samples. You can create both AWS and Azure accounts free of charge. Both of these platforms provide a free quota/credits to evaluate the platform and the same can be used to try out sample codes. Both platforms provide CLI tools to interact with the cloud. The download and installation instructions for the CLI tool are given in the corresponding section.

You can find the code files for this chapter at `https://github.com/ PacktPublishing/Cloud-Native-Applications-with-Ballerina/tree/ master/Chapter06`

The Code in Action video for the chapter can be found here: `https://bit.ly/3zGGpds`

Introducing serverless architecture

When you are building a web service, you need to think about where it should be deployed and how it should be deployed. We have already discussed **Infrastructure as a Service (IaaS)** – the cloud platform that provides physical or **Virtual Machines (VM)** to host services. With serverless architecture, you do not need to worry about infrastructure implementation and maintenance and you can focus on only writing code that serves your customers. The traditional way of deploying cloud applications uses the following infrastructure options to host the service:

- Runs on bare metal
- Runs on a VM
- Runs on a container

There are multiple problems and challenges associated with deploying applications with the aforementioned methods. The following are some of those:

- **Management and maintenance**: Infrastructure management tasks such as updating and patching and so on bring an additional overhead for developers. They need to consider infrastructure-level implementation details while thinking about the business logic.

- **Security**: Aside from maintenance, developers need to consider the security and network connectivity of the system. This increases the complexity of the overall application and the chances of having possible vulnerabilities.

- **Scalability and availability**: Developers need to consider how and when to scale the system by doing proper capacity planning during the development stage. This requires the developer to install multiple dependencies within the system, such as Kubernetes.

- **Paying for the server even if it's idle**: Most of the time, we need to keep application servers running continuously even if there are no active users. Therefore, it is a waste of money and computing power and there is no way of paying for actual consumption/usage.

In serverless architecture, developers no longer need to be concerned about infrastructure-related tasks, rather the functions that contain the business logic. All of the aforementioned challenges are handled by the cloud platform provider. They will manage servers, secure the services, and automatically scale servers based on the traffic. You do not need to pay for long-running servers. Instead, you need to pay for the actual resource consumption.

FaaS is one of the leading types of serverless architectures. AWS Lambda, Azure Functions, and Google Cloud Functions are common serverless service providers. Each of these platforms has different types of implementations. The price of the serverless platform depends on the number of requests that are handled and the duration of the execution. You can allocate memory for each function and scale each function individually.

There are both benefits and drawbacks associated with serverless architecture. Here are some of the advantages of using serverless architecture:

- **Low cost**: As mentioned previously, developers do not need to worry about the underlying infrastructure implementation. This significantly reduces the complexity of the application from the developer's point of view. Cloud providers are responsible for the operation and maintenance of the underlying deployment infrastructure. Organizations do not need to spend money for DevOps engineers to support and maintain the infrastructures. There is zero CAPEX and reasonable Operating Expenses (OPEX) based on actual consumption only.

- **Fast deployment**: Since an application is a collection of functions, developers can build and deploy functions separately. Small changes to a function don't require the entire application to be restarted.

- **Scalable**: Serverless is scalable based on incoming traffic. Resources are allocated on demand, where organizations do not need to pay for idle servers. The software developers or operations team do not have to worry about implementing and managing scalability as automatic scaling and built-in high availability features are baked into the platform.

There are also disadvantages attached with the serverless architecture as follows:

- **Vendor lock-in**: Functions are specific to the cloud provider. At some point, if you need to switch from one cloud provider to another, there would be a huge migration process. Even though the configurations, tools, and languages used to program are supported in both platforms, the migration process will become an overload due to platform-specific features.

- **Cold start problem**: Serverless functions remain active when traffic comes to a function. When a function is not invoked for a while, it goes to an idle state. If you try to invoke an idle function, it will take some time to reactivate.

- **Limitation by vendor**: There could be some limitations specified by the cloud service provider to operate functions. There can be limitations on the memory size, execution time, the number of concurrent executions, and code size.

If you plan to use serverless functions to create your applications, always consider the limitation of using serverless. However, due to its simplicity and low maintenance features, organizations consider serverless as an option and try to move to a serverless architecture. In the next section, we will discuss how to develop Ballerina applications with Azure Functions.

Developing Ballerina applications with Azure Functions

Azure is a leading cloud application development platform for hosting cloud applications. Azure provides Azure Functions as a serverless solution to build **Function as a Service (FaaS)** applications. Ballerina provides built-in libraries to create Azure functions. In this section, we will learn how to implement a serverless application with the Ballerina language. We begin by learning about the Azure platform that provides different cloud solutions.

Introducing Azure cloud features

Azure is a cloud service offered by Microsoft that has distributed data centers around the world. Azure provides computing, storage, and network solutions to build cloud applications. Azure pricing is based on the resources you have consumed, in other words, a pay-as-you-go approach. Azure gives you some free credits with the trial subscription to try out their cloud platform.

Azure Portal is the management portal where you can manage Azure Cloud services. You can build cloud applications on the Azure platform using either the Azure management web portal or the Azure CLI, which is a command-line tool used to interact with the Azure platform.

Azure compute provides a different range of computation solutions to build cloud applications. You can create VMs with different operating systems hosted in the Azure cloud. You can install software on VMs and perform some computing tasks on them. Azure also supports containers, which are much lighter than VMs. Azure provides **Azure Container Instance** (**ACI**) to run Docker containers on the Azure platform. This is a PaaS solution provided by the Azure platform. On the other hand, Azure provides **Azure Kubernetes Service** (**AKS**) as a container orchestration solution to deploy microservices. You can simply deploy a Kubernetes cluster with the Azure platform.

Azure Virtual Network (**Azure VNet**) provides networking solutions on the Azure platform. This helps developers to build secure applications connected over the internet or on-premises. The virtual network is scoped to a single region. You can use virtual network peering to connect services that are running in different regions. On the other hand, Azure Application Gateway gives you a way to handle HTTP traffic on your system. Azure provides the **Azure Domain Name System** (**Azure DNS**) as the DNS solution, Traffic Manager to handle traffic across the regions, and **Content Delivery Network** (**CDN**) to deliver content to the user with caching support.

Azure Storage Services provides storage solutions to build cloud applications. Azure Blob storage lets you host large and small unstructured binary objects on the Azure cloud. Queue storage is used to hold smaller messages that need to be kept in a queue. File storage can be used to store data as files. Azure provides Cosmos DB as a NoSQL database solution, while Azure SQL Database is a relational database solution.

Aside from these key features, Azure includes hundreds of features offered to cloud native application developers. These include secure vaults, key-value storage, Traffic Manager, machine learning tools, IoT solutions, data analytics tools, data streaming solutions, and so on. In this chapter, we will focus on building serverless applications. Azure provides Azure Functions to provide serverless solutions. In the next section, we will learn what Azure Functions is and how to create a simple Azure function.

Building serverless applications with the Azure cloud

Azure provides various infrastructure capabilities and tools to build a cloud application easily. Azure Functions provide a FaaS solution to build serverless applications. You can use Azure Functions to build web applications, IoT systems, data streams, and message queues. This section includes a sample implementation of a simple Azure function.

Azure triggers and bindings provide a way to invoke Azure functions with input arguments. **Triggers** are used to define how the Azure function should be triggered. **Bindings** are used to connect input and output components to an Azure function. Azure functions can have only one trigger. Binding can be either an input or output binding. Input bindings are used to pass arguments to the function, and output bindings are used to respond to the function result or trigger another resource. You can use multiple combinations of bindings to combine different types of services with a single function. Ballerina provides multiple triggers and binding types to work with Azure Functions, including the following triggers and bindings:

- **HTTP trigger and bindings**: Invoke Azure Functions using an HTTP request.

- **Queue trigger and bindings**: Trigger Azure Functions with an Azure Storage queue.

- **Blob trigger and bindings**: Trigger Azure Functions with blob storage.

- **Cosmos DB trigger and bindings**: Trigger Azure Functions with Cosmos DB operations.

- **Timer trigger**: Trigger Azure Functions at a predefined time.

- **Twilio output bindings**: Send an SMS with Twilio output bindings.

You can choose one of the triggers and multiple bindings to create a function. For example, an HTTP trigger can be used to trigger a function with an HTTP request, while the function publishes messages to a queue with a Queue binding and sends an SMS with Twilio bindings.

To get started with Azure Functions, perform the following steps:

1. First, you need to create an account in the Azure portal.

2. Next, you need to configure the Azure CLI tool to interact with the Azure cloud via the CLI interface. You can download and install the Azure CLI client from `https://docs.microsoft.com/en-us/cli/azure/install-azure-cli` (the Azure web page) depending on your operating system. Once you download and install the CLI tool, you can verify the installation by issuing an `az version` command in the command-line tool.

3. The next step is to authenticate the Azure CLI tool with the Azure cloud. You can use the `az login` command to do this. This command will open a web page on your web browser and request login information.

4. Provide the login information and complete the CLI initialization.

The Azure management web portal contains a UI to easily manage your cloud deployments. The examples and instructions mentioned in this book use both the web portal and the CLI tool to interact with the Azure Cloud. To find any component or resources on the Azure web portal, you can use the search options on the top of the Azure portal. With the Azure portal, you can organize frequently used resources and tools on the left menu bar panel, as shown here:

Figure 6.1 – Azure portal search bar and menu bar

> **Tip**
> The Azure management portal goes through user interface changes from time to time. The instructions given in this book might not be the same with new interface changes. Refer to the Azure documentation guide in case you require further details on using Azure services.

To create a function, first, you need to create an Azure artifact called a resource group. A resource group is used to group multiple related Azure resources. You can use this resource group to manage all allocated resources for this Azure function. Follow these steps to create a resource group:

1. To create a new resource group, go to **Resource groups** either by searching using the search bar, or by clicking on the **Resource groups** link on the left side menu bar.

2. On the **Resource groups** page, click the **New** button to create a new resource group.

3. Select the subscription that you want to bill for the service under the project details page. In the **Resource group** input box, provide a name for the resource group. Then you can select the region where you need to deploy the resource group.

4. Finally, click on **Review + create** to create the resource group. On the review page, verify the details entered and then click on the **Create** button to create the resource group.

Let's now start creating the Azure function with the Ballerina language that gets triggered with an HTTP request and responds to the client with HTTP bindings. Check out the following example of creating an HTTP trigger with HTTP input and output bindings:

```
import ballerinax/azure_functions as af;
@af:Function
public function hello(@af:HTTPTrigger { authLevel: "anonymous"
    } string input) returns @af:HTTPOutput string|error {
    return "Response from Azure, " + input;
}
```

This function uses the Azure Functions Ballerina library to implement the function. Here we have defined a function, `hello`, along with the `@af:Function` annotations. This annotation is used to mark this function as an Azure function. In function input arguments, we can set triggers and bindings. This function has an HTTP trigger with HTTP input and output bindings. The `af:HTTPTrigger` annotation is used to mark the given argument as the trigger and it has string-type input bindings. Here, we also specified the function to not have any authentication scheme by setting the `authLevel` field to `anonymous`.

You can have various types of authentication methods to authenticate the incoming request to a function. The output binding of the HTTP trigger is given as the `return` statement. The `return` statement contains an output type as a string type with `af:HTTPOutput` annotations to label output as an HTTP response type. If there are any errors found in the function, it will return an error response to the client application with the `error` data type. In the function content, it appends the incoming request body to a string and returns it. Build the project with the Ballerina `build` command and generate the Azure function artifacts.

Now we need to create a **Function App** to store functions on the Azure Cloud. The function app is capable of grouping and keeping multiple Azure functions together. Follow these steps to create a function app:

1. Go to the **Function App** page on the Azure portal and click **New** to create a new function app.

2. Select the billing subscription on the **Subscription** input filed. Select the resource group that you have created on the **Resource Group** input. Give the function app a name in the **Function App name** field under the **Instance Details** section.

3. Select **Java** as the **Runtime stack** and the Java version as **11.0**. Also, select the region where you need to deploy the function in the **Region** selection box:

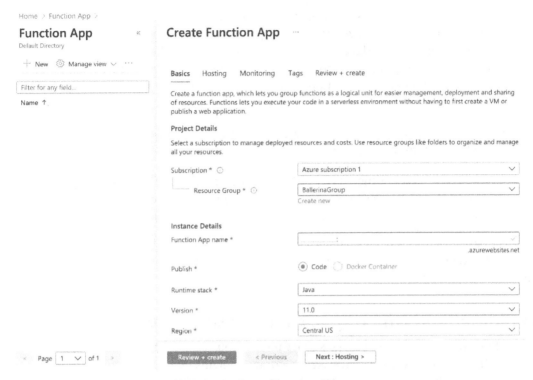

Figure 6.2 – Create Function App page

4. You can configure the Azure function with the host operating system, storage account, and monitoring with the **Hosting** and **Monitoring** tabs. For this example, we will stick with the default values. Once you complete all the fields, click on **Review + create** to continue creating the function app. This will bring up a review page and then click on the **Create** button to create the function app. It may take around 1 minute to complete the function app creation and the Azure portal provides you with a notification once it has been completed.

The next step is to upload the function by using the Azure CLI tool. The command used to upload the function artifacts can also be found on the build command output as well. The following is the template of the Azure `create` function command to upload artifacts to the Azure Cloud:

```
az functionapp deployment source config-zip -g <resource_group_
name> -n <function_app_name> --src <build_artifact_location>
```

In this command, you need to specify the resource group name you have just created along with the artifact location, which is `<project_home>/target/bin/azure-functions.zip`. Make sure to log in with the Azure CLI tool before using this command. Once you enter this command, the Azure CLI uploads artifacts to the Azure Cloud and creates a new function. Once the function gets deployed, we can test the function by using the following `curl` command, which invokes the function:

```
curl -d "Hello World" https://<function_app_name>.
azurewebsites.net/api/hello
```

You can find the corresponding service URL from the Azure portal's **Function App** page. The URL for the function is created with the function app name and the function name. Once you invoke this service, you will get the concatenated response from the backend.

This is a basic example that used an HTTP trigger along with an HTTP binding to read and send a response back to the client. In the next section, we will learn to connect multiple Azure functions by using Azure Queue Storage.

Building Ballerina applications with Azure Queue

Azure Storage offers a pay-as-you-go service model. You can store as much data as you want with the Azure Storage service. Azure Storage contains two types of storage services – Standard and Premium. The standard storage account type includes the following:

- Blob
- Table
- File
- Queue

The premium storage account includes data storage on **Solid State Drives** (**SSDs**) to provide faster data access. Azure Queue is a queue that can be used to send messages over the Azure platform. It acts as storage as it can store messages even after you consume them. Therefore, you can use Azure Queue as a backlog of work. As we have described in the previous chapter, we can use a queue to connect multiple services in an asynchronous manner. For this example, we will build an order validation scenario by using queues.

In our order management system, the customer has a credit limit with which to purchase items. When the order confirmation request comes to the system, the system checks that the order does not exceed the given credit limit. We will create a function to read the order ID from the incoming request, a function to validate the order credit limit, and another two functions to send the validation result back to the customer. Here are the steps:

1. To simplify the example, we will read order details from a Ballerina table data type variable that is already initialized with dummy values instead of using a database connection. We will use the following data structure to hold the order details:

```
type Order record {|
    readonly string orderId;
    string customerEmail;
    float totalPrice;
|};
type OrderTable table<Order> key(orderId);
```

2. Now we can use this data structure to create dummy values with the following Ballerina code:

```
OrderTable orders = table [{
    orderId: "343232",
    customerEmail: "customer@email.com",
    totalPrice: 50
}];
```

3. We can use Ballerina SMTP libraries to send emails. The following utility method, sendEmail, enables us to send an email to a specified address:

```
function sendEmail(string to, string subject, string
    body) returns error?{
        email:SmtpClient smtpClient = check new ("<SMTP
            client URL>",
        "<SMTP client username>" , "<SMTP client
            password>");
    email:Message email = {
        to: [to],
        subject: subject,
        body: body,
        'from: "<sender_email_address>"
    };
    check smtpClient->sendMessage(email);
}
```

4. Change your SMTP client URL, username, and password according to the SMTP gateway. Also, provide an email that you will use to send the email to the `sender_email_address` placeholder. Make sure to read these values from the environment variables rather than hardcoding them to the source code. You can set environment variables from the Azure portal.

5. Now we will implement the function that gets triggered with an HTTP request. The following function contains an HTTP trigger with a queue output binding. This function simply reads the incoming HTTP request and submits the body to the Azure queue:

```
@af:Function
public function submitOrder(
    af:Context ctx,
    @af:HTTPTrigger { authLevel: "anonymous" }
      af:HTTPRequest req,
    @af:QueueOutput { queueName: "order-queue" }
      af:StringOutputBinding msg)
    returns @af:HTTPOutput af:HTTPBinding {
    msg.value = req.query.get("orderId");
    return { statusCode: 200, payload: "Submitted
      successfully" };
}
```

Here, we set the authentication level to `anonymous`. The queue name is set as `order-queue`, which holds a queue of order validation messages. We can send a message to the queue indicated by output binding by assigning a value to the `msg` variable. Here we set the message as the order ID. The order ID is sent from the HTTP request with a query parameter. This function reads the order ID and forwards it to the `order-queue` queue.

6. Now we need to build the function that reads the queue message and does the credit limit validation. The following function reads the messages from the `order-queue` queue and performs the credit limit validation:

```
@af:Function
public function validateOrder(af:Context ctx,
    @af:QueueTrigger { queueName: "order-queue" } string
      inputMessage,
    @af:QueueOutput { queueName: "order-success-queue" }
      af:StringOutputBinding outSuccessMsg,
```

```
    @af:QueueOutput { queueName: "order-fail-queue" }
        af:StringOutputBinding outFailedMsg) {
    Order orderItem = orders.get(inputMessage);
    if orderItem.totalPrice > maxCreditLimit {
        outFailedMsg.value = inputMessage;
    } else {
        outSuccessMsg.value = inputMessage;
    }
}
```

This function has a queue trigger and two queue output bindings. If validation is a success, it publishes a message on the order-success-queue queue. Otherwise, it publishes a message on the order-fail-queue queue. Once the validation process is complete, this function publishes order id either to order-success-queue or order-fail-queue as the message content.

7. Next, we can create two functions to read each of the queues and send an email to the user. The following is the function that listens to the order-success-queue queue:

```
@af:Function
public function validateOrderSuccess(af:Context ctx, @
    af:QueueTrigger { queueName: "order-success-queue" }
        string inputMessage) returns error?{
    Order orderItem = orders.get(inputMessage);
    check sendEmail(orderItem.customerEmail, "Order
        validation success", "Order validation successful
            for order ID: " + orderItem.orderId);
}
```

This function has an input binding to read messages from the order-success-queue queue and sends an order validation success email to the customer.
The following is the function listening to the order-fail-queue queue and sends a validation failure email to the customer:

```
@af:Function
public function validateOrderFail(af:Context ctx, @
    af:QueueTrigger { queueName: "order-fail-queue" }
        string inputMessage) returns error? {
    Order orderItem = orders.get(inputMessage);
    check sendEmail(orderItem.customerEmail, "Order
        validation failed", "Order validation failed for
            the order ID: " + orderItem.orderId);
}
```

You can try out this example code by sending a `GET` request to the `submitOrder` function with the query parameter `orderId`. Since the price of the order is lower than the credit limit, you will get an email saying that the order validation was successful. This is an example of where you can use Azure Functions with queue and HTTP bindings to build a system that has multiple functions. You can combine multiple output bindings to build an Azure function. By using and combining other libraries and components provided by Ballerina and Azure, you can build more complex and meaningful applications.

In this section, we used Azure Functions to build and deploy a serverless system. Similarly, the AWS Cloud provides AWS Lambda to implement a serverless system. In the next section, we will discuss what AWS Lambda functions are and how to implement them with the Ballerina language.

Developing Ballerina applications with AWS Lambda functions

AWS is a leading cloud service provider that hosts web services on spatially distributed data centers all over the world. AWS provides many cloud solutions that you can use to host different types of applications. AWS Lambda is the FaaS solution provided by the AWS platform. You can easily create a Lambda function by using both the CLI tool and the AWS Web Console. The Ballerina language supports the creation of AWS Lambda functions with the support of Ballerina AWS Lambda layers. In this section, you will learn how to set up the AWS CLI tool, how to create a simple Lambda function, and how to improve it to provide a complex solution by using AWS Step Functions and the AWS API Gateway trigger.

Understanding AWS Cloud services

Amazon Web Services (**AWS**) provide a complete range of tools for implementing cloud-based solutions. Similar to other cloud vendors, AWS also provides cloud computing solutions at affordable rates. AWS pricing is also based on a pay-as-you-go scheme. The cost depends on the resource you have consumed. AWS provides computing, storage, database, and networking workloads for cloud developers to develop cloud-based applications. Developers can leverage these components to build applications that are running on the cloud.

The AWS computing solution includes **Amazon Elastic Computing Cloud** (**Amazon EC2**) resizable VMs. You can install software on these machines and perform various tasks. AWS provides Lambda functions for developing serverless-based solutions. AWS **Elastic Kubernetes Service** (**EKS**) offers ways to create and deploy on Kubernetes clusters easily. AWS **Elastic Container Service** (**ECS**) also provides a scalable container orchestration service for developers.

AWS provides a broad range of storage and database solutions to store and manage data. AWS **Simple Storage Service (S3)** stores data as objects on the AWS Cloud. Amazon **Elastic Block Storage (EBS)** provides a storage service to store data. EBS storage is commonly used with Amazon EC2 VMs as it is internal storage. Compared to S3, EBS is faster and also expensive than S3. Amazon database solutions include Amazon Aurora, which is a high-speed, cost-effective relational database solution. Amazon RDS is another popular relational database provided by the AWS Cloud. Amazon Dynamo DB is a NoSQL database provided by the AWS platform.

AWS Lambda functions were introduced in 2014 and act as a serverless solution for the AWS Cloud platform. The AWS Lambda pricing model is based on the number of requests the function receives and the duration to execute the function. AWS calculates the time duration by measuring the round-trip time for the incoming request. If the request failed while executing, it takes the execution time for a given request. With AWS Lambda, you can configure the amount of memory allocated for a function. The CPU allocation is also increased or decreased based on the amount of memory allocated for a function. To start developing with AWS Lambda, first, we will set up the AWS CLI on your computer.

Configuring the AWS CLI and AWS Console

To start developing AWS Lambda functions, first, you need to install the CLI tool that is used to manage the AWS Cloud. This tool can be downloaded as an installer and installed on your system. The following are the download and installation instruction web pages to set up the AWS CLI tool. Download and install the tool based on the operating system that you use:

- **Linux**: https://docs.aws.amazon.com/cli/latest/userguide/install-cliv2-linux.html.

- **MacOS**: https://docs.aws.amazon.com/cli/latest/userguide/install-cliv2-mac.html.

- **Windows**: https://docs.aws.amazon.com/cli/latest/userguide/install-cliv2-windows.html.

After you have installed the CLI, you can verify the installation with the aws -version command. This command prints the currently installed AWS version. Here, we used AWS CLI version 2 to demonstrate the examples. If you are using version 1, make sure to use the relevant CLI commands.

To start using the AWS CLI, perform the following steps:

1. First, you need to authenticate the AWS CLI to access the AWS Cloud. To do this, you need to create a user and generate permissions to work with AWS Lambda functions. To create a new user, click on **My Security Credentials** under your username on the top menu bar of the AWS Console as follows:

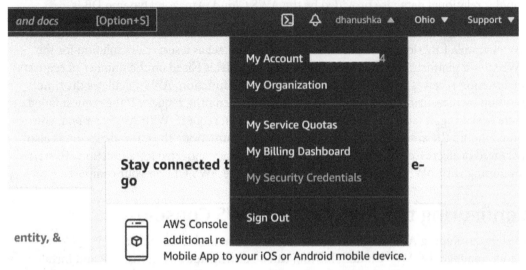

Figure 6.3 – AWS Console – My Security Credentials link

2. This will lead you to a new window called **Identity and Access Management (IAM)**. This window lets you create new users and roles for authentication and authorization on the AWS environment. From this window, select **Access management | Users** on the left menu. This menu lists available users in AWS where you can control access for each user.

3. Create a new user by clicking on the **Add user** button on the top left of the page:

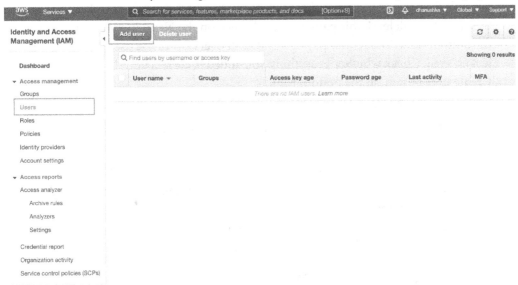

Figure 6.4 – User management window in the AWS Console

4. When you click on the **Add user** button, it will show the **Add user** page. Here you need to provide the username to identify the user. Make sure that you select the **Programmatic access** option for **Access Type** under the **Select AWS access type** section.

5. Next, click on the **Next: Permissions** button.

    In the next view, you need to set the permission for the user that you are going to create. For this example, let's give all the Lambda function permissions to the newly created user.

6. Under the **Set permissions** section, select the **Attach existing policies directly** button to show existing policies. This will give you a list of predefined policies. Search for AWSLambdaFullAccess to give full access to the new user. Click the **AWSLambdaFullAccess** tick button on the left and proceed to the next window by clicking on the **Next: Tags** button on the bottom right of the page.

7. In the next window, you can set a tag for the user you are creating. This is an optional field and you can use these tags to organize, track, and control access for this user. We will leave this page as it is and click on the **Next: Review** button on the bottom right. The review page gives you an overview of the user details that we just entered. Check whether everything you added is correct and then click on the **Create user** button on the bottom right.

    In the next window, you can see that it successfully created the user and is listed in the list. In this list, you can find the **Access key ID** and the **Secret access key** keys that can be used to log in as this user. Save these two keys and close the page.

8. The next step is to configure the AWS CLI tool to use these keys to log in with the user we have just created. Open a terminal in your computer and execute the `aws configure` command. This command asks you for the AWS access key ID and the AWS secret access key. Enter those values copied from the user creation process. Leave the other optional fields blank and complete the configuration process.

Once you have completed this step, you can use the AWS CLI tool to work with AWS Lambda functions.

In the same way that we have created a user, you need to create a role to deploy Lambda functions as well. To create a new role, perform the following steps:

1. Click on **Roles** on **Identity and Access Management (IAM)** just under the **Users** button. This window lists all the available AWS roles.

2. Click on the **Create role** button on this page to create a new role.

3. The next window lets you create a new role with multiple options. Select **AWS service** under **Select type of trusted entity** and then select **Lambda** under the **Choose a use case** section.

4. Click **Next: Permissions** to add permission for the Lambda function. Here, we will also add **AWSLambdaFullAccess** permission to the new role. Search for `AWSLambdaFullAccess` and select it.

5. Then, click on the **Next: Tags** button on the bottom right. In the same way as for users, you can create a tag for the roles as well.

6. Leave this page as it is and click on the **Next: Review** button. Under the **Review** window, you need to give a role name to identify the role and complete the **Role description** field – a brief description regarding the role.

7. Click on the **Create role** button to create the role.

Now we have both a user and a role that we can use to deploy our AWS Lambda function. In this section, we have focused on connecting the AWS CLI with the AWS Cloud by creating a new user. In the next section, let's discuss how to write a simple AWS Lambda function with the Ballerina language.

Creating your first Ballerina AWS Lambda function

Ballerina supports the creation of an AWS Lambda function with the Ballerina built-in AWS Lambda libraries. You can create and deploy functions on AWS Lambda easily with Ballerina and the AWS CLI tool. We will demonstrate the AWS Lambda function's functionality by building a simple total order amount calculation function that we have already discussed in *Chapter 4, Inter-Process Communication and Messaging*, with a gRPC sample. This function reads an order item that has an item ID and quantity. The Lambda function gets the price of the item by the item ID and calculates the total price. In this scenario, instead of the gRPC service, we will use the Lambda function to implement the functionality. A response with the total price is sent back to the caller as a JSON response. You can follow these steps to create a simple Lambda function that runs on the AWS Cloud:

1. First, we will create the following data structure to decode an incoming order item JSON request:

    ```
    public type OrderItem record {|
        string itemId = "";
        int quantity = 0;
    |};
    ```

2. We will create the following `InventoryItemTable` table data type to hold all available types of order items and an `itemList` variable to hold all item types:

    ```
    type InventoryItem record {|
        readonly string itemId;
        float price;
    |};
    type InventoryItemTable table<InventoryItem> key(itemId);
    InventoryItemTable itemList = table [
        {
            itemId: "item1",
            price: 120
        },{
    ```

```
            itemId: "item2",
            price: 20
    }
];
```

3. Now we can implement the AWS Lambda function as follows with the Ballerina language to calculate the total price of the incoming order item request:

```
@awslambda:Function
public function totalPrice(awslambda:Context ctx,
    OrderItem item) returns json|error {
    InventoryItem? inventoryItem = itemList[item.itemId];
    if (inventoryItem is InventoryItem) {
        return { "totalRes" : inventoryItem.price *
            <float>item.quantity};
    } else {
        return error("Error while retrieving table data");
    }
}
```

Here we used the `@awslambda:Function` annotation to mark this function as an AWS Lambda function. The name of the function is set as `totalPrice`, which reads context and the order item as the input argument. This function returns either a JSON response or an error to the caller. This function simply gets the price of the item from the `itemList` table and multiplies the price by the quantity. The response is sent back as a JSON response with a key of `totalRes`.

The next step is to deploy this function on AWS Lambda. Perform the following steps to deploy the function:

1. First, build the Ballerina project with the Ballerina `build` command. You can use either the CLI tool or the AWS Console to upload the function. To upload the function with the AWS CLI, use the following terminal command. Make sure to replace the placeholder with the appropriate arguments. A sample AWS CLI deployment command can also be found in the Ballerina `build` command output:

```
aws lambda create-function --function-name <function_
name> --zip-file fileb://<relative_file_path> --handler
<project_name>.<function_name> --runtime provided
--role <role_arn> --layers arn:aws:lambda:us-west-
1:134633749276:layer:ballerina-jre11:1 --memory-size 512
--timeout 10 --region us-west-1
```

For this example, the function name would be `totalPrice`. `role_arn` is the ARN value that you can find on the role description in the AWS Console. To get the ARN value, go to the **Identity and Access Management (IAM)** page from the AWS Console and select the role that we have just created. On the **Summary** page, you can find the **Role ARN** value.

In this command, you can see the `--layers` option passed to the CLI command. AWS currently supports the Java, Go, PowerShell, NodeJS, C#, Python, and Ruby programming languages natively to implement the underlying Lambda functions. Therefore, we need to use the Ballerina wrapper layer to support AWS Lambda to work with Ballerina-based functions. AWS layers are a way of separating library util functions from the main business logic. A layer can contain libraries, a custom runtime, and other dependencies. Here we used the Ballerina JRE wrapper to deploy the Ballerina functions on the AWS server.

2. Once you execute this command, the AWS CLI deploys this function on the AWS Cloud. If you need to redeploy the function, build the Ballerina program again and use the following command to update the function:

```
aws lambda update-function-code --function-name
<function_name>--zip-file fileb://<relative_file_path>
--region us-west-1
```

Make sure you replace the function name and the file path placeholders. If you log in to the AWS Console and list the available functions, the function that we have just uploaded is listed on the function list.

3. You can trigger this Lambda function by using the AWS CLI tool with the following command:

```
aws lambda invoke --function-name totalPrice --payload
'{"itemId": "item1", "quantity": 5}' response.json
--region us-west-1 --cli-binary-format raw-in-base64-out
```

`payload` is set as JSON content and the `--cli-binary-format raw-in-base64-out` option is set to mark the `payload` given in raw format. Otherwise, you need to provide `payload` as a `base64`-encoded string. This will send the payload to the Lambda function and write the response to the `response.json` file on the current directory. If you check the content of the `response.json` file on the current directory, you can find the response received from the Lambda function.

You can do the same deployment and testing in the AWS UI Console as well. You can create a new function and upload the Ballerina artifact .zip file to the function with the AWS Console. You can also test the function with the **Test** button provided in the top left of the corresponding function page. On the **Test** page, you can specify a payload and invoke the function.

Adding triggers to invoke a Lambda function

We have discussed how to build a Lambda function and how to invoke it with the AWS CLI tool. In a production system, we need to specify a way of invoking this function. AWS Lambda functions provide triggers to invoke Lambda functions. These triggers include API gateways, message brokers, databases, and streaming interfaces. You can define the type of the trigger and invoke Lambda functions using that trigger.
For example, we will discuss how to add the API gateway trigger to a Lambda function.

To add an API gateway trigger, perform the following steps:

1.   Navigate to the AWS Management Console and select the AWS Lambda function.

2.   Click on the **Add trigger** button on the **Function** overview page for the selected Lambda function.

3.   Next, it will redirect you to the **Add trigger** page. On this page, in the Trigger configuration section, select the **Select a trigger** dropdown and then select **API Gateway** from the list.

4.   Once you select the **API Gateway** option, it asks for an API. On this drop-down list, select the **Create an API** option.

5.   Then it asks to select either **HTTP API** or **REST API**. Select **REST API** as the API type and then select **Open** from the **Security** dropdown for now. When you are building a production-grade program, make sure to include a secure API invocation method rather than using the **Open** type.

6. Then, click on **Add** to create a new API Gateway trigger with a new API:

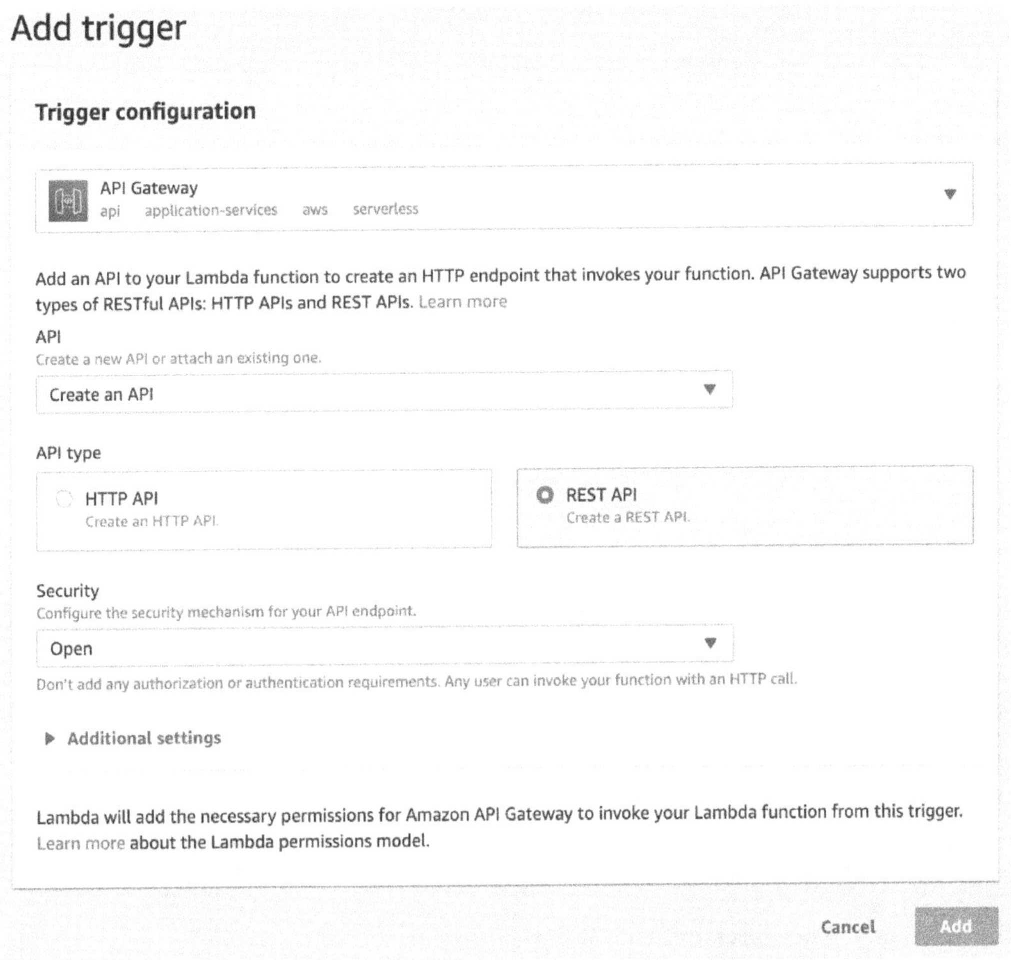

Add trigger

Trigger configuration

API Gateway
api   application-services   aws   serverless

Add an API to your Lambda function to create an HTTP endpoint that invokes your function. API Gateway supports two types of RESTful APIs: HTTP APIs and REST APIs. Learn more

API
Create a new API or attach an existing one.

Create an API

API type

HTTP API	● REST API
Create an HTTP API.	Create a REST API.

Security
Configure the security mechanism for your API endpoint.

Open

Don't add any authorization or authentication requirements. Any user can invoke your function with an HTTP call.

▶ Additional settings

Lambda will add the necessary permissions for Amazon API Gateway to invoke your Lambda function from this trigger. Learn more about the Lambda permissions model.

Cancel   Add

Figure 6.5 – Creating a new API Gateway for your Lambda function

Now you can see that the new API Gateway trigger that we have just created is listed on the Lambda function details page. If you go to the API gateway list page, `https://us-west-1.console.aws.amazon.com/apigateway/main/apis?region=us-west-1`, you can see that the API gateway is also listed here (you need to change the preceding URL based on the region where you have to create resources). From the list, select the API gateway that we have just created:

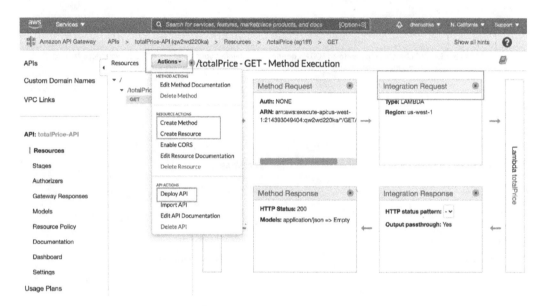

Figure 6.6 – Creating an API page

Initially, this page does not contain any resources. First, we need to create a resource that accepts an HTTP connection. To create a new resource, perform the following steps:

1. Click on the **Actions** button in the top-left corner.

2. Select the **Create Resource** option on the drop-down list. This will give you a **New Child Resource** page.

3. Add the resource name where the resource name should be in the **Resource Name** input box. For this example, we will name this resource `totalPrice`. **Resource Path** also gets changed when you enter the resource name. We will keep that value as it is.

4. Then, click the **Create Resource** button to complete the creation of the resource. The resource will then be listed on the resource list in the **Resources** tree view.

Now, we need to add an HTTP method to handle incoming HTTP requests. Follow these instructions to add resources:

1. Select the resource that we have just created on the **Resources** list tree view and then click the **Actions** button.

2. From the drop-down list, select **Create Method** to add a new method for a resource. This will reveal a small drop-down box just below the resource that we have just created on the **Resources** tree view.

3. From the drop-down list, select the **GET** method. Optionally, you can also select other methods such as **POST** and **PUT**.

4. Then, click on the checkmark button to create the method. Once you click the checkmark, it will reveal the **/totalPrice – GET – Setup** page to configure the method invocation details. This will give you multiple integration types.

5. From the list of options for the integration type, select **Lambda function**. **Lambda Region** is the region where you have created the Lambda function.

6. Then, select the Lambda function that you need to integrate from the **Lambda Function** input box.

7. Then, click on **Save** to save the changes. The console asks you to permit the API gateway to invoke the Lambda function.

8. Click **OK** to grant permission. Now it will show you a page, **/totalPrice – GET – Method Execution**, that visually represents the API gateway workflow as follows:

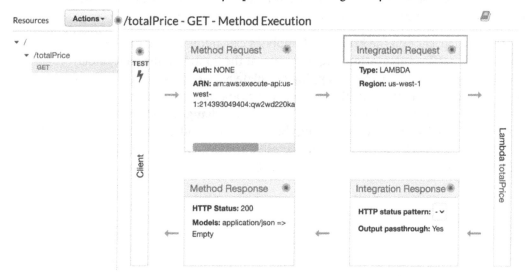

Figure 6.7 – totalPrice Method Execution page

In the sample application, we have specified that the Lambda function caller needs to provide `itemId` and `quantity` to calculate the total price. Since we have created an API gateway using the `GET` method, we need to map incoming request data to the Lambda function input. Therefore, we need to perform a mapping function on the API gateway.

The AWS API gateway contains four phases to handle a request. It contains incoming and response message flows. Each of these two message flows is separated into two phases – Method and Integration. The following are the four phases of the API gateway that handle request and response flow:

- **Method Request**
- **Integration Request**
- **Method Response**
- **Integration Response**

Method handles incoming and outgoing messages. Integration integrates the request input and the output. We need to add the `GET` request to the Lambda function mappings in the **Integration Request** phase.

To add a request mapping, click on **Integration Request** on the **/totalPrice – GET – Method Execution** page, as shown in *Figure 6.7*. This will reveal a similar page, **/totalPrice – GET – Integration Request**, which is the same as the method creation window. In this view, you can create a Lambda mapping in multiple different ways. Here you can use path parameters, query parameters, and message headers to map `GET` request content to the Lambda function input. For this example, we will use query parameters to submit the request.

To create a map, click on the **Mapping Templates** expanding button on the **/totalPrice – GET – Integration Request** view. At the bottom of the page under **Content-Type**, click on the **Add mapping template** button. This will reveal a textbox for adding the content type. Set the content type as **application/json** and click on the checkmark button. This will reveal a pop-up window, **Change passthrough behavior**, and then click on the **No, use current settings** button. Then it will add an application/JSON to the **Content-Type** list. Click on the newly created content-type application/JSON to add the mapping logic. This will show you an input box to enter the mapping logic:

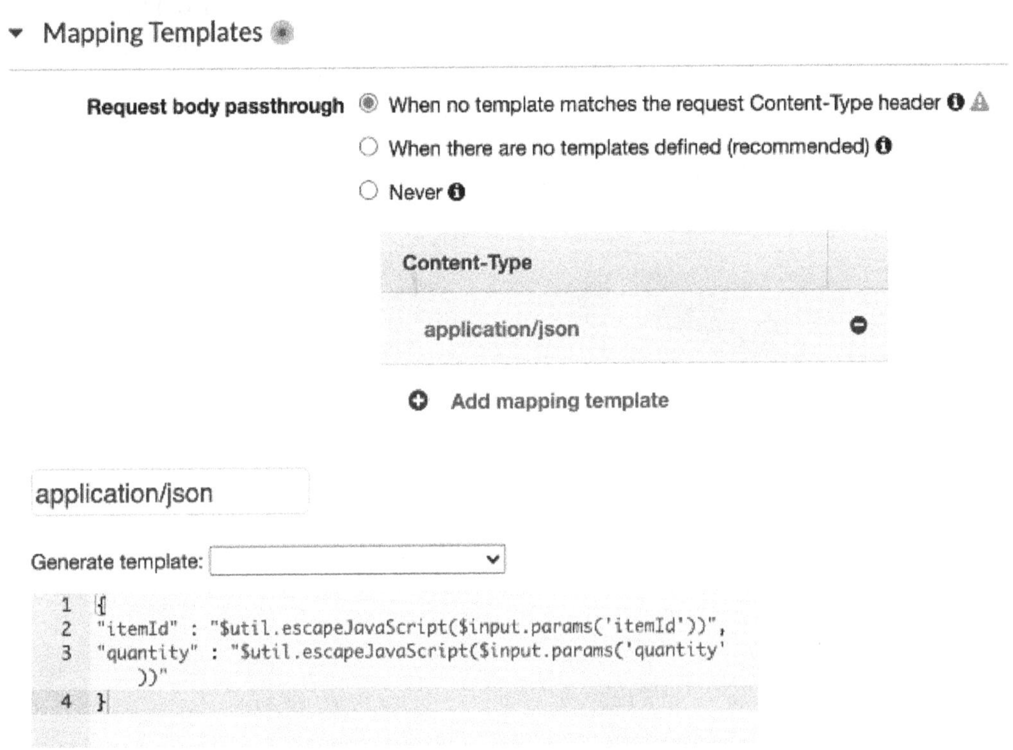

▼ Mapping Templates ◉

Request body passthrough   ◉ When no template matches the request Content-Type header ⓘ ⚠

○ When there are no templates defined (recommended) ⓘ

○ Never ⓘ

Content-Type	
application/json	⊖

⊕   Add mapping template

application/json

Generate template: [▾]

```
1  {
2   "itemId" : "$util.escapeJavaScript($input.params('itemId'))",
3   "quantity" : "$util.escapeJavaScript($input.params('quantity'
        ))"
4  }
```

Figure 6.8 – Adding API gateway mapping logic

Add the following content to the mapping details to map query parameters to the Lambda function:

```
{
"itemId" : "$util.escapeJavaScript($input.params('itemId'))",
"quantity" : "$util.escapeJavaScript($input.
    params('quantity'))"
}
```

This JSON definition creates two keys, `itemId` and `quantity`, extracted from the URL parameter `itemId` and `quantity`. Then, click on the **Save** button to save the mapping data. Once everything has been done, click on the **Deploy API** button on the **Actions** dropdown. On the **Deploy API** view, create a new deployment stage if this has not already been created. Then, click **Deploy** to deploy the API gateway. Now you will be automatically redirected to the **Stages** page, which lists all of the deployed APIs. Click on the resource method you need to invoke on the API **Stages** tree view to show the HTTP endpoint address:

Figure 6.9 – GET endpoint address from the deployed API

Get the API URL and invoke it with the query parameters. For this example, the sample `GET` request is available at `https://<aws_api_domain>.execute-api.us-west-1.amazonaws.com/default/totalPrice?itemId=item1&quantity=3`. This `GET` request returns a JSON response of the total price.

This is a simple example that demonstrates Lambda functions along with an HTTP API gateway as a trigger. To build more complicated FaaS applications, AWS provides a Step function that lets you store the state of the system in a state machine. In the next section, we will discuss AWS Step Functions and how to use them to build more complicated business requirements.

Using AWS Step Functions with an AWS Lambda function

AWS provides a Step function feature that lets you design complex business logic with a state transition. You can integrate different AWS services, including Lambda and API gateways, to build workflows. The Step function contains a series of steps with input and output at each step. For each request that starts a Step function, it keeps a separate state machine. Lambda functions are designed to work in a stateless manner, while Step functions give you a way to keep the states and use the Lambda function to move between states.

AWS Step Functions is written in Amazon States Language. This is a JSON-based domain-specific language. AWS provides a visualization over the given state machine JSON definition. You can also visualize the current state of the state machine in real time. In this section, we will demonstrate how to use AWS Step Functions with the Ballerina AWS Lambda function. For this example, we will use the order delivery scenario where a delivery agent delivers the item and sends the order delivery confirmation email to the customer.

In this example, the customer creates an order, pays for the order, and submits it to the delivery service. The delivery service sends an email to the delivery agent once the customer submits the order. Then, the delivery agent picks the item and ships the product. Once the delivery process has been completed, the delivery agent confirms that the item is delivered to the given address. Then, the delivery service sends an email to the customer confirming that the order has been delivered. The workflow of this implementation can be visualized in the following diagram:

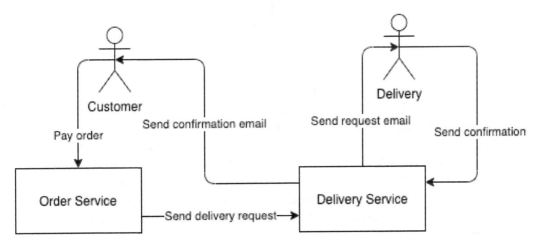

Figure 6.10 – Order delivery workflow

This workflow uses emails to communicate with different parties. For this example, we will assume that the order service and the payment service do the order validation and payment processing part and send a request to the delivery service to send the confirmation email. You can use the same Step function to implement these components as well. For this instance, we will focus on the implementation of the delivery service. When the order service sends an order request to the delivery service, it sends an email to the delivery agent that contains a GET request URL to confirm the order. The delivery agent can click on the URL once the delivery process has been completed.

You can create Step functions in AWS to implement this scenario. We need to create two API gateways to handle the order service delivery request and the agent order confirmation request. These two APIs trigger the Step function and invoke the relevant Lambda functions to send emails. The overall architecture of the API gateway and the Lambda functions can be represented in the following diagram:

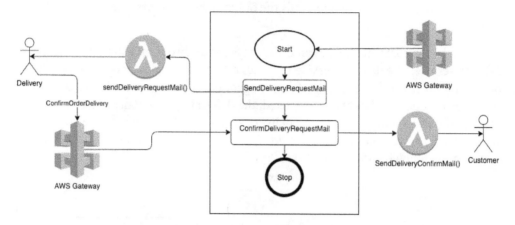

Figure 6.11 – AWS component architecture of the delivery service

The Step function gets started with the incoming request coming from the order service through the API gateway, which contains the order ID. First, the Step function comes to the `SendDeliveryRequestMail` state that triggers the `sendDeliveryRequestMail` Lambda function. This Lambda function generates an email and sends it to the delivery agent. This email contains the order ID and the task token generated by the Step function. This task token is used to find which Step function execution this order belongs to. As described earlier, each Step function creates a new execution state machine for each request. The task token is used to identify each of the executions.

Once the order is delivered by the delivery agent, the agent can click on the email link and confirm the delivery. This link again triggers an API gateway and calls the Step function to mark the confirmation of delivery. Then, the SendDeliveryRequestMail step becomes successful and moves to the next step – ConfirmDeliveryRequestMail. This step calls the `sendDeliveryConfirmMail` Lambda function. This Lambda function sends an email to the customer saying that the order was delivered successfully.

Building the Lambda function with Ballerina

First, we will create two Lambda functions to send emails to the delivery agent and the customer. To do this, we will use the Ballerina Lambda library along with the email library. We need to read the order ID from the request and load the order details from the database. To simplify this, we will create the following data structure with the sample values to mock the database services:

```
type Order record {|
    readonly string orderId;
    string customerEmail;
    string inventoryEmail;
    string address;
|};
type OrderTable table<Order> key(orderId);
OrderTable orders = table [{
    orderId: "43234234",
    customerEmail: "customer@email.com,
    inventoryEmail: "inventory@email.com",
    address: "215, New York"
}];
```

The `Order` record holds the details of the order and the `OrderTable` table is used to hold all available orders in a table data type. Next, we need to initialize the `orders` table with sample orders. We will use this variable to access order data in both of the Lambda functions that we are going to create. The sample Ballerina Lambda function code to send the delivery request email is as follows:

```
@awslambda:Function
public function sendDeliveryRequestMail(awslambda:Context ctx,
    json item) returns json|error {
    string orderId = <string> check item.req.orderId;
    string taskToken = <string> check item.taskToken;
    string encodedTaskToken = check url:encode(taskToken,
        "UTF-8");
    Order orderToSend = orders.get(orderId);
    string deliveryAddress = orderToSend.inventoryEmail;
    string mailToSend = string 'New order received with ID
        ${orderToSend.orderId} to address ${orderToSend.address}
```

```
        Click following link to approve the delivery.
        https://abcdef.execute-api.us-west-1.amazonaws.com/staging/
            confirmdeliveryrequest?orderId=${orderId}
                &taskToken=${encodedTaskToken}"';
        check sendEmail(deliveryAddress, "Order confirmation",
            mailToSend);
        return { "status" : "success"};
    }
```

The @awslambda:Function annotation at the beginning indicates that this function is an AWS Lambda function. This function reads a JSON input. From this JSON input, we will extract the order ID and the task token. Both of these values are passed from the Step function. We will discuss how these values are generated later. Since we have used query parameters to pass a task token, we need to URL encode the task token since it contains special characters that cannot be sent through the REST protocol with query parameters. The Ballerina URL library's encode function is used here to generate the URL-encoded string.

Now we can get the order details with the given orderId variable. From the orderToSend variable, we can infer the delivery person's email address. The mailToSend variable is used to generate the message that we are going to send. It includes a description of the order and a link to confirm the order once it has been delivered. This link contains the API gateway address, the orderId, and the encoded taskToken as query parameters. Finally, we have used the sendEmail utility method to send the email.

The next Lambda function that we need to implement is the function that sends the confirmation email to the customer. This function is also defined as a Lambda function as follows:

```
@awslambda:Function
public function sendDeliveryConfirmMail(awslambda:Context ctx,
    json item) returns json|error {
    string orderId = <string> check item.orderId;
    Order orderToSend = orders.get(orderId);
    string customerAddress = orderToSend.customerEmail;
    string mailToSend = string 'Your order with ID
        ${orderToSend.orderId} sent';
    check sendEmail(customerAddress, "Order delivered",
        mailToSend);
    return { "status": "success"};
}
```

This function simply reads the order ID from the incoming call from the Step function. The `orderId` variable is used to take the order details from the order list. Then, the customer address is taken from the order. The `mailToSend` variable contains the message that needs to be sent to the customer. Then, the message is sent using the `sendEmail` utility function.

Creating Step functions in the AWS Console

Step functions are used to hold the current state of each execution. Each request made to the API gateway begins with Step function execution. To create a Step function, follow these steps:

1. You can log in to the AWS management console and create a Step function on the **Step Functions** page. Go to the **Function overview** page, select the **Configuration** tab, and then select **State machines** from the left navigation. From the **State machines** section, click on the **Create state machine** button to create a new state machine. This will redirect you to the **Define state machine** page.

2. Select **Author with code snippets** to create a Step function from scratch.

3. You can select either **Standard** or **Express** as the type based on the number of transactions you need to perform and the pricing schema. For this example, we will select the **Standard** type.

4. Next, you need to define the Step function in JSON format. In the **Definition** section, you can find a text area to insert the JSON definition. You can create a Step function with pre-designed code snippets. These snippets contain basic blocks that you can use to design a Step function. Here, we will create a Step function with the following JSON definition:

```
{
    "Comment": "OrderDeliveryWorkflow",
    "StartAt": "SendDeliveryRequestMail",
    "States": {
        "SendDeliveryRequestMail": {
            "Type": "Task",
            "Resource": "arn:aws:states:::lambda:invoke.
              waitForTaskToken",
            "Parameters": {
                "FunctionName": "<sendDeliveryRequestMail_ARN_
                  name>",
                "Payload": {
```

```
          "req.$": "$",
          "taskToken.$": "$$.Task.Token"
        }
      },
      "Next": "ConfirmDeliveryRequestMail"
    },
    "ConfirmDeliveryRequestMail": {
      "Type": "Task",
      "Resource": "arn:aws:states:::lambda:invoke",
      "Parameters": {
        "FunctionName": "<sendDeliveryConfirmMail_ARN_
          name>",
        "Payload": {
          "orderId.$": "$.orderId"
        }
      },
      "End": true
    }
  }
}
```

In this Step function definition, you will see that there are two steps, defined
as `SendDeliveryRequestMail` and `ConfirmDeliveryRequestMail`.
The `StartAt` field specifies that this Step function begins execution
from the `SendDeliveryRequestMail` Step function. The
`SendDeliveryRequestMail` Step function has `Task` for the type and
`waitForTaskToken` as the resource. This ensures that this task waits until it gets
a success message to mark this step as passed.

Inside `Parameters`, we can define the Lambda function that we need to invoke
when this step started. Give your function ARN name to the `FunctionARN` field.
This ARN can be found in the Step function details view. The `Payload` section
defines the arguments that need to be sent to the Lambda function. Here, we read
the request body and assign it to the JSON key, `req`. The autogenerated task token
used to identify this execution is also passed as an input parameter.

In the Lambda function, we can read this as a JSON value. As described earlier,
this step invokes the Lambda function and it waits for a success message from the
delivery agent to complete the `SendDeliveryRequestMail` step. The `Next`
parameter defines the next step, once this step succeeds.

The `ConfirmDeliveryRequestMail` step gets executed after the delivery agent confirms the order delivery. This step invokes the `sendDeliveryConfirmMail` Lambda function to send a notification to the customer. This Step function invokes the `sendDeliveryConfirmMail` function with the payload parameters as specified in the `Payload` section. Here, we used `orderId` to pass order details to the Lambda function. Once this Step function invokes the Lambda function, its execution ends.

5. Once you copy and paste this definition into the AWS Console, it will show a visual representation of the Step function as follows:

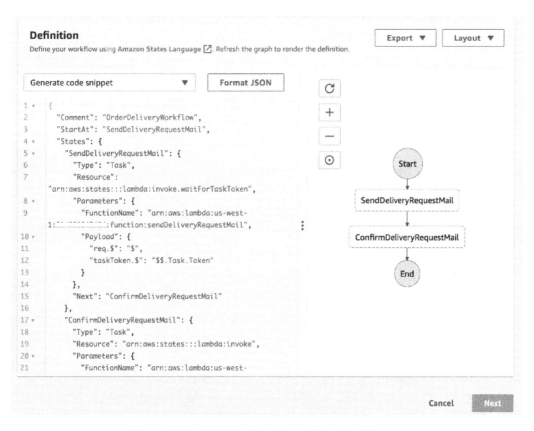

Figure 6.12 – AWS Step Functions with a state diagram

6. Click on the **Next** button to proceed with creating the Step function.

7. On the next page, **Specify details**, you need to provide a name for the state machine in the **State machine name** input box.

8. Under the **Permissions** section, you can either create a new role or use an existing role that we have created. Make sure to give the required access for the created role. For this example, you can try out the **AWSStepFunctionsFullAccess** access level. Under the **Logging** section, you can specify the log level for the Step function.

9. Then, click on **Create state machine** on the bottom right of the page to create the Step function.

This is a simple Step function that has two steps. You can have more steps to design complex business use cases. Now we need a way to invoke these step functions. AWS supports many types of triggers to invoke Step functions. An API gateway is one such way of triggering Step functions. In the next section, we will discuss invoking Step functions with an API gateway in further detail.

Creating an API gateway to invoke Step functions

The next step is to create an API gateway to invoke the Step function. We will create a new API gateway to access this Step function with two different resources to invoke each step in the Step function. Here are the steps to achieve it:

1. Search for `API gateway` on the AWS management console and go to the API gateway home page.

2. Click on **Create API** to create a new API. From **Choose an API type**, select **REST API**, and then click the **Build** button. This will redirect you to another page – **Create new API**.

3. Select **New API** from the option list and give the API a name under the **Settings** section. Keep **Endpoint Type** as **Regional**.

4. Click on **Create API** to create the API. In the API details page, create two new resources with the context `startdeliveryprocess` and `/confirmdeliveryrequest`.

5. Create two `GET` methods for each resource you have created. The `startdeliveryprocess` resource is used to start the Step function, while the `confirmdeliveryrequest` resource is used to confirm the order.

6. In the method creation page of the `startdeliveryprocess` resource's `GET` method, you need to define the endpoint details that these methods are connected to. Under **Integration type**, select **AWS Service**. This will give you a list of details to fill in.

7. Select the AWS region where your services are deployed. Select **Step Functions** as the AWS service. Set the **HTTP method** as POST, which sends a POST request to the Step function. For **Action Type**, select **User action name**. Set the **Action** field to StartExecution to indicate that this API initiates execution of the Step function.

8. For **Execution role**, provide the ARN of the execution role. Make sure to give permission to the role to invoke services. For this example, you can add the **AmazonAPIGatewayInvokeFullAccess** permission to the role. Keep **Content Handling** as **Passthrough**.

9. Click on the **Save** button to complete the method creation, as shown in the following screenshot:

Figure 6.13 – API creation page for the startdeliveryprocess resource's GET method

10. For the `confirmdeliveryrequest` resource also, create a new GET method. **Integration type**, **AWS Region**, **HTTP method**, **Action Type**, **Execution role**, and **Content Handling** should be the same as the `startdeliveryprocess` resource.

11. For the **Action** field, set the value as **SendTaskSuccess** to invoke the Step function as a successful task. Click the **Save** button to create this resource method as well.

12. The next step involves creating request mapping for the Step function. For this step, click on the **GET** method on the **startdeliveryprocess** resource from the **Resources** list tree view. Then, click on the **Integration Request** phase. From the **Mapping Templates** section, add a mapping template with the application type `application/json`, the same as for the previous example. Add the following mapping definition to the template view:

```
{
    "input": "{\"orderId\":\"$input.params('orderId')\"}",
    "stateMachineArn": "<State_machine_ARN>"
}
```

In this mapping definition, we provide Step function input parameters with the `input` field. This field takes a string value to pass data to the Step function. Here, we create a serialized JSON content that has the `orderId` field on it. The `orderId` field is extracted from the request query parameters. Here, we also define the Step function ARN with the `stateMachineArn` field. You can find this ARN on the **Step Function** details page.

13. Once the changes are complete, click on the **Save** button to save it.

14. Just as you created an API for `startdeliveryprocess`, `confirmdeliveryrequest` should also have a mapping definition. For that, add the following mapping definition to the **Integration Request** phase:

```
{
    "output": "{\"orderId\":\"$input.params('orderId')\"}",
    "taskToken": "$input.params('taskToken')"
}
```

This mapping configuration has `output` mapping to send the `orderId` variable taken from the query parameters. The task token is read from the query parameter and is sent with the `taskToken` key. When a customer clicks on the email, it will invoke this API and complete the `SendDeliveryRequestMail` step in the Step function.

15. Then you can deploy your API to get the API endpoint. Once you deploy the API, it will give you an endpoint address to invoke the service. In this phase, make sure that the endpoint URL that you have specified in the `confirmdeliveryrequest` resource is correctly set in the `sendDeliveryRequestMail` Lambda function.

This is the URL that the delivery agent is redirected to after clicking **Confirm URL** in the email. You can test the implementation by taking the URL for the `startdeliveryprocess` API and invoke it with a `GET` request along with the `orderId` query parameter. A sample `GET` request is `https://<aws_api_domain>.execute-api.us-west-1.amazonaws.com/staging/startdeliveryprocess?orderId=43234234`.

When you invoke the system using this URL with a `GET` request, the delivery agent receives an email. Once the delivery agent clicks on the email link, it completes the first step of the Step function and goes to the next step. In this step, the Step function calls the `sendDeliveryConfirmMail` Lambda function and sends the confirmation mail to the customer.

You can use Step functions to build complex cloud applications with serverless architecture. Since FaaS applications are charged based on the resources consumed, it is cost-effective and there are no complex deployment procedures. However, always keep in mind the problems that are associated with FaaS platforms, including vendor lock-in, the lack of features, and cold start.

Summary

In this chapter, we discussed serverless architecture and how to use the Ballerina language to implement a serverless function. Serverless functions are largely used to build a stateless application that has a shorter execution time. We discussed why and when to use serverless functions. We primarily focused on two main cloud vendors that provide FaaS solutions for cloud developers – AWS and Azure.

The Azure Cloud platform is widely used by cloud application developers to build cloud applications. We discussed the services provided by the Azure Cloud platform. Then, we discussed a simple Azure function implementation with an HTTP trigger. We also discussed a complex scenario where multiple Azure functions are connected with Azure Queue triggers and bindings.

Next, we discussed the AWS Cloud platform, which provides different types of cloud solutions for developers. AWS provides AWS Lambda as the serverless solution. In that section, we discussed how to create a simple AWS Lambda function. Furthermore, we learned to use triggers to invoke Lambda functions. To build more complex systems with Lambda functions, we used Step functions in conjunction with Lambda functions.

When you are creating a cloud native application, decide when to use FaaS in your application. Then, select the appropriate cloud vendor to host the system based on features, limitations, and cost. FaaS provides an elegant way of building a cloud application that has lower maintenance costs and low complexity.

In the next chapter, we will focus on another important aspect that you should always give high priority to, which is security. Since we are discussing cloud native applications, security is a key aspect of building cloud applications as these applications are more vulnerable to threats. We will discuss security features provided by the Ballerina language, along with tools that you can use to implement security with Ballerina applications.

Questions

1. What is the difference between PaaS and FaaS?

2. When should you use serverless architecture?

3. How is the Step function invoked from the Ballerina AWS Lambda function?

Further reading

- *Learning AWS: Design, Build, and Deploy Responsive Applications Using AWS Cloud Components*, by Aurobindo Sarkar and Amit Shah, available at `https://www.packtpub.com/product/learning-aws/9781784394639`

- *Hands-On Azure for Developers: Implement Rich Azure PaaS Ecosystems Using Containers, Serverless Services, and Storage Solutions*, by Kamil Mrzygłód, available at `https://www.packtpub.com/product/hands-on-azure-for-developers/9781789340624`

- *Serverless Design Patterns and Best Practices: Build, Secure, and Deploy Enterprise-Ready Serverless Applications with AWS to Improve Developer Productivity*, by Brian Zambrano, available at `https://www.packtpub.com/free-ebook/serverless-design-patterns-and-best-practices/9781788620642`

- *Learn AWS Serverless Computing: A Beginner's Guide to Using AWS Lambda, Amazon API Gateway, and Services from Amazon Web Services*, by Scott Patterson, available at `https://www.packtpub.com/product/learn-aws-serverless-computing/9781789958355`

Answers

1. PaaS provides a platform to deploy and run cloud-based applications, while FaaS provides a platform to deploy and run individual functions. Both PaaS and FaaS platforms do not need to manage internal infrastructure. PaaS is more about deploying services as a single unit, while FaaS is focused on deploying applications that are separated into smaller units called functions. PaaS application scaling is done as a whole, whereas the FaaS application can be scaled by each function. FaaS functions can be scaled automatically based on the traffic that needs to be handled.

2. Serverless architecture can especially be used in cases where you have a small number of functions that need to host. If your application is complex, then you need to design your application such that it is compatible with serverless architecture. However, it is not recommended to build functions that are long-running using serverless architecture. Serverless is effective in building low computation-intensive, highly responsive services.

3. When you need to invoke the Step function from the AWS Lambda function, you can use the AWS API (`https://docs.aws.amazon.com/step-functions/latest/apireference/Welcome.html`). The AWS API contains a REST interface to invoke the Step function with the required arguments. You can also find Ballerina libraries on Ballerina Central to invoke Step functions.

7
Securing the Ballerina Cloud Platform

Modern web-based software applications are a collection of distributed components rather than a single application. Integrating different components is a key factor that exposes software applications to threats. Therefore, securing cloud applications is critical as it has multiple components connected over a network. Building secure cloud applications includes securing communication over a network, authenticating users, authorizing users in different access levels, and so on.

In this chapter, we will go over how to create a Ballerina cloud application securely by using Ballerina's security features. We will go through several multiple free open source tools that can be used with the Ballerina language to develop secure Ballerina cloud applications.

The following is the list of topics that we are going to discuss in this chapter:

- Managing certificates in Ballerina applications
- Authenticating and authorizing with **Lightweight Directory Access Protocol (LDAP)** user stores
- Token-based authorization methods

By the end of this chapter, you should be able to understand the various aspects of cloud native application security and how to implement them using the Ballerina language.

Technical requirements

In this chapter, we will use multiple freely available tools to improve the functionalities of the Ballerina security feature. We will use the Keytool tool to work with certificates. **Keytool** is a certificate management tool that comes with the JDK. We also use Apache Directory Studio to build an LDAP server and manage it. All the download and installation instructions for Apache Directory Studio can be found in the *Setting up Apache Directory Studio* section.

For this chapter's examples, we used WSO2 Identity Server to manage users and roles and to provide different authentication and authorization methods. Download and running instructions for WSO2 Identity Server can be found in the upcoming sections. To run WSO2 Identity Server, you must have Java 11 or later installed.

The code files for this chapter can be found on GitHub at `https://github.com/PacktPublishing/Cloud-Native-Applications-with-Ballerina/tree/master/Chapter07`.

The Code in Action video for the chapter can be found here: `https://bit.ly/3x8ZFi7`

Managing certificates in Ballerina applications

When a client application needs to access a service over a network, there should be a way to verify that the service that is attempting to access is the service it claims to be. This validation can be done with the use of certificates. Certificates use public-key cryptography to validate endpoints that we need to connect. In this section, we will learn how to use a certificate in Ballerina to secure services. We will discuss creating SSL-enabled services, clients, and mutual SSL to validate both parties.

Securing the Ballerina service with HTTPS

Certificates are often used in computer security to verify a claim given by a service. Public key cryptography uses a private key and a public key to manage certificates. A **private key** is a key that is only known to the component that generates messages and claims. The private key is used to generate a signature from the content that it needs to send. A **public key** is a key that can be shared with anyone who needs to verify the sender. This public key is used by the listener to check that the messages were sent by the sender who has the private key.

Securing communication is an essential part of building a microservices application. Services running on the microservice architecture need to validate other services. In a client-server architecture, you might need to validate the service that the client needs to access. Mutual validation, on the other hand, can be used to add extra security by allowing both the client and the server to verify one another. You can use service mesh to build a secure cloud application on microservice architecture to let services communicate using HTTPS. Ballerina supports HTTPS, allowing microservices to be built securely.

Ballerina has built-in support to implement HTTPS-enabled web services. To configure the HTTPS service endpoint, simply configure the Ballerina service with key and certificate information. With a few configurations, you can also build a Ballerina client that uses the HTTPS protocol to invoke a remote endpoint. The key is the private key that is used to generate a digital signature, while a certificate is a public key that is used to verify the signature.

HTTPS provides a secure connection over an HTTP connection. To start creating an HTTPS server, we must first create a keystore that contains a private key that needs to generate the certificate. This keystore can be generated with a `keytool` command provided by the JDK. `keytool` is a certificate management tool that can be used to create and manage certificates. For this example, we will create a keystore with `keytool` with the following command:

```
keytool -genkeypair -alias ballerina -keyalg RSA -keysize 2048
-storetype PKCS12 -keystore ballerinaKeystore.p12 -validity
3650
```

This command will ask you for some details, including the keystore password that is used to access the certificate store. For this example, we will use `ballerina` as the password. It will again ask you to re-enter the same password. Then it asks for the first name and the last name, organization unit, organization, city, state, or province, and the two-letter country code. Once you provide this information, it will create a keystore file in the current directory with the name of `ballerinaKeystore.p12`. Here, we used `RSA` as the key generation algorithm, `PKCS12` as the `keystore` type, and set the `validity` of the keystore to `3650` days.

> **Note**
>
> When you are trying this sample on localhost, call it `localhost` for the **What is your first and last name?** field. Since Ballerina validates domain names as well, the program might fail if the domain does not match.

Now we need to generate a certificate file from the keystore that we have just created. The following command can be executed with the `openssl` command-line tool to generate a certificate file from the given keystore file:

```
openssl pkcs12 -in ballerinaKeystore.p12 -nokeys -out
ballerinaServer.crt
```

Once you execute this command, it will generate a `ballerinaServer.crt` certificate file. This is your public certificate file that can be shared with the client application. The client can use this certificate file to validate the server.

Next, we will generate a key file with the following `openssl` command:

```
openssl pkcs12 -in ballerinaKeystore.p12  -nodes -nocerts -out
ballerinaServer.key
```

This command generates a `ballerinaServer.key` file that we can use as the private key to generate signatures. We can keep these key files and certificate files in a folder and use them on a Ballerina program to secure services with HTTPS. The following HTTPS service starts an HTTPS service on port `9093` with a single resource function definition that responds to the client with `Hello World!`:

```
http:ListenerConfiguration epConfig = {
    secureSocket: {
        key: {
            certFile: "resources/ballerinaServer.crt",
            keyFile: "resources/ballerinaServer.key"
```

```
            }
        }
    };
    listener http:Listener httpsListener = new (9093, epConfig);
    service /testHTTPS on httpsListener {
        resource function get .() returns string{
            return "Hello World!";
        }
    }
```

Here, we need to define this service to use HTTPS by using
`ListenerConfigurations` configs. The certificate file path should be provided
with the `certFile` parameter, and the private key file path should be provided to the
`keyFile` parameter in the `secureSocket` field. The Ballerina service definition reads
the key file and creates a secure service endpoint in port `9093`. The resource function
defined in this example simply responds with a text to the caller.

You can try out this example by running this Ballerina code and accessing this service
through the browser. The URL to access the service is `https://localhost:9093/`
`testHTTPS`. You may notice that instead of the `HTTP` protocol, here, we specified the
protocol as `HTTPS` in the URL.

In the browser, you will get the response `Hello World!` on a web page with an
HTTPS-enabled connection.

> **Note**
>
> Since this certificate is not included in the browser-trusted certificate list, the
> browser indicates to you that your connection is not a private page. Click on
> the **Advanced** button and click on the **Proceed to localhost (unsafe)** link to
> access the service. Otherwise, you can generate the trust certificate from
> the keystore and add it to the browser certificate list to trust this Ballerina
> service endpoint.

We can now create a secure service with the Ballerina platform. Next, we need a way
to call secure services from another Ballerina client application. In the next section,
we will discuss how to create a Ballerina client to invoke the HTTPS endpoint and
retrieve a response.

Calling an HTTPS endpoint with Ballerina

To use HTTPS for client-side communication, the client application must have a certificate that includes the public key of the backend service. The following Ballerina program creates a client endpoint to the service that we have created and sends a GET request:

```
http:ClientConfiguration epClientConfig =  {
    secureSocket:{
        cert: "resources/ballerinaServer.crt"
    }
};
http:Client securedEP = check new("https://localhost:9093",
  epClientConfig);
string response = check securedEP->get("/testHTTPS");
io:println(response);
```

In this sample, we have defined the certificate file that needs to connect with the backend service in the epClientConfig variable. Aside from the certificate file configuration, this example is exactly the same as a simple HTTP client example. This Ballerina code uses the certificate to verify the backend service and execute a GET request.

In the next section, we will discuss how to expand HTTPS communication to validate both the client and server certificate with mutual SSL instead of just validating the server certificate.

Securing Ballerina interservice communication with mutual SSL

HTTPS is a method for establishing a secure HTTP connection between the client and the server. In certain cases, SSL validation for both the client and the server may be needed. Other than validating the server certificate by the client, mutual SSL forces the server to certify the client. Therefore, both the client and the server need to have two different keys and certificate files to verify identity. This is useful in microservice architecture when two services need to trust one another.

Mutual SSL certificate support is provided by Ballerina to enforce mutual SSL client-server validation. We will go through a simple validation scenario in which a client application tries to communicate with a server that has mutual SSL enabled.

To demonstrate these samples, perform the following steps:

1. You need to create two keys and certificates with the `keytool` and `openssl` commands. Here are the names of the key files and certificate files to run this example:

    a) **Server key file**: `ballerinaServer.key`

    b) **Client key file**: `ballerinaClient.key`

    c) **Server cert file**: `ballerinaServer.crt`

    d) **Client cert file**: `ballerinaClient.crt`.

    We can use the previous certificate and key file for the server and we need to generate another certificate and a key for the client application. You can use the same previous `keytool` and `openssl` commands to generate those key and certificate files.

2. The Ballerina service and client endpoint provide a simple way of handling mutual SSL by only giving the required parameters to the service definition and the client endpoint definition.

    The sample code for the Ballerina HTTP service, which accepts HTTP requests with mutual SSL enabled, is shown as follows. Let's start with the service endpoint configuration definition:

```
http:ListenerConfiguration mutualSSLServiceEPConfig = {
    secureSocket: {
        key: {
            certFile: "resources/ballerinaServer.crt",
            keyFile: "resources/ballerinaServer.key"
        },
        mutualSsl: {
            verifyClient: http:REQUIRE,
            cert: "resources/ballerinaClient.crt"
        },
        protocol: {
            name: http:TLS,
            versions: ["TLSv1.2", "TLSv1.1"]
        },
        ciphers:["TLS_ECDHE_RSA_WITH_AES_128_CBC_SHA"]
    }
};
```

Unlike previous HTTPS service implementations, here we need to define
two certificate files. Under the `key` configurations in the `secureSocket`
configurations, we need to define the key file and the certificate file of the service.
Since we need to enable mutual SSL, we need to define another parameter, called
`mutualSsl`, under the `secureSocket` configurations. On this configuration,
we need to provide a certificate location with the `cert` parameter and the
`verifyClient` parameter as `http:REQURE` to mark this endpoint as a mutual
SSL-enabled endpoint. You can define the SSL protocol and version information
under the `protocol` parameter. The service definition for this endpoint is very
similar to the previous HTTPS service example except for the fact that the server
also requests a client certificate to validate the client application. Refer to the
following service listener:

```
listener http:Listener serviceEP = new (9093,
    mutualSSLServiceEPConfig);
service /testMutualSSL on serviceEP {
    resource function get .(http:Caller caller,
        http:Request req) returns string {
            return "Hello, World!";
    }
}
```

Now we can create the `ballerina` client to access this service. This client
application calls the endpoint with a `GET` request. The client endpoint configuration
to call the mutual SSL-enabled service endpoint is as follows:

```
http:ClientConfiguration clientEPConfig = {
    secureSocket: {
        key: {
            certFile: "resources/ballerinaClient.crt",
            keyFile: "resources/ballerinaClient.key"
        },
        cert: "resources/ballerinaServer.crt",
        protocol: {
            name: http:TLS
        },
```

```
     ciphers:
              ["TLS_ECDHE_RSA_WITH_AES_128_CBC_SHA"]
  }
}
```

In the same way as the Ballerina service configurations, here you need to provide the client key location and certificate location along with the server's certificate location.

3. To certify the server and the client, we must include protocol and cipher types, just as we did with the Ballerina service configurations. The following is the client code that uses this configuration to invoke the Ballerina service:

```
http:Client securedEP = check
  new("https://localhost:9093",clientEPConfig);
string response = check securedEP->get("/testMutualSSL");
io:println(response);
```

This sample uses the client endpoint configuration to create a mutual SSL-enabled client endpoint. Then, it sends a GET request to the mutual SSL-enabled service and prints the result to the terminal if it is a success.

Mutual SSL is an important step in building microservices applications to provide stronger security for both ends. Ballerina makes it easy to manage mutual SSL connections between services by simply providing key and certificate configurations for both the client and server.

In a microservice architecture, a service mesh can also be used to establish an SSL connection between the client and the server. Depending on the architectural requirements of the system and maintenance cost, you can choose either option.

Certificate validation is an important step in building a web application to improve security and avoid *man-in-the-middle* attacks. Always make sure to validate endpoints if your application is running on the cloud as it is vulnerable to threats.

Another requirement for building a cloud application is to authenticate and authorize users. This is a common requirement as most cloud applications are required to have a list of users and authorization levels. In the next section, we will discuss how to maintain a list of users with an LDAP registry.

Authenticating and authorizing with LDAP user stores

Authentication and authorization are key aspects of building an application. **Authentication** is the act of identifying the user, and **authorization** is giving the authority to perform an action. In a larger software system, there are requirements to give different levels of access levels to different users. Different types of access controlling methods, such as **Mandatory Access Control (MAC)**, **Role-Based Access Control (RBAC)**, and **Discretionary Access Control (DAC)**, are available for handling access levels. The system architect should select the most appropriate authentication method and access control method based on system requirements.

In this section, we will discuss maintaining user authentication and authorization details by using the most popular directory service, which is LDAP. Here, we have provided some tools that you can use to build an LDAP user store and how to integrate it with the Ballerina application.

Authenticating and authorizing users with the LDAP server

There are multiple methods of authenticating a user into a system. These methods include passwords, multi-factor login, certificates, and tokens. Password-based authentication is the simplest authentication method that can be used to implement authentication. Password-based authentication methods use a unique username to identify the user and a password to verify the user that the person claims. The disadvantage of using a username and password is that it makes you vulnerable to phishing attacks.

To store usernames and passwords, password-based authentication methods require the use of a database. Relational databases or LDAP servers can be used to accomplish this. When you are building applications that use relational databases, make sure to use the proper mechanism to store passwords securely. Hashing passwords with a salt value is a standard practice for storing passwords. Almost all of the enterprise databases provide special functionality to store passwords. You can use those functions directly to create a column to store the password as a hashed value. You can also implement a relational database user store to keep user information with the help of the Ballerina database accessing libraries.

Another common way of creating a user store is by using an **LDAP server**. LDAP is primarily used to store user details and membership privileges in a directory-like structure. An identity server can be connected to an LDAP server to store and read user data. We will discuss Identity Server in more detail in the forthcoming sections. LDAP is not only used to store usernames and passwords, but it can also store information, including email addresses, contact information, and so on.

When it comes to storing user information, there are advantages and disadvantages to using LDAP over a relational database. LDAP is more optimized for reading entries than writing entries. If the read-to-write ratio is *greater than 10,000:1*, it is standard practice to choose LDAP over relational databases. On the other hand, you should not use directories to store transactional data that is frequently changing since directories do not provide good performance when it comes to writing data. In addition, directories are not strongly supported transactions in concurrent execution. As a result, monetary transaction information should not be stored in LDAP servers. LDAP also does not provide support for complex queries the same way as SQL does. It just supports a few basic search options.

LDAP was created with the purpose of establishing a user store with built-in security features for storing user information. If you are building an application with a relational database, you need to create all of the functionalities from scratch. Another advantage of using LDAP is that it has a simple design in access control due to the hierarchical nature of data storage. Furthermore, rather than building everything from scratch with a relational database, you can incorporate several third-party applications in LDAP directly to manage user details.

Directory services store data in a key-value format. The entries in an LDAP database are used to store attributes and values. For example, a given employee entry can have the attributes of name, email, and phone number. You can define what attributes are available in an entry by specifying a schema for it. Entries in the LDAP server can be represented with **LDAP Data Interchange Format** (**LDIF**). The LDIF format can be used to define and interchange LDAP entries. Take a look at the LDIF definition of a user entry in the following example:

```
dn: cn=John Doe,ou=Users,dc=ballerina,dc=io
objectClass: inetOrgPerson
objectClass: organizationalPerson
objectClass: person
objectClass: top
cn: John Doe
givenName: John
sn: Doe
telephoneNumber: +94 111 222 3333
telephoneNumber: +94 111 222 3334
ou: Users
mail: john@mail.com
```

An entry is made up of a collection of attributes and values. Each of these entries has a unique identifier known as **Distinguished Name (DN)**.

In the given LDIF definition example, dn: cn=John Doe,dc=ballerina,dc=io specifies the DN for the entry. This entry contains John Doe as the **Common Name (CN)** and ballerina and io as **Domain Components (DC)**.

Here, cn=John Doe is the **Relative Distinguished Name (RDN)** and dc=ballerina,dc=io is the DN of the parent entry. **Organization Unit (OU)** is used to represent which group this entry belongs to. Since LDAP stores data in a hierarchical structure, you can allocate entries to OUs. For this example, this entry is regarded as a user entry that stores user information.

The next few lines of the LDIF example contain object classes that are used to define the schema of the entry. The list of given object classes provides a set of optional and required attributes that should be defined in the entry. Some of these attributes can be ignored if they are optional attributes. It is compulsory to add required attributes. For example, the person object class is used to define attributes for a user. This object class contains the required field of **Surname (SN)** and the optional field of **Description**.

The next few lines of the LDIF definition contain attributes and values that belong to the entry. This includes John Doe as the CN, John as the given name, Doe as the SN, two telephone numbers, and the email address. All of these LDIF definitions, including object classes, attributes, and values, are represented as a single entry.

Apache Directory Studio provides a set of tools to create and manage LDAP servers. In the next section, we will discuss how to use Apache Directory Studio and build a simple LDAP server to store user login and authorization information.

Setting up Apache Directory Studio

Apache Directory Studio (Apache DS) comes with an LDAP server as well as other tools, including a schema editor, LDIF file editor, and LDAP browser. We will use this tool to build a simple LDAP server to implement a user authentication and authorization scenario with the Ballerina language. Alternatively, you can use OpenLDAP, 389 Directory Server, and other **Directory as a Service (DaaS)** platforms to build an LDAP server.

Apache DS can be downloaded from the official download page (https://directory. apache.org/studio/downloads.html) for all three major operating systems. You can download the installer from there and install Apache DS on your computer.

Since it is an Eclipse plugin, you can update the existing Eclipse IDE with the Apache DS plugin as well. When you install Apache DS, it provides you with a user interface that you can use to develop and manage LDAP servers. Refer to the following screenshot:

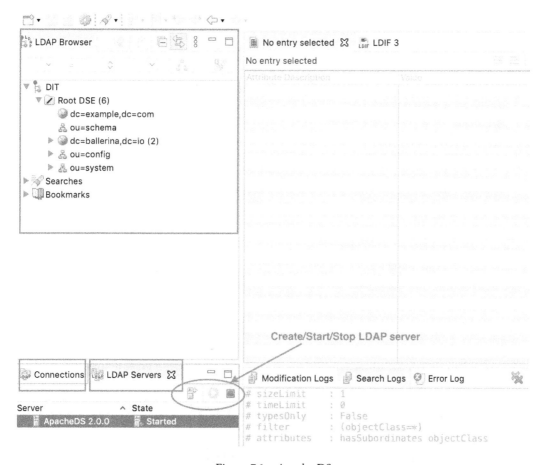

Figure 7.1 – Apache DS

The next step is to create a new LDAP server on Apache DS to store user information:

1. In the Apache DS IDE, click on the **LDAP Servers** tab on the bottom left of the screen and click on **New Server** to create a new LDAP server. This will bring up the **Create an LDAP Server** window.

2. Select the server type as **Apache DS 2.0.0** from the **Apache Software Foundation** drop-down list. Then, click **Finish** to create the server. This will create a new LDAP server and it will be listed in the **LDAP Servers** tab with the state as **Stopped**.

3. Now, we must start the server by selecting the server from the server list and clicking the **Run** button on the **LDAP Servers** tab. After a while, the state of the server will show as **Started**.

4. Next, we will create a connection to the LDAP server to access it through the Apache DS. *Right-click* on the running server in the **LDAP Servers** tab server list and click **Create a Connection**. This will create a new connection and it will be listed in the **Connections** tab on the left panel.

5. Once you select the connection that you have just created from the connection list on the **Connections** tab, Apache DS will list the LDAP content on the **LDAP Browser** tree view on the top left. In this panel, you can find the entries listed in the LDAP server under **DIT | Root DSE** on the entries tree view.

Now we can start adding entries to the LDAP server. This can be done either by manually entering each entry using the interface or by importing the LDIF file directly. Let's understand each in the following sections.

Adding entries by importing the LDIF file

If you want to import LDIF, you can right-click on the **Root DSE** entry root on the LDAP Browser panel and click **Import | LDIF Import**. Provide the LDIF file location and import entries into the LDAP server.

On the other hand, you can create a new LDIF file by clicking on **File | New** on the menu bar, select **LDIF File** from the **LDAP Browser** list, and then click **Finish**. Then, copy and paste the following content into the LDIF file and click on **Execute LDIF**, which is the green play button in the top-right corner of the LDIF file editor window.

The LDIF content to generate sample LDAP entries is as follows. We will go over each LDAP entry in turn:

1. Copy all of the following LDIF content into the LDIF editor and execute it:

```
dn: dc=ballerina,dc=io
objectClass: extensibleObject
objectClass: domain
objectClass: top
dc: ballerina
```

This entry definition is used to define the root domain component of the LDAP server. The distinguished name for this entry is dc=ballerina,dc=io.

2. In the next step, we will add two OUs to handle users and user groups, as follows:

```
dn: ou=Users,dc=ballerina,dc=io
objectClass: organizationalUnit
objectClass: top
ou: Users
dn: ou=Groups,dc=ballerina,dc=io
objectClass: organizationalUnit
objectClass: top
ou: Groups
```

The user `ou` is used to store and handle `Users` in the LDAP server. The group `ou` is used to store access levels for each user. Therefore, users work as the authentication part, and the group works as the authorization part of the LDAP server.

3. The next step is to allocate users and group entries to the LDAP server. We will create two user accounts in the LDAP server. The first user account definition is as follows, with `delivery` as the user ID, `uid`:

```
dn: uid=delivery,ou=Users,dc=ballerina,dc=io
objectClass: organizationalPerson
objectClass: person
objectClass: extensibleObject
objectClass: uidObject
objectClass: inetOrgPerson
objectClass: top
cn: Delivery User
givenName: delivery
sn: Delivery
uid: delivery
mail: delivery@mail.com
ou: Users
userpassword: delivery
```

This user account has `uid=delivery,ou=Users,dc=ballerina,dc=io` as the DN to uniquely identify this particular user. Here, we used a few object classes to define the schema that is used to store user details. Object details should be filled with attributes and values. Here, the password of the user account is given in plain text, which will be hashed by the LDAP server by itself when inserted into the server.

4. Let's add another user entry to the LDAP server with the following LDAP definition:

```
dn: uid=customer,ou=Users,dc=ballerina,dc=io
objectClass: organizationalPerson
objectClass: person
objectClass: extensibleObject
objectClass: uidObject
objectClass: inetOrgPerson
objectClass: top
cn: Customer User
givenName: Customer
sn: User
uid: customer
mail: customer@email.io
ou: Users
userpassword: customer
```

This user account is created to represent a customer who logs into the system. In the same way as the delivery user account, the customer account contains the same set of fields to initialize attributes and values. Similarly, you can create any number of users in the LDAP server.

5. Next, we need to add the group entries to handle the authorization of each user. We will use the following LDIF definition to create two groups to allocate these user accounts. Use the following LDIF to insert a delivery group:

```
dn: cn=delivery,ou=Groups,dc=ballerina,dc=io
objectClass: groupOfNames
objectClass: top
cn: delivery
member: uid=delivery,ou=Users,dc=ballerina,dc=io
```

The delivery group is used to store delivery-level access to the application. Here, we have allocated the delivery user account to the delivery group to give delivery privileges to the delivery account with the `member` attribute. You can add as many members as you wish to the delivery account to add delivery privileges for the user account.

6. We will create another group with customer privilege with the following LDIF definition:

```
dn: cn=customer,ou=Groups,dc=ballerina,dc=io
objectClass: groupOfNames
objectClass: top
cn: customer
member: uid=delivery,ou=Users,dc=ballerina,dc=io
member: uid=customer,ou=Users,dc=ballerina,dc=io
```

The customer group contains the privilege to handle customers. Here, we have allocated both the `delivery` and `customer` user accounts to the customer groups to give both customer-level access. You can create more user accounts and allocate users as customers by adding more members to the customer group.

Adding entries manually to Apache DS LDAP server

If you are creating new entries manually, you can start adding entries from the root to child objects. Here is how it is done:

You need to start creating entries by right-clicking on the DIT Root DSE and click **New | New Entry**. This will open up the **Entry Creation Method** dialog box.

Select **Create entry from scratch** and then click **Next >**.

Then, search for the object class required to add the entry schema. Once you click and add the required object classes to the **Selected object** classes list from the **Available object** classes with the **Add** button, click **Next >**.

In the next window, you need to add the attributes and values required for each object class. You can find all of these object classes, attributes, and values from the LDIF definitions that we discussed earlier. Once attributes and values are filled, click **Next >**.

The next window gives you a summary of the entry. Click **Finish** after reviewing the details. You need to repeat this process until all the entries are added to the LDAP server.

By connecting to the LDAP server and querying entries, you can now verify the LDAP entries you have entered. We will use the `ldapsearch` tool to access the LDAP server and query details. Apache DS starts the LDAP server on default port `10389`. We can use the following search query to list all the entries we have added to the LDAP server:

```
ldapsearch -h localhost:10389 -b "dc=ballerina,dc=io" -x
```

This command connects to the LDAP server running on port `10389` in localhost and lists all the entries starting from the base entry, `dc=ballerina,dc=io`. Once entries are verified, we can start developing the Ballerina application to query user details from the LDAP server and authenticate and authorize users.

In the next section, we will create a Ballerina application that will authenticate and authorize users with the LDAP store we have created.

Authenticating and authorizing Ballerina services with LDAP

Ballerina service interfaces provide built-in support for LDAP integration. You can start an LDAP server and use that server as the user detail database to authenticate and authorize users.

To demonstrate how Ballerina connects to an LDAP server, we will use the previously built LDAP server along with the entries that we have added:

1. First, you need to import the `ballerina/auth` module to the Ballerina project to use the LDAP auth server.

2. Then, you need to create the following LDAP configuration to connect with the LDAP service:

```
auth:LdapUserStoreConfig ldapAuthProviderConfig = {
    domainName: "ballerina.io",
    connectionUrl: "ldap://localhost:10389",
    connectionName: "uid=admin,ou=system",
    connectionPassword: "secret",
    userSearchBase: "ou=Users,dc=ballerina,dc=io",
    userEntryObjectClass: "identityPerson",
    userNameAttribute: "uid",
    userNameSearchFilter:
        "(&(objectClass=person)(uid=?))",
    userNameListFilter: "(objectClass=person)",
    groupSearchBase: ["ou=Groups,dc=ballerina,dc=io"],
    groupEntryObjectClass: "groupOfNames",
    groupNameAttribute: "cn",
```

```
    groupNameSearchFilter:
      "(&(objectClass=groupOfNames)(cn=?))",
    groupNameListFilter: "(objectClass=groupOfNames)",
    membershipAttribute: "member",
    userRolesCacheEnabled: true,
    connectionPoolingEnabled: false,
    connectionTimeout: 5,
    readTimeout: 60
};
```

The first nine attributes in the configuration are about the LDAP server connection details Ballerina uses to connect to the LDAP server. `domainName` is a unique name for identifying the user store. `connectionUrl` is the connection URL that is used to connect to the LDAP server. Since Apache DS starts the LDAP at default port `10389`, you can specify the URL as `ldap://localhost:10389`. `connectionName` is the user that we used to connect to the LDAP server. By default, Apache DS creates a user, `admin`, with the password `secret` to connect to the LDAP server.

3. You can connect to the LDAP service with the `uid=admin,ou=system` connection name, and the password `secret` to connect to the LDAP server.

   `userSearchBase` is the root DN where the directory search begins. When Ballerina looks for a given user, it starts to search from the `ou=Users,dc=ballerina,dc=io` DN and gets the user details. Here, we need to provide some information about how to access user entries with the following configurations:

   a) `userEntryObjectClass`: The `Object` class used to create the user entries

   b) `userNameAttribute`: The attribute to identify the user

   c) `userNameSearchFilter`: The filter query to filter the entries by username

   d) `userNameListFilter`: Filtering criteria for the user entries object class

   The same set of configurations are repeated for the group as well. Here, you also need to define the user search base, entry object class, name attribute, name search filters, and filtering criteria. Additionally, you need to provide `membershipAttribute` to identify the member allocation attribute in the group entry, caching roles, and LDAP connection details.

4. Now, we need to create the Ballerina service definition to start a new HTTP service. The following code sample creates a Ballerina service listening to port 9090 with LDAP configurations:

```
listener http:Listener securedEP = new(9090);
@http:ServiceConfig {
    auth: [{ ldapUserStoreConfig:
        ldapAuthProviderConfig }]
}
service /oms on securedEP { }
```

This service starts at port 9090 and uses LDAP to authenticate and authorize users. The service has the root context as /oms.

5. In the next step, we will implement resource functions to handle REST methods. The following resource function is built to authorize only admin user groups:

```
@http:ResourceConfig {
auth: [{
    ldapUserStoreConfig: ldapAuthProviderConfig,
    scopes: ["delivery"]
    }]
}
resource function get deliveryAccess() returns string {
    return "Hello, Delivery!";
}
```

This resource definition uses delivery as the scope, which only allows users in the delivery group to have access to this particular resource.

6. Now we can run this sample and try to invoke this resource function by means of an HTTP request. Since this is basic authentication, we need to send the authorization method as Basic along with the authorization header on the HTTP request headers. The following is the curl command to invoke this resource function with the delivery user credentials:

```
curl http://localhost:9090/oms/deliveryAccess -H
"Authorization: Basic ZGVsaXZlcnk6ZGVsaXZlcnk="
```

Base64 encoding is used to encrypt the credentials. Since the username and the password of the delivery account are `delivery` and `delivery`, the authorization header is the base64 encode of `delivery: delivery`. If you give an incorrect username or password, the Ballerina service gives you a `401` unauthorized response. In the same way, if you try to invoke this service with a customer account, you will receive a `403` forbidden error. The `curl` command to invoke the service with customer credentials is as follows:

```
curl http://localhost:9090/oms/deliveryAccess -H
"Authorization: Basic Y3VzdG9tZXI6Y3VzdG9tZXI="
```

In the same way, you can add `customer` as the scope to another resource function to give customer-level access to the resource function. If so, this function should be able to handle both delivery and customer users. Check the following sample code that allows access for a customer group that contains both the customer and the delivery user account:

```
@http:ResourceConfig {
auth: [{
    ldapUserStoreConfig: ldapAuthProviderConfig,
    scopes: ["customer"]
    }]
}
resource function get customerAccess() returns string {
    return "Hello, Customer!";
}
```

This resource function allows both delivery and customer users to access this service. You can extend this concept of using the LDAP user store to provide authentication and authorization for users in Ballerina services. Other third-party software can help you to handle users and groups in an LDAP server. Compared to using a database, you can easily authenticate and authorize users with LDAP servers without building a user store from scratch.

In this section, we have discussed building a Ballerina application that uses basic authentication with an LDAP server. In some cases, you might also need to access basic auth-enabled service endpoints. In this scenario, you need to send authentication credentials to an endpoint with basic auth enabled. These details are sent along with the HTTP request headers. Ballerina HTTP client endpoints provide a way to send HTTP requests along with the auth header by defining auth details on the client endpoint definition.

Check the following example Ballerina code, which tries to invoke the `deliveryAccess` resource function that we created in the previous example:

```
http:Client adminEP = check new("http://localhost:9090", {
auth: {
        username: "delivery",
        password: "delivery"
      }
});
string response = check adminEP->get("/oms/deliveryAccess");
```

This endpoint definition contains the username and password required to access the service. Ballerina will take this username and password and then encode them in base64 and attach them to the request's authorization headers. Then, it sends a GET request to the target endpoint and gets the response to the `response` variable. In the same way, you can use other HTTP methods to access endpoints that require basic authentication with these client endpoint configurations.

In this chapter, we are mainly focusing on creating Ballerina applications that support authentication and authorization with an LDAP server. Optionally, you can also create your own user store with a relational database or NoSQL database. However, managing the LDAP user store is easier and you can find many tools to integrate it with. Here, we used the Basic authentication method to access the backend Ballerina service. Basic authorization is the simplest authorization method that can be used to implement a service. In many use cases, however, using a plain text password is not secure enough as it is more vulnerable to hackers.

In the next section, we will discuss token-based authorization methods, which will make the platform more secure.

Token-based authorization with Ballerina

In the previous section, we discussed the Basic authentication method, which is the simplest way to authenticate and authorize a user. Having Basic authentication is not practical in a larger distributed system that has separate components connected over a network. Every time a user needs to access a resource, the resource function needs to verify the user by calling an authorization server.

Instead of calling an auth server for each request, a token-based authentication method uses a token to access services. This allows the system to avoid using the user's username and password and use a token instead. Token are temporary keys that can be validated with a public key which is published by the token generator. Using the username and password is more vulnerable since these remain unchanged for a long time. Tokens are comparatively short-lived and can revoke access in case of a security threat.

In this section, we will discuss different methods of building token-based authentication and authorization platforms with the Ballerina language.

Securing Ballerina services with a JSON Web Token

JSON Web Token (JWT) is a type of token that is used to represent a claim between two parties. The JWT protocol is built on the RFC 7519 standard that defined the structure of the JWT payload and format in 2015. JWT uses a signature of the content to verify that the content is not altered during transmission. Rather than using a username and password for authentication and authorization, JWT provides a way to build a more distributed authentication and authorization mechanism. A JWT issuer can authenticate the user and issue a token that contains the access-level information. Client applications or services can use this token to access the required service.

A token is valid for a certain amount of time. Once it expires, the client application or service needs to generate another JWT and access the required service. Refer to the following diagram, which shows a sample communication flow for accessing a service secured with JWT:

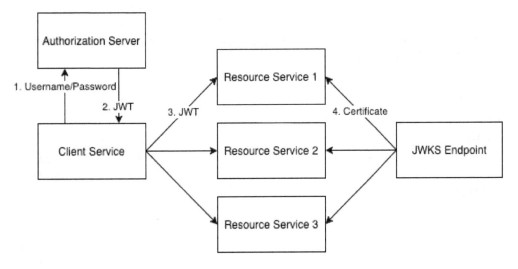

Figure 7.2 – JWT authentication flow

When a **Client Service** needs to access a resource, first it sends authentication details to an **Authorization Server**. This server authenticates the user with the given credentials. Once a user is authenticated, the authorization server generates a **JWT** and sends it to the client service. The client service can use this JWT before it expires. The JWT can now be used by the client service to access the target resource. The **Resource Service** validates the token with a public certificate exposed by the authorization server. If the given JWT is valid, then the resource service can identify the authority given to the user from the token claims and allow the resource function to be invoked.

HTTP is a stateless protocol that does not keep track of the client states. When you are building a web application, the user application needs a way of storing the identity of the client in the browser itself. Cookies are used to solve this problem by adding a unique ID for the client and storing it in the user's web browser. Each time the client sends a request to the server, it validates the user by cookies. The server also has to maintain a map of those cookie IDs and mapping details. However, using this technique to create applications in a microservice architecture is challenging. By design, services in a microservice architecture are stateless. When a service maintains session information for a given user in a service, the next request might deliver to another service that does not have that particular session information. Then, it recognizes the request as unauthorized.

JWT, on the other hand, does not need to maintain a session to authenticate users. It validates client requests using certificates. A JWT is included in the authentication header of each request. The token is validated using the **JSON Web Key Set (JWKS)** endpoint certificate where it keeps all the public certificates of the authentication server. JWT claims can be used to add additional client details. In a microservices environment, each service can independently use JWT to authenticate and authorize the client. On the other hand, a single JWT can be used to access all of the authorized services rather than having a separate authentication and authorization mechanism for each service. In the next section, let's discuss the structure of the JWT.

Structure of the JWT

A JWT is made up of three sections separated by a dot. Each of these three parts is a Base64-encoded JSON string. You can see the data sent over the JWT by decoding these three components separately. When a user application requests a JWT, the token generation endpoint generates a JWT along with claims that it can be used to authorize services that trust the token issuer.

A JWT contains the following structural components:

- Header
- Payload
- Signature

Refer to the following diagram, which illustrates three parts of the JWT:

eyJhbGciOiJIUzI1NiJ9.eyJzdWIiOilxMjM0NTY3ODkwIn0.ErO0fS1yKjr73zJmeYqazVauy8z4Xwuhebs9fXVr3u4

Header   Payload   Signature

Figure 7.3 – JWT structure

The header contains details of the JWT metadata. It includes the token type, typ, as the JWT and a signing algorithm, alg, such as HMAC, SHA256, or RSA. Optionally, a JWT header can contain a kid header claim when there are multiple keys used to sign JWT tokens. You can use this kid claim to identify the key used to sign the token when you need to verify the token.

A sample JWT header claim is set as follows:

```
{
  "alg": "RS256",
  "typ": "JWT",
  "kid": "MzMxM2UxNDUyZDg3MTQ1YjM0MzEzODI0YWI4NDNlZDU1ODQzZW
   FjMQ=="
}
```

A JWT payload contains a set of claims issued by the token generation endpoint. The payload contains a set of claims as key-value attributes in JSON format. This payload contains both custom claims and reserved claims. Let's see what these claims are:

- **Reserved claims**: A set of optional claims that are already reserved in JWT format. Reserved claims are commonly used in JWT as these contain the primary details of the token. These claims include iss, sub, aud, and so on.
- **Custom claims**: You can create your own custom claims inside the JWT. You need to make sure to have namespaces to avoid name collision.

The following is an example JWT payload claim set:

```
{
    "iss": "ballerina",
    "sub": "admin",
    "aud": "vEwzbcasJVQm1jVYHUHCjhxZ4tYa",
    "jti": "3559be28-cf2f-4131-b80b-efc3bf5d2641",
    "exp": 1615268093,
    "nbf": 1615264493,
    "iat": 1615264493
}
```

This JWT payload claim set contains multiple reserved claims that are associated with the JWT token. The following are some of those reserved claims generated by the token generation endpoint:

- iss (**the issuer**): The issuer of the JWT.

- sub (**the subject**): The subject of the JWT.

- aud (**the audience**): This claim is used to define who is the recipient of the JWT token.

- jti (**JWT ID**): This is a unique identifier for each JWT token. This claim is used to avoid replaying the same JWT.

- exp (**the expiration time**): This specifies the expiration time of the JWT. Once this time has elapsed, the JWT is no longer valid.

- nbf (**not before time**): This claim is used to define to not accept the JWT until this given time.

- iat (**issued at**): This specifies the time that the token was issued.

Finally, the signature is used to sign the JWT. The signature is generated by hashing the header, payload, and secret with a hashing algorithm. The secret is the secret that is held by the JWT issuer to generate tokens. The following is a sample function that uses HMAC SHA256 to show how the signature calculates from the header, payload, and the secret:

```
HMACSHA256(base64UrlEncode(header) + "." +
base64UrlEncode(payload), secret)
```

The JWT generator combines all of these three parameters that are separated by a dot to generate the JWT. When you need to invoke an HTTP service that requires a JWT, you can simply send the JWT with authorization headers with the `Bearer` type. The request header of the request with the JWT will be as follows:

```
Authorization: Bearer <token>
```

This request is sent to the backend server, which authorizes the request using the token value. We'll create a JWT using the Ballerina language and validate it in the next section. In later sections, we will discuss more on sending authorization headers in the Ballerina HTTP request.

Generating and validating a JWT with Ballerina

Ballerina provides built-in JWT generation support to issue new JWTs. First, you need to create a keystore to generate a signature for the JWT. In this example, we will also use the certificate that we created in the previous sample involving securing the Ballerina service with HTTPS. The following is the sample code to create a JWT token using the Ballerina language:

```
jwt:IssuerConfig issuerConfig = {
    username: "admin",
    issuer: "ballerina",
    audience: "aVe3D33DWUseSeisADe33DYsl3",
    keyId:"MzMxM2UxNDUyZDg3MTQ1YjM0MzEzODI0YWI4NDNlZDU1ODQzZ
      WFjMQ==",
    expTime: 3600,
    customClaims: { "foo": "bar" },
    signatureConfig: {
        config: {
            keyStore: {
                path: "resources/ballerinaKeystore.p12",
                password: "ballerina"
            },
            keyAlias: "ballerina",
            keyPassword: "ballerina"
        }
    }
};
string|jwt:Error jwtToke = jwt:issue(issuerConfig);
```

The Ballerina JWT library contains the `issue` function that generates JWTs. You need to provide all the requisite parameters that need to generate the token as a `jwt:IssuerConfig` record type. This configuration contains the username, which is `sub`, the issuer, which is `iss`, and the audience, which is `aud`, in the JWT payload. The `keyId` field is `kid` in the JWT header, which represents the certificates used to sign the request. The `expireTime` field specifies how long the JWT token is valid for in seconds. You can add custom claims to the JWT with the `customClaims` field as key-value pairs.

The `signatureConfig` field is used to define the keystore configurations that are used to sign the token. You can give either the absolute path or relative path where the keystore is stored. Here, we have reused the existing keystore that was generated in the previous example. You need to provide the keystore password with the `password` field.

The `jwt:issue` function returns the JWT in a string data type. It uses type checking to see whether the token generation function throws any errors before using the `jwtToken` variable. For authorization and authentication purposes, the created token can be shared with the client application. You can check what content is available in the JWT you created by decoding it. If you add a customer claim to the token, then you can read those claims on the JWT token payload section. We will discuss using custom claims to authorize users in the *Authorizing Ballerina services with WSO2 IS-generated JWT claims* section.

You can validate the token generated in the Ballerina program itself. We will use the token that we have just generated to validate whether it is a valid token. Check the following example code, which validates a given JWT and reads available claims:

```
jwt:ValidatorConfig jwtValidatorConfig = {
    issuer: "ballerina",
    audience: "aVe3D33DWUseSeisADe33DYsl3",
    clockSkew: 60,
    signatureConfig: {
        trustStoreConfig: {
            certAlias: "ballerina",
            trustStore: {
                path: "resources/ballerinaTruststore.p12",
                password: "ballerina"
            }
        }
    }
};
jwt:Payload|jwt:Error result = jwt:validate(check jwtToken,
    jwtValidatorConfig);
```

The `jwt:validate` function is used to validate a given JWT. The `jwt:ValidatorConfig` record type is used to provide the token validation configurations. `issuer` is an optional field that is used to define the JWT issuer. The `audience` field defines the audience that the JWT is issued to. To prevent token validation failures due to clock synchronization issues, you can use `clockSkew`. In the `trustoreConfig` field, add the certificate details, such as the trust store alias, path, and password.

Token validation can fail either due to an invalid JWT or due to JWT expiration. By validating the signature with the trust store public certificate, the JWT validator determines whether the JWT is a valid token. This validation can fail due to an invalid signature or on account of expiration. This JWT can be sent to the client who needs access to the resource. Then, the client can use this JWT to access the resource service where the resource service can validate whether this is a valid token by validating JWT with the JWT issuer public certificate.

In this section, we have focused on generating and validating JWT with a Ballerina program. JWT is important in access controlling in a distributed application. In the next section, we will discuss how to generate a JWT with Identity Server rather than generating it from the Ballerina and use that JWT to validate a Ballerina service. As the identity server, we will use WSO2 Identity Server to manage users, roles, and JWT.

Generating JWT with WSO2 Identity Server

WSO2 Identity Server (**WSO2 IS**) is an open source identity service provider that has many identity and access management features. It supports **Single Sign-On** (**SSO**), OpenID Connect, OAuth2, among other things. You can use WSO2 IS to manage LDAP as well. In the previous LDAP example, we have manually created all the entries that are required for the LDAP service example. With WSO2 IS, you can integrate an LDAP server and create users and groups by using the WSO2 IS management console itself. This is important where you need to manage the users and assign roles for each user by using a web-based user interface. The instructions on how to use WSO2 IS can be found on the WSO2 IS web page at `https://is.docs.wso2.com/en/latest/`.

WSO2 IS provides multiple features to secure your cloud native applications. For this example, we will check how to use the WSO2 IS JWKS endpoint to validate a JWT.

WSO2 IS allows you to manage multiple certificates and expose public certificates over a JWKS endpoint. You can download the WSO2 IS installer from the `https://wso2.com/identity-and-access-management/` website and install it on your computer. The installation locations for WSO2 IS on your computer are listed as follows, depending on the operating system you're using:

- **macOS**: `/Library/WSO2/IdentityServer/`
- **Windows**: `C:\Program Files\WSO2\IdentityServer\`
- **Ubuntu**: `/usr/lib/wso2/IdentityServer/`
- **CentOS**: `/usr/lib64/IdentityServer/`

Now we need to copy the keystore file that we have generated in the previous example to the WSO2 IS. A default keystore file is included with WSO2 IS. But we will add another keystore file to generate JWTs. You can copy the `ballerinaKeystore.p12` keystore file into the keystore location in WSO2 IS, which is located in the `<WSO2_IS_HOME>/repository/resources/security/` directory. This directory contains the keystores and trust stores that are used in the WSO2 IS to generate and validate certificates.

Now we need to provide the newly added keystore location to the WSO2 IS by giving the keystore filename to the WSO2 configuration. The configuration file is located in the `<WSO2_IS_HOME>/repository/conf/deployment.toml` file. This file contains the WSO2 IS configurations in TOML format. Now you need to change the keystore location configuration as follows in the TOML configuration file:

```
[keystore.primary]
file_name = "ballerinaKeystore.p12"
type = "PKCS12"
password = "ballerina"
alias = "ballerina"
key_password = "ballerina"
```

Here, we specify the filename of the keystore file as `ballerinaKeystore.p12`, and the keystore type as `PKCS12`. You need to provide keystore information such as the keystore `password`, `alias`, and `key_password` as well in this TOML configuration.

Now you can start up the WSO2 IS server. Get a new terminal and change the current directory to the `<WSO2_IS_HOME>/bin` directory. Execute the `./wso2server.sh` command on Linux and Mac, or `wso2server.bat` in Windows, to start the WSO2 IS server. This will take a few seconds to start up the server. Once it completes the start up process, it will show you a few links that you can use to access the web-based management console. With default configurations, you can access the management console through the `https://localhost:9443/carbon/` URL.

When you access the management console through the web browser, it gives you a login window to give the username and the password. The default admin privilege user account can be used to access the management console. Provide `admin` as both the username and password and click **Sign-in** to log in to the console. You can use the management console to configure the WSO2 IS to secure and manage your system. To build our sample, let's start creating the required users and roles to authenticate and authorize users.

WSO2 IS uses the built-in LDAP store to maintain users and roles in the local environment. As previously mentioned, you can connect LDAP to store these users and roles. For this example, we will use the embedded LDAP user store to build this sample.

In the management console, the left panel contains four tabs to navigate through various configurations. By default, you will get the **Main** tab, which contains frequently changing configurations. On the left navigation panel, click **Users and Roles | Add** to add new users and roles. This takes you to **Add Users and Roles** to select either **Add New User** or **Add New Role**.

Click on **Add New Role** to create a new role. This brings up the following page, where the **Domain** and **Role Name** fields need to be completed. Keep **Domain** as **PRIMARY** and provide a name for **Role Name**. For this example, we will create a new role as `customer`, shown as follows. Click **Next >** to continue to add a new role:

Figure 7.4 – Add New Role page to the WSO2 IS management console

On the next page, you can select different levels of permission to assign to the newly created role. For this example, we will add all permissions by marking the tick in **All Permissions**. Then, click on the **Finish** button to complete creation of the user. In the same way, you can create new users from the **Add User and Roles** page. Instead of roles, click on **Add New User** to create a new user. On the **Add New User** page, provide **PRIMARY** as the domain, along with the username of the user, a password, and password verification details. For this example, let's create a user with the name `customer` and `customer` as the password. Then, click **Next >** to select roles for the user. In the **Select Roles** field of the **User** step page, mark the role as **customer**, which is the role that we created earlier. Then, click **Finish** to complete creation of the user.

Now we need to create a service provider that provides different methods of authentication and authorization services to our system. A service provider provides web services and relies on **Identity Provider (IdP)** to provide authentication and authorization. We will create a service provider that provides a JWT-based authentication service to the system users.

To add a new service provider, click on **Service Providers | Add** on the **Main** tab on the left panel. This will take you to the **Add New Service Provider** page, which includes some basic information about the service provider. Select **Manual Configuration** from the **Select Mode** section and provide a name for the service provider in the **Service Provider Name** input box. Then, click **Register** to register a new service provider. This will bring up a new page – **Service Providers**.

Now, we will add a JWT-based inbound authentication method to the service provider that we have just created. Click on the **Inbound Authentication Configuration** expanding list and click on the **OAuth2/OpenID Connect Configuration** drop-down button. Then, click on the **Configure** button to configure a JWT inbound authentication method. Refer to the following screenshot:

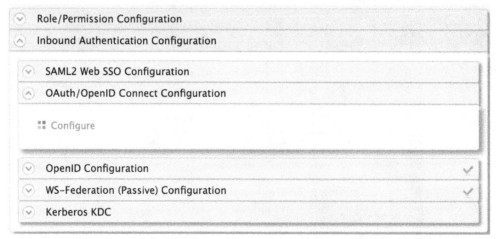

Figure 7.5 – Configuring the service provider section

This will take you to another page called **Register New Application**. Then, perform the following steps:

1. From this form, remove the tick marks for the **Code** and **Implicit Auth** grant types. Make sure to keep the tick mark of the **Password** grant type.

2. On the **Token Issuer** option list, select **JWT**.

3. Then, click on the **Add** button to add configurations. This will take you back to the **Service Providers** page. You will see that a new entry is added under the **Inbound Authentication Configuration | OAuth/OpenID Connect Configuration** section in the table.

4. This table contains **OAuth Client Key** and **OAuth Client Secret** fields to access this service provider. You can use the **Show** button to show the hidden client secret that can be used to invoke this service provider.

Now we can use this service provider and the user account to generate a JWT from the WSO2 IS. Use the following `curl` command to generate a new JWT by invoking the WSO2 IS oauth2 endpoint:

```
curl -u <Client_key>:<Client_secret> -k -d "grant_
type=password&username=<Username_of_user>&password=<Password_
of_user>" -H "Content-Type:application/x-www-form-urlencoded"
https://localhost:9443/oauth2/token
```

You need to replace the `Client_key` and `Client_secret` placeholders with `OAuth Client Key` and `OAuth Client Secret`. Also, replace `Username_of_user` and `Password_of_user` with the username and password information of the user that we have just created. A sample command for this CURL is as follows:

```
curl -u NfFYJG740zelDUlVUY2kCV5fbbca:vnULYoZaC24AlwtS4v9jbvVj
XJ4a -k -d "grant_type=password&username=customer&password
=customer" -H "Content-Type:application/x-www-form-urlencoded"
https://localhost:9443/oauth2/token
```

This command will give you the following JSON response that contains a JWT:

```
{
    "access_token":"xxxxxxx",
    "refresh_token":"xxxxxxxx",
    "token_type":"Bearer",
    "expires_in":3600
}
```

Here, `access_token` is the JWT generated from the WSO2 IS server. We will use this access token and validate whether this token is valid. The refresh token is a token used to generate a new access token after the existing token expires. The token we have to generate contains a default life span of `3600` hours that can be configured in WSO2 IS.

Now we have the JWT generation endpoint that can generate a JWT for client applications. In the next section, we will use this JWT to secure the Ballerina service.

Authorizing Ballerina services with JWT

A Ballerina service can have a JWT-based authorization method to allow access for a resource function. Instead of validating the JWT with a locally available certificate, we can use the JWKS certificate endpoint that holds the certificate publicly to authorize requests. **JWKS** is a read-only HTTP endpoint that provides a list of certificates issued by a given identity server. This certificate's details are exposed in JSON format. A sample JWKS endpoint response is as follows:

```
{
  "keys":[{
  "kty":"RSA",
  "e":"AQAB",
  "n":"XXXXXXXXXX",
  "use":"sig",
  "kid":"XXXXXXXXX_RS256",
  "alg":"RS256"
}]
}
```

This JSON response contains an array of keys for each certificate hosted in WSO2 IS. For each certificate, there is `kty`, to identify the cryptographic algorithm, `e` and `n` to represent the RSA encryption exponent and the modulus, `use` to specify the intention of the public key, `kid` to identify the certificate from multiple certificates, and `alg` to identify the encryption algorithm. You can access the WSO2 IS JWKS list from the `https://localhost:9443/oauth2/jwks` URL. This will give you a WSO2 IS registered JWKS public certificate list. We will use this JWKS endpoint to validate JWTs that are generated from the WSO2 IS.

The following is the source code to validate a JWT by using the WSO2 IS JWKS endpoint:

```
jwt:ValidatorConfig validatorConfig = {
    issuer: "https://localhost:9443/oauth2/token",
    audience: "uWnIibMbqvtMcURlORd2y44sX9ka",
    signatureConfig: {
        jwksConfig: {
            url: "https://localhost:9443/oauth2/jwks",
            clientConfig: {
                secureSocket: {
                    cert: {
                        path: "resources/
                            ballerinaTruststore.p12",
                        password: "ballerina"
                    }
                }
            }
        }
    }
};
string jwt = "xxxx";
jwt:Payload|jwt:Error result = jwt:validate(jwt,
    validatorConfig);
```

You need to provide the JWT string that you have generated from the `curl` command in the `jwt` variable. These Ballerina JWT configurations are much similar to the previous example. Here, `issuer` is the token issuer endpoint that generates the JWT. `audience` is the client key that we received for the service provider. The JWKS endpoint URL is specified in the `jwksConfig` field under the `signatureConfig` attribute. Here, we need to define the JWKS endpoint certificate to certify the JWKS endpoint. Ballerina uses certificate details just to create an SSL connection to the JWKS endpoint rather than validating the JWKS from the trust store certificate. In this example, we validate a JWT that was issued by an identity provider. Next, we will expand the idea of using JWT claims to authorize users for different services.

Authorizing Ballerina services with WSO2 IS-generated JWT claims

As we have already discussed so far, a JWT can be used to prove that the user is indeed who they claim to be. In addition to user information, it can also carry authorization details. For example, an identity server is able to give some authorization to a user with a claim on the token. The identity server can transfer this authorization information to the resource service through the JWT claims. The identity server can add claims regarding authorization levels to the JWT and give the token to the client application. Then, the client application can send this JWT to the resource service and authorize it.

WSO2 IS allows you to add claims to the JWT generated for each user. We will go over how to create a custom claim for a user and how to allow a user to limit access to certain services.

In the first step, we will create a local claim and then we will expose that local claim as an external claim with the existing scope. Let's see how this happens:

1. First, we begin by creating a local claim that will be used to claim the user authorization level. On the management console, click on **Claims | Add** from the left navigation panel to create a new claim.

2. This will take you to a new page, **Add New Dialect/Claim**, to select what type of claim to add. Click on **Add Local Claim** to add a new local claim.

On the **Add Local Claim** page, you need to fill in some information about the new claim. For **Claim URI**, give a unique URI to identify the claim. We will provide `http://wso2.org/claims/authlevel` as the **Claim URI** to identify the user authorization level. Provide a name in the **Display Name** field to identify the claim. We will provide `AuthLevel` as the **Display Name** attribute. Also, add a description relating to the claim in the **Description** field. In the **Mapped Attribute (s)** field, you must provide the mapped attribute. Give a value of **role** to the **Mapped Attribute** column in the **Mapped Attribute (s)** field. Also, make sure to mark the **Supported by Default** tick button. Refer to the following screenshot:

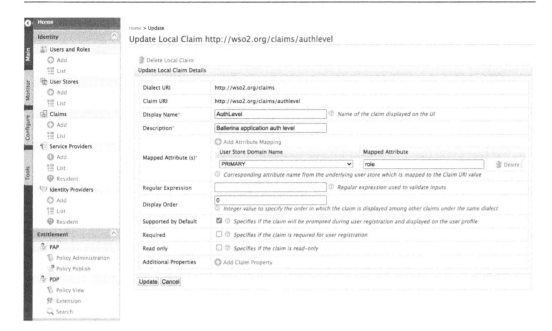

Figure 7.6 – Adding local claim page configurations

Once the information is filled in, click **Add** to add the local claim. This will create a new local claim in the WSO2 IS. Local claims are local to the WSO2 IS.

Now, we need to create an external claim from the local claim to add claims to the JWT. Here's how:

1. To add a new external claim, click on **Claims | Add** on the left navigation panel.

2. Click on **Add External Claim** from the **Add New Dialect/Claim** page. This brings up a new page, **Add External Claim** that has a mapping to expose local claims to external claims. For the **Dialect URI** field, select `http://wso2.org/oidc/claim` from the drop-down list. For the **External Claim URI** field, give the name of the claim as `authlevel`. For the **Mapped Local Claim** input, select the local claim, `http://wso2.org/claims/authlevel`, that we have created from the drop-down list. Then, click the **Add** button to add the external claim.

Now we need to add this claim to a scope. With WSO2 IS, you can create your own scopes and access claims. Otherwise, you can add the claim to an existing scope. For this example, we will create a new scope and add the claim that we have just created to that scope:

1. To create a new scope, click on **OIDC Scopes | Add** on the left navigation panel. This will bring up the **Add New Scopes and Claims** page.

2. For the scope name, provide a name to identify the scope. We will provide `appscope` as the scope name.

3. From **Select OIDC Claims**, click **Add OIDC Claim** to add the claim to the scope. In the **Enter OIDC Claims** column, select the claim that we have just created.

4. Click **Finish** to complete the scope creation.

5. If you list down all the available scopes by clicking on **OIDC Scopes | List**, you can see that the scope we have just created is listed here.

6. On the scope list, click on the **Update** button in the **Action** column on the relevant newly created scope.

7. This will take you to a page, **View claims appscope,** with a list of OIDC claims. Mark the **AuthLevel** tick to attach the claim we have created to this scope. Click **Finish** to complete the addition of a claim to the new scope.

Now, we can assign different auth level claims to different users:

1. To attach claims for each user, click on **Users and Roles | List** on the left navigation panel. Then, click on **Users** to list all users on the system.

2. Now, click on the **User Profile** button in the **Actions** column for the **customer** user that we have just created. This will bring up a new page named **Update Profile : customer** where some details need to be filled in.

3. On this page, you can see that the claim configuration, **AuthLevel**, is also available to fill with an input box. Let's add `customer` as the **AuthLevel** to indicate to this user to issue a JWT with the claim as the customer access level.

4. Click **Update** to update the user information.

> **Note**
> If you filling out this information for the first time, then make sure to fill out all the required fields, including **First Name**, **Last Name**, and **Email**.
> Otherwise, it will give you an error message saying you need to complete all the mandatory fields.

In the final step, you need to configure the service provider to expose the claims. Here is how we do it:

1. To configure, go to **Service Providers | List** and click on **Edit** to edit the service provider.

2. In the **Service Provider** edit view, click on **Claim Configurations** to set up the claims.

3. Click on **Add Claim URI** to add a new claim.

4. In the **Local Claim** column, select the claim, `http://wso2.org/claims/authlevel`, that we created earlier. Then, click **Update** to update the configurations.

Now we can access the service provider and get a JWT token that can be used to authenticate and authorize a user. In the same way as the previous `curl` command, let's send a request to the `oauth` endpoint and get a JWT. This time, we will send the scope along with the request as follows:

```
curl -u NfFYJG740zelDU1VUY2kCV5fbbca:vnULYoZaC24AlwtS4v9
jbvVjXJ4a -k -d "grant_type=password&username=customer&password
=customer&scope=appscope" -H "Content-Type:application/
x-www-form-urlencoded" https://localhost:9443/oauth2/token
```

In this `curl` command, we set the scope as `appscope`. This will give you a JWT that contains the claim that we have added to the customer user account. If you decode the payload of the JWT, you can check that it contains a custom claim as `"authlevel"`: `"customer"`. This is the custom claim that we have added to the WSO2 IS. Here, the key is the claim that we created and the value is the value that we set on the user's user profile. In the next section, we will build a Ballerina application that reads a JWT and authorizes a user to use a particular resource function.

Building a Ballerina service with JWT authorization

JWT claim validation can be used to authenticate and authorize users in the Ballerina service. We will use WSO2 IS as the identity provider to generate a JWT. The Ballerina service can read the request authentication headers and allow access to authorize resources. The Ballerina service can validate a JWT with the JWKS endpoint exposed by the WSO2 IS. For this example, we will use a previously configured WSO2 IS service provider that generates a JWT with authentication-level claims.

The following is the Ballerina source code for the simple service implementation of using a JWT claim to authorize users. This service starts at port 9090 and allows only users with customer-level permission to access the resource function. If the `authlevel` claim is `customer`, this will allow access to the resource:

```ballerina
listener http:Listener ep= new(9090);
@http:ServiceConfig {
    auth: [{
            jwtValidatorConfig: {
                issuer: "https://localhost:9443/oauth2/token",
                audience: "NfFYJG740zelDUlVUY2kCV5fbbca",
                signatureConfig: {
                    jwksConfig: {
                        url: "https://localhost:9443/oauth2/
                            jwks",
                        clientConfig: {
                            secureSocket: {
                                cert: {
                                    path: "resources
                                        /ballerinaTruststore.p12",
                                    password: "ballerina"
                                }
                            }
                        }
                    }
                },
                scopeKey: "authlevel"
            },
            scopes: ["customer"]
        }]
}
service /hello on ep{
    resource function get .() returns string {
        return "Hello, World!";
    }
}
```

This Ballerina service contains a single service with a resource function that responds to an HTTP request with the payload `Hello, World!`. This service can only be accessed if the JWT argument includes the `authlevel` claim with the value of `customer`. The `scopeKey` attribute in `jwtValidatorConfig` is used to identify which claim is supposed to be used to authorize user. The `scopes` attribute is an array of allowed claim values that allow you to access the resource function. In this scenario, the `authlevel` claim should be `customer`. You can add multiple claim values such that the Ballerina service authorizes a request with those claims.

The same as the previous JWKS validation sample, we used a JWKS endpoint to validate the JWT signature. The Ballerina service uses a JWKS endpoint to obtain the public certificate of the identity provider. Here also, we need to provide certificate details to connect to the JWKS endpoint to validate a JWKS endpoint.

You can invoke this service endpoint with the token that we have generated. Create a new token if it has already expired. You can invoke this service endpoint with the following `curl` command. Make sure to replace the `JWT_token` placeholder with the JWT you received:

```
curl --location --request GET 'http://localhost:9090/hello'
--header 'Content-Type: application/x-www-form-urlencoded'
--header 'Authorization: Bearer <JWT_token>'
```

Other than invoking the endpoint with the `curl` command, you can create a Ballerina client application to invoke the Ballerina service. You can use the Ballerina password type `oauth2` client to access WSO2 IS, get the JWT token, and send a request to the Ballerina service with a few lines of codes. Check out the following example implementation of calling the Ballerina service that we have created:

```
http:Client ep check new("http://localhost:9090", {
    auth: {
        tokenUrl: "https://localhost:9443/oauth2/token",
        username: "customer",
        password: "customer",
        clientId: "NfFYJG740zelDUlVUY2kCV5fbbca",
        clientSecret: "vnULYoZaC24A1wtS4v9jbvVjXJ4a",
        scopes: ["appscope"],
        clientConfig: {
            secureSocket: {
                cert: {
                    path: "resources/
```

```
                        ballerinaTruststore.p12",
                password: "ballerina"
            }
        }
    }
  }
});
public function main() {
    http:Response response = check ep->get("/hello");
    io:println(response.statusCode.toString());
}
```

This client endpoint example calls the Ballerina service in port `9090` with the given auth configurations. `tokenUrl` is set as the WSO2 IS `oauth2` token generation endpoint. `username` and `password` are the client logging credentials for the identity provider. We need to provide `clientId` and `clientSecret`, which are the OAuth client key and OAuth client password, respectively. Here, too, you need to provide trust store certificate details to the WSO2 IS server to secure the `oauth2` token generation. When you run this application, it will print the response code to the terminal output.

In this section, we have learned about JWT-based authentication to authorize users. We can easily manage users and roles in Identity Server with the WSO2 IS to give different levels of access levels for the Ballerina application. In the next section, we will learn how to use OAuth2 authorization to authorize users in Ballerina applications.

OAuth2 authentication and authorization with WSO2 IS

The OAuth2 protocol is a standard authorization protocol that is commonly used in systems nowadays to expose different services to users. If you ever came up with a web application that lets you use Google, Facebook, or a GitHub account to log in to a web application or allows a third-party web application to use your Google, Facebook, or GitHub resources, then you have experienced the OAuth2 protocol. This protocol lets an identity provider share resources among third-party applications. The following are the major components in the OAuth2 service:

- **Authorization server**: This is the component that provides identity to the users in the system. Users can use the authorization server to generate tokens that can be used to access other third-party resources.

- **Resource owner**: A resource owner is a user who has the authorization to use resources. A resource owner can log in to the authorization server to get tokens or let third-party applications give the authorization to use the user's resources.

- **Resource server**: This entity contains the resource owner's data that is required to expose the third-party application. This endpoint lets the data interact with third-party applications if it has a valid token issued by the authorization server on behalf of the resource owner.

- **Client**: The client application is the third-party entity that is trying to access the resource owner's data. The client should have a valid token to access the resource server that was issued by the authorization server.

OAuth2 provides different solutions to implement the authorization of resources. Based on how the OAuth2 protocol is implemented, we can categorize OAuth2 deployment types into the following categories:

- **Authorization code**: This method is commonly used in web browsers where the resource owner needs to permit the resource server to use a client application. Client applications use redirections to the authorization server to grant permission to the client and issue a token to the latter. This is the method that you might use when you use Google, Facebook, or GitHub to grant permissions to its resources.

- **Implicit**: This grant type is commonly used in building mobile applications and web applications. Unlike the previous method, this authorization server sends the access token to the user agent (to a mobile application or web browser) to forward it to the client. This method is not regarded as being very secure as it exposes access tokens to the intermediate user agent.

- **Client credentials**: Client credentials are used to generate tokens for machine-to-machine communication scenarios, such as CLI tools and daemons, rather than generating tokens on behalf of users.

- **Password grant**: In this scenario, the user shares the username and password to generate a token. Therefore, this grant type should not be used with third-party applications. This grant type can be used where the client trusts the resource owner.

For this example, let's create a WSO2 IS auth server, generate a token with the password grant type, and invoke a Ballerina service with a Ballerina client program. Let's now see how to go about this:

1. To create an OAuth2 authorization server, click on **Service Providers | Add** to add a new service provider. As with the previous service provider, give a name to the service provider in the **Service Provider Name** input box. We will name the service provider BallerinaOAuth2. Click **Register** to register it:

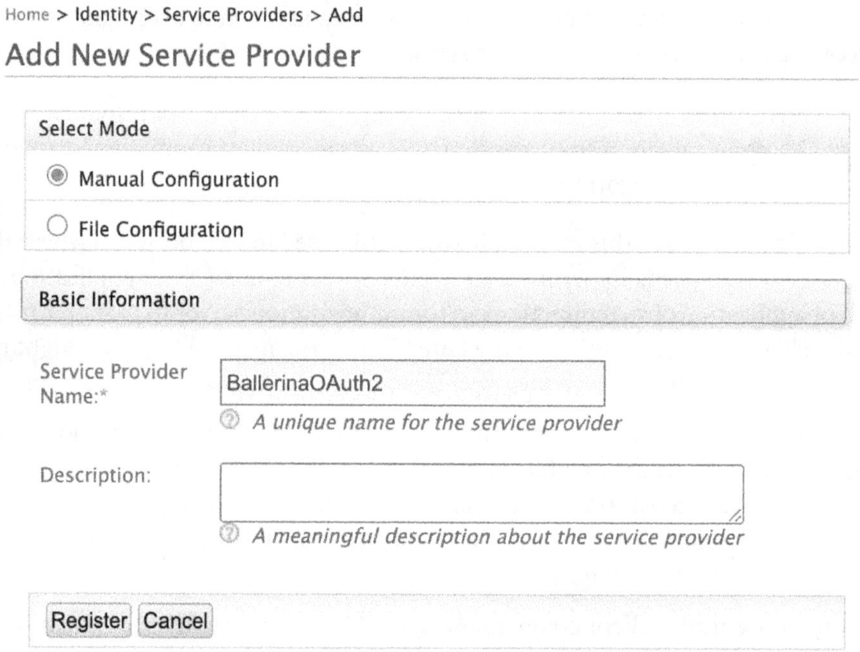

Figure 7.7 – Service provider configuration view

2. On the next **Service Providers** page, you need to configure a service provider to create inbound authentication with the OAuth2 service. Click on the **Inbound Authentication Configuration** drop-down button and then click on the **OAuth/OpenID Connect Configuration** dropdown to configure the OAuth2 configurations.

3. Next, click on the **Configure** button. On the next **Register New Application** page, remove the tick mark for **Code** and **Implicit OAuth2** methods. Make sure to check the box for the role-based scope validator in the **Scope Validators** selection. For this service provider, keep the token issuer as the **Default** option selected:

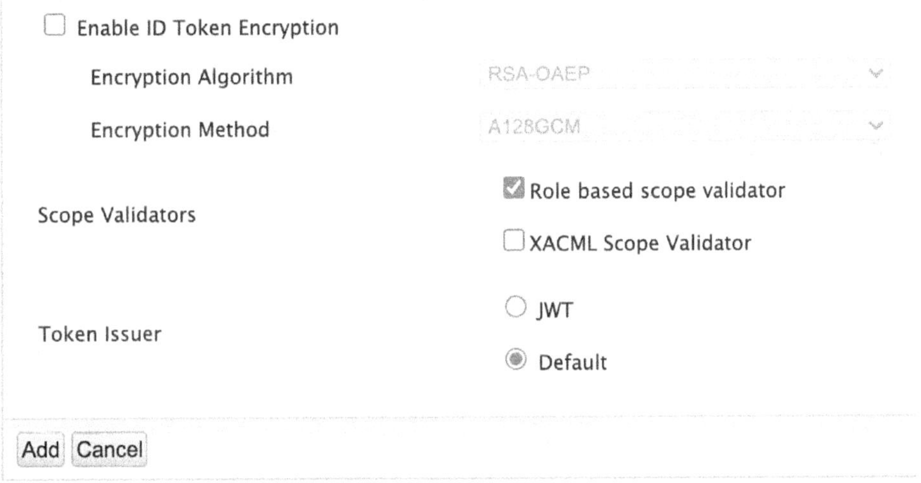

> ☐ Enable ID Token Encryption
>
>   Encryption Algorithm   RSA-OAEP   ⌄
>
>   Encryption Method   A128GCM   ⌄
>
> Scope Validators   ☑ Role based scope validator
>
>   ☐ XACML Scope Validator
>
> Token Issuer   ○ JWT
>
>   ◉ Default
>
> [Add] [Cancel]

Figure 7.8 – OAuth/OpenID Connect Configuration view

4. Then, click **Add** to add the inbound authentication method. This will give you the OAuth client key and the client secret.

Let's now create two roles that are assigned to two separate users to provide a role-based authorization method. Let's give different levels of permissions for two roles and use those permissions to set different authorization levels for a Ballerina application. Create a role and a user to represent the delivery role. We will name both the role and the user as `delivery`. Give the delivery user the delivery role type once the delivery role has been created.

WSO2 IS provides role-based access with OAuth2 scopes. We can associate different levels of scopes with different roles. You can generate a token that only allows the resource service to use it if it is available for the allowed role.

For this example, let's create a scope for the delivery role to add new products. Let's name this scope `add_product`. WSO2 IS provides a REST API to interact with scopes. For this example, we will add a scope to a given role. You can try out all the different types of REST methods provided in WSO2 IS from the `https://is.docs.wso2.com/en/5.12.0/develop/oauth2-scope-management-rest-apis/` URL. The following is the `curl` command to generate a new scope, `add_product`, that has a delivery role associated with it:

```
curl --location --request POST 'https://localhost:9443/api/
identity/oauth2/v1.0/scopes' --header 'Authorization: Basic
YWRtaW46YWRtaW4=' --header 'Content-Type: application/json'
--data-raw '{
    "name": "add_product",
    "displayName": "add_product",
    "description": "add_product",
    "bindings": [
        "delivery"
    ]
}'
```

This `curl` command sends an HTTP POST request to the WSO2 IS OAuth2 scope management API and creates a new scope with the name of `add_product`. You can give a list of binding roles that the scope belongs to. For this example, we allocate this scope to the delivery role. Here, you need to provide the username and password of the user who has permission to change scopes. Here, we used an admin account that has `admin` as both the username and the password. `YWRtaW46YWRtaW4=` is the base64-encoded value of `admin:admin`.

Now we can generate a token from WSO2 IS to access services that have the `add_product` scope. We can generate a token from the WSO2 IS token generation endpoint with the following example `curl` command:

```
curl -u nobQxoLIp3BfoG4Anncf82szJnga:wSSdSy_
eyfXIISCfJSLeyVz5VlAa -k -d "grant_
type=password&username=delivery&password=delivery&scope=add_
product" -H "Content-Type:application/x-www-form-urlencoded"
https://localhost:9443/oauth2/token
```

Here, you need to provide the client key and the client secret as the authorization details. We set `password` as the grant type, along with a username, client password, and the scope details of the resource. This `curl` command will give you a JSON response of the access key token that can be used to invoke the OAuth2-enabled service. This `curl` command also gives you a refresh token that can be used to generate a new access token if it has expired. Since the refresh token is valid for a longer period, make sure to store the refresh token in a secure way in your client application.

The token generated can be validated against the introspect endpoint provided by the WSO2 IS. You can provide the token and user credentials to validate the token with the following `curl` command:

```
curl -k -u admin:admin -H 'Content-Type: application/x-www-
form-urlencoded' -X POST --data 'token=<oauth2_token>' https://
localhost:9443/oauth2/introspect
```

Replace the `oauth2_token` placeholder with the token you received and replace the user credential with a user who has sufficient access to use introspection. This command gives you the validity of the token in JSON format. If the `active` field is false in the JSON response, this means that the token is no longer valid.

Now we can build the Ballerina resource service that reads authorization information from the request headers and validates the request based on the user role and scope. The following is the source code implementation to authorize users with OAuth2 scopes in the Ballerina service:

```
listener http:Listener ep = new(9090);
@http:ServiceConfig {
    auth: [{
            oauth2IntrospectionConfig: {
                url: "https://localhost:9443/oauth2/introspect",
                tokenTypeHint: "access_token",
                scopeKey: "scope",
                clientConfig: {
                    customHeaders: {"Authorization":
                        "Basic YWRtaW46YWRtaW4="},
                    secureSocket: {
                        cert: {
                            path: "resources/
                                ballerinaTruststore.p12",
                            password: "ballerina"
```

```
                    }
                  }
                }
              },
              scopes: ["add_product"]
            }
        ]}
service /hello on ep {
    resource function get .() returns string {
        return "Hello, World!";
    }
}
```

Here, the Ballerina server starts on port 9090. We need to provide OAuth2 authorization configurations for the service endpoint to configure OAuth2 scope-based authorization. Here, we need to specify the introspect URL for the url field. scopeKey is the key that is used for scope-based authorization. In customHeaders, you need to provide admin credentials to access the identity service and validate the OAuth2 token. As with the previous examples, you need to provide the trust store configurations to verify the introspect URL.

The resource function defined in this example sends a plain text response to the caller. The scopes field defined in the auth configurations only allows access tokens with the correct access role permissions. The Ballerina program only allows access to this service if the authorization token has a user who has role support for the add_product scope. For this example, this service only allows users who have the role of delivery, which has the scope of add_product.

We can use a curl command to access this resource function with the OAuth2 token that we have generated. If a token is expired, make sure to generate a new token. The following curl command can be used to invoke the resource function with a valid OAuth2 token:

```
curl --location --request GET 'http://localhost:9090/hello'
--header 'Content-Type: application/x-www-form-urlencoded'
--header 'Authorization: Bearer <oauth2_token>'
```

Replace the oauth2_token placeholder with the token you received from the WSO2 IS key generation endpoint. When you invoke the Ballerina service with this command, it will give you the Hello World! plain text response.

We can create a client application to access this service with the Ballerina language. Instead of generating tokens separately, let's use the Ballerina client endpoint that will call the WSO2 IS, get an OAuth2 token, and access the Ballerina service with that client. The following is the sample code that accesses the Ballerina OAuth2 protected service with the OAuth2 grant type password:

```ballerina
http:Client ep = check new("http://localhost:9090", {
    auth: {
        tokenUrl: "https://localhost:9443/oauth2/token",
        username: "delivery",
        password: "delivery",
        clientId: "nobQxoLIp3BfoG4Anncf82szJnga",
        clientSecret: "wSSdSy_eyfXIISCfJSLeyVz5VlAa",
        scopes: ["add_product"],
        clientConfig: {
            secureSocket: {
                cert: {
                    path: "resources
                          /ballerinaTruststore.p12",
                    password: "ballerina"
                }
            }
        }
    }
});
public function main() {

    http:Response response = check ep->get("/hello");
    io:println(response.statusCode.toString());
}
```

This Ballerina client application sends a request to the localhost port 9090 with the OAuth2 credentials. The OAuth2 token is generated by calling the WSO2 IS OAuth2 token generation endpoint. You need to provide the username and password of the client, along with clientId, clientSecret, and scope. Since we set up the delivery user to have the delivery role that has add_product scope, this request will succeed. If you try this client application with another user who is not associated with the scope, this will give you an unauthorized error response.

You can select different types of token-based authentication and authorization methods to provide access control in your system. The method should be selected based on the requirement of the application. Using LDAP to hold user and role information, as well as using OAuth2-based authorization methods, provides a good combination for building cloud applications rapidly. Optionally, you can build the system from scratch using relational databases to store users and roles. Ballerina gives you support to create a user store with relational databases and build everything from scratch. The decision of selecting the most appropriate access control system should be decided based on security requirements and cost.

In this section, we have focused on building Ballerina applications with the OAuth2 protocol. The OAuth2 protocol provides multiple ways to authorize users. We also focused on building a Ballerina application that has an OAuth2 password grant type. Token-based authorization methods provide strong security features. You can combine security options provided by the Ballerina language to provide strong security.

Summary

In this chapter, we have focused on various security aspects of developing a cloud native application. In the first section, we covered certificate management in a Ballerina application. We discussed securing Ballerina services with HTTPS and mutual SSL. When you are building a cloud application, make sure to validate endpoints with a proper certificate management system. Certificate validation prevents man-in-the-middle attacks and secures cloud applications.

Next, we discussed the different types of authentication and authorization methods for a Ballerina application. We learned what LDAP is and how to set up an LDAP server using Apache DS. The sample program demonstrated how to implement a simple scenario by means of which different users with different access levels access a Ballerina resource function. For this example, we manually added the user information to the LDAP server, but you can have a third-party application to manage the LDAP server.

In the last section, we focused on token-based authorization methods to build a highly scalable cloud application. First, we discussed JWT-based user authorization. Here, we first learned how to generate and validate JWT with the Ballerina language. We used WSO2 IS as the JWT generation endpoint and used custom claims to give users a different access level for applications. Then we extended the token-based authorization methods to include the OAuth2 method, which has a variety of solutions in building cloud native applications.

The security of a cloud native application is a wide topic with many aspects that you have to consider. This chapter only covered some basic and critical aspects of building cloud native applications. The topics covered explain how to communicate securely with two endpoints in a Ballerina application, authenticate and authorize users with LDAP, and token-based methods such as JWT and OAuth 2. Aside from the topics discussed in this chapter, you need to focus on different types of vulnerabilities, such as injection attacks, man-in-the-middle attacks, and denial-of-service attacks. Always follow security best practices to build a cloud application and use Ballerina security features such as data encryption, certificates, and access controlling to secure your application.

In the next chapter, we will learn how to monitor Ballerina language-based services with different freely available tools. There, we will discuss observability methods, including logging, tracing, and metrics. These methods allow you to monitor, analyze, and maintain your system.

Questions

1. Why are dynamic secrets important?

Further reading

- You can refer to *Learn Kubernetes Security*, by Kaizhe Huang, available at `https://www.packtpub.com/product/learn-kubernetes-security/9781839216503`

Answers

1. Dynamic secrets are important since secrets in the system can be exposed to a third party. If secrets are permanent values, then the possibility of exposing and using those credentials by a hacker is much higher. Secret values can be accidentally exposed outside with system logs and source codes. Also, a hacker might try to steal secrets from a user or by logging in to the system. If secrets are dynamic and change after a period of time, then they are valid for a particular time range. You can use Kubernetes secrets to manage secret values on a Kubernetes cluster. Third-party applications such as HashiCorp Vault can be easily integrated into the Kubernetes platform, where you can dynamically change secret values.

8
Monitoring Cloud Native Applications

There are no programs without bugs and issues, and there is no code that is perfect. Failures can occur at any time, and maintainers should be able to identify and fix them as quickly as possible. Unlike monolithic applications, monitoring cloud native applications is challenging in a number of ways. In this chapter, we will focus on building applications that can be monitored to identify possible failures and debugging the flow in the case of failure.

In this chapter, we will discuss the following topics:

- Understanding the importance of observability and monitoring cloud applications
- Using logs to analyze system behaviors
- Using traces to identify the request flow over a distributed application
- Using metrics to collect important values to monitor the system

By the end of this chapter, you should understand the importance of observing and monitoring a **cloud native system**. The sample tools described in this chapter will help you to build an observable system that lets **DevOps** engineers maintain the system easily.

Technical requirements

This chapter contains **Ballerina** examples along with different types of observability tools such as **Prometheus, Grafana, Logstash, Jaeger,** and **Elasticsearch**. All of this aforementioned software is open source and can be downloaded onto your local computer or the target platform. Make sure that you select the correct **operating system (OS)** and install any prerequisites if needed.

You can find the code files for this chapter at `https://github.com/ PacktPublishing/Cloud-Native-Applications-with-Ballerina/tree/ master/Chapter08`.

The Code in Action video for the chapter can be found here: `https://bit.ly/2VbdVJU`

Applications that we have mentioned here can also be used as a **Docker container**. In *containerized* platforms such as **Kubernetes**, you can use Docker containers instead of using executables. Each section in this chapter, contains **Docker pull commands** for all of the Docker images that we have used.

Introduction to observability and monitoring

Programs can fail at any time. These failures might be due to issues with the machines that we used to build the infrastructure, or they may be due to human error. Machines are reliable to some extent, but there can be hardware failures while running the program. On the other hand, programmers may also create code-faulty programs or unresponsive programs due to human error. In either case, the system should be able to recover from failure and continue to operate. To achieve this, the system should be observable to find issues. In this section, we will discuss the techniques that we can use to observe and monitor a distributed system.

Observability versus monitoring

Observability and **monitoring** are common terms that you will have heard in programming. The concept of observability and monitoring is not only applicable to cloud native applications but also an important part of developing any software system. A simple statement to understand observability and monitoring is that if it is observable, then it can be monitored. Observable systems can collect data that can be further used for monitoring and analyzing. Observable systems reveal the data to the monitoring system and the monitoring system analyzes the data.

To further understand a system with observability, we can analyze the system with three different methods known as the **three pillars of observability**. The following are the three methods that can be used to collect data on a distributed system:

- **Logs**
- **Traces**
- **Metrics**

Now, in the following sections, we will discuss each of these pillars and what each of these methods is able to do in a distributed system.

Logs

Logs are the easiest way of generating useful insights about program execution. Logs can also be treated as an immutable sequence of events. Each log contains a timestamp associated with it that can be used to identify when the program flow happened. Logs are generally plain text strings. But they can represent structured data with formats such as **JSON** and **XML**. Depending on the severity of the current status of the running program, the log level can also be changed.

For example, `Error` is a log level that represents an issue in the system, while debug logs are informative logs to track the program flow. If you are monitoring a system, error logs are an indication of a serious issue in the system. If you need to drill down to the issue, you might also need debug logs as they contain more details about how the program flowed into the error state.

Traces

Traces are specially developed to analyze distributed systems. A microservices application can have multiple services that work together to perform some task. A request that comes into a microservices system can be passed along various paths. Traces are used to track how the request travels through various services. By looking at the traces, we can find out how the request flows in the system and what the failures and performance bottlenecks are on each request flow.

OpenTracing and OpenTelemetry are standard specifications that are used to collect traces. We will discuss tracing further in upcoming sections. Tracing is a solution that lets developers debug the system in a distributed system. The following are situations where you can use traces to track issues:

- **Distributed transaction monitoring**: You can monitor how each transaction travels through a distributed system and analyze it in case of transaction failures.

- **Performance and latency optimization**: You can analyze what the components are that request travel in a microservices application and what kinds of bottlenecks there are in the request flow. Based on the traces collected on each service, we can find the performance bottlenecks and improve those components.

- **Root cause analysis**: You can find the issue on a system in the case of system failure. It is hard to debug an issue in a microservices application by just using logs. Traces give an elegant way of debugging and finding the root cause in microservices applications.

- **Service dependency analysis**: You can visualize how the overall system interacts with each service together by using traces. Based on how the request flows between services, we can infer the service dependencies.

In the next section, we'll learn about the metrics method.

Metrics

The third pillar of observability is **metrics**. Metrics are values that can be used to represent the various characteristics of the system. Metric values can be aggregated and visualized to analyze the system. Unlike the aforementioned methods, metrics consume a constant number of resources in the host machine to collect metrics. Therefore, metrics do not affect the system performance as much as other approaches. Metrics are mainly important in analyzing the following aspects:

- **Latency**: This is the amount of time that was taken to reply.

- **Saturation**: This property can represent how system resources such as CPU, memory, and network resources are allocated for the services.

- **Traffic**: Traffic measures the number of requests that are coming into the system.

- **Error**: Errors represent the number of errors in the system.

In this section, we learned about observability and monitoring. We also learned about the three methods used to collect data in a distributed system. In the upcoming sections, we will discuss how we can practically implement each of these methods by using the Ballerina programming language. In the next section, we will discuss the simplest way of observing software, which is logging.

Ballerina logging

Logging is the simplest way to collect data from a running Ballerina instance. Ballerina's built-in log module provides different functionalities to work with logs. In this section, we will discuss the features provided by Ballerina and the tools that can be used to collect and analyze logs.

Printing logs with Ballerina

Ballerina's log module provides you with a simple way of printing logs to the standard output of the terminal. Ballerina offers four log levels to log messages:

- DEBUG log: These logs are for debugging the flow of the code. This log level is widely used when issues occur and you need to debug the program flow.

- INFO log: This is used to log information-level logs. These logs are used to log normal behavior of the program's execution flow.

- WARN log: These are possible failures that might cause the program to go in an erroneous flow. Warning logs can be used to give a warning about the system before there is a failure.

- ERROR log: This, on the other hand, is used to log errors in the program flow. Error logs indicate a serious problem in Ballerina's program execution flow.

All of these log functionalities are provided by the ballerina/log module. The following code is an example of using Ballerina's log module for printing log messages:

```
log:printDebug("debug log");
log:printInfo("info log");
log:printWarn("warn log");
log:printError("error log");
```

Ballerina also supports filtering logs at four different levels. Those are also DEBUG, INFO, WARN, and ERROR. If you filter a log with the DEBUG state, the Ballerina program logs all the log messages. If the log filter level is INFO, then it logs INFO, WARNING, and ERROR logs. We can represent this log level and log filter level in the following table:

	DEBUG	INFO	WARN	ERROR
printDebug()	Yes	No	No	No
printInfo()	Yes	Yes	No	No
printWarn()	Yes	Yes	Yes	No
printError()	Yes	Yes	Yes	Yes

Table 8.1 – Log level hierarchy

The first column represents the Ballerina log level and the first row represents the log filter level. As you can see in the table, by increasing the log filter level, we can log only the critical logs, and by decreasing the log filter level, we can have more context from the logs. We can set the log filter level with the Config.toml file and by setting the [ballerina.log] value. By default, Ballerina has the INFO log level. If you need to change it to another log level, you can specify it in Config.toml as follows:

```
[ballerina.log]
level = "<log_level>"
```

By changing log_level, you can fine-tune the log level required to your production, deployment, and debug sessions.

Next, we will explore the structure and content of the log messages. Ballerina's log function generates log message in the following format:

```
time = <time_stamp> level = <log_level>  module =
[<organization name>/<module_name>] message = <log_message>
```

A sample output of the Ballerina program is as follows:

```
time = 2021-07-07T14:02:29.138+05:30 level = DEBUG module =
user/logging_basic message = "debug log"
time = 2021-07-07T14:02:29.148+05:30 level = ERROR module =
user/logging_basic message = "error log"
```

This log format is known as the logfmt format, which is widely used in collecting logs. This format is human-readable such that the debugger can easily go through the logs and debug the program flow. Optionally, you can use JSON format to log in to a Ballerina program. You can set format = "json" under [ballerina.log] in the Config. toml file configuration to log messages in JSON format. Even though this format is not very readable, you can easily use it with an analytics tool to analyze logs. If you are using the default logfmt format, then you can use a log message parser to read log content. In the next section, we will discuss Logstash and how we can use it to parse logfmt-formatted log messages.

Using Logstash to collect logs

Logstash is a popular log collector tool that can be used to collect logs produced by the Ballerina application. You can download Logstash onto your computer from the Logstash website (https://www.elastic.co/downloads/logstash), or you can use a Docker container on a containerized production system. If you download Logstash executables, you need to unzip the files first. Logstash mainly does the following things to collect and publish logs:

- **Input**: It reads logs from different sources such as the file system, stdin Kafka events, and so on.

- **Filter**: It applies filters on collected logs before sending them to the backend.

- **Output**: It publishes logs to backend services such as Elasticsearch, the file system, MongoDB, and message brokers such as Kafka and RabbitMQ.

The following diagram represents the different types of input and output methods provided by Logstash:

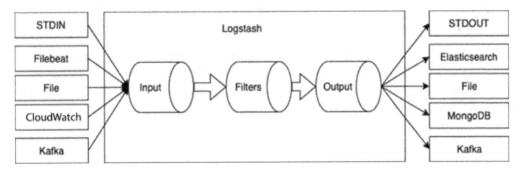

Figure 8.1 – Logstash components and plugins

Logstash is capable of collecting data from different sources. Logstash provides multiple plugins for data collection. It can also publish collected log messages for persistence or further processing. Logstash output plugging also supports multiple protocols for sending output data. To learn how Logstash input, filter, and output elements work, we will first discuss the simplest way of reading and writing logs, which is using the `stdin` and `stdout` plugins.

When you are starting up Logstash, you need to provide input, filter, and output definitions in JSON-like format. The Logstash configuration language is very similar to the JSON format. You can provide configuration as a file or as command-line arguments.

For example, you can try out Logstash by starting it with the `./logstash -e 'input { stdin { } } output { stdout {} }'` command. The `-e` flag signifies that we provide Logstash configurations from the command line. Otherwise, you can set the `-f` flag to read configuration from a file.

Here, we set the input as the `stdin` terminal and the output as `stdout`. If you try to write something into the terminal and press the *Enter* key, the phrase you just entered will be displayed in JSON format as follows:

```
{
    "message" => "hello",
    "@timestamp" => 2020-11-20T14:48:07.025Z,
    "host" => "user.local",
    "@version" => "1"
}
```

The preceding message contains the log that has just been collected, the timestamp, the host to identify the node, and the version of the message. The Elasticsearch backend can be used to publish these messages so that we can use this data to analyze collected logs using the **Kibana** dashboard interface.

In Ballerina, you may realize that there are a few parts that make up the log message. It includes the timestamp when a message is created, the log level, the module name, and the log message. These are plain text messages. But if we can convert this message format into a separate representation, we can analyze the log details further. To do this, we can use filters that split this log message and publish it as multiple fields.

We used the **Logstash Grok filter** to parse Ballerina log messages. The Grok filters are used to parse unstructured log messages to structured messages such that you can analyze the log messages in much more detail. To add this filter, you need to add filter configurations to the Logstash configuration file. Here, we create the following file, which can be passed as a -f flag to load configuration from a file:

```
input {
  stdin { }
}
filter {
grok{
    match => {
    "message" =>
      "time%{SPACE}=%{SPACE}%{TIMESTAMP_ISO8601:date}
        %{SPACE}level%{SPACE}=%{SPACE}%{WORD:logLevel}
          %{SPACE}module%{SPACE}=%{SPACE}%
            {GREEDYDATA:package}%{SPACE}message%
              {SPACE}=%{SPACE}\"%{GREEDYDATA:logMessage}\""
    }
  }
}
output {
  stdout {}
}
```

Inside the `filter` section, we added a Grok filter that matches keywords for the given format. When the matching format has been found, it restructures the response according to the values found. Here, we set templates with the `%{PATTERN:FIELD_NAME}` format.

If a matching pattern is found, then it creates a new attribute with the name of the field and updates the response message. TIMESTAMP is used to match the timestamp on the log message and WORD is a single word separated with a space or character. A space is used to show single or multiple spaces, and GREEDYDATA is a wildcard that can represent **regex** as .*. Check the following log line that has been inserted into the Logstash terminal:

```
time = 2021-07-07T14:31:42.982+05:30 level = INFO module =
user/logging_basic message = "This is sample info log"
```

This log message generates the following Logstash message, which contains parsed information of the log message:

```
{
    "package" => "user/logging_basic ",
    "message" => "'time = 2021-07-07T14:31:42.982+05:30
        level = INFO module = user/logging_basic message =
            \"This is sample info log\"'",
    "@version" => "1",
    "@timestamp" => 2021-07-07T18:22:11.443Z,
    "logLevel" => "INFO",
    "host" => "user.local",
    "date" => "2021-07-07T14:31:42.982+05:30",
    "logMessage" => "This is sample info log"
}
```

If the given log message does not fit the Grok expression that we provided, it adds another tag to the message indicating that there is a Grok parsing failure. If you need to add more Grok parsers if there are multiple patterns of publishing logs, you can add more Grok expressions as an array to the `message` field. However, Ballerina publishes log messages in the previously mentioned way; you don't need to add more Grok filters unless you use your own message formatting.

You can also use an `if` condition to check log message values and change the resulting message. For example, the following configuration block after the Grok block adds another attribute to the message if the log level is `INFO`:

```
if [logLevel] == "INFO" {
    mutate { add_field => {"tagType" => "info log"} }
}
```

So far, we have been discussing filtering and mutating the output message. For the preceding example, we used `stdin` as input and `stdout` as the output of the messages.

To connect Logstash with the Ballerina application, we need to configure Logstash to collect logs from a file. There are multiple plugins that Logstash supports to collect logs. Using **file plugging** is the simplest way of collecting logs. You can start Ballerina with the relevant log level and point the log result into a file. The Ballerina console logs can be directed into a file with the following command:

```
bal run logging_basic/  2> /var/log/logging_basic.log
```

The input configurations of Logstash can be changed as follows to collect Logstash data from a file:

```
input {
  file {
    path => "/var/log/logging_basic.log "
  }
}
```

We can use a `file` collector to collect data from the filesystem if Logstash is also running in the same instance. In a container environment, to collect data from Ballerina log files, we need to run a Logstash server parallel to the Ballerina service. But Logstash is a heavyweight process that requires lots of processing power and memory. In a container environment, we can use a lightweight log collector tool, **Filebeat**, that collects logs from the host node and publishes them to Logstash.

Using Logstash with Filebeat in container environments

We can use Docker images in the container environment to implement log collection scenarios. Here, we use Filebeat to collect logs from Ballerina and publish them to Logstash. Filebeat should be running on the same host that the Ballerina application is running. We can mount the Ballerina log files to the Filebeat container and send updates whenever the file is updated.

You can use Filebeat directly to send logs to Elasticsearch as well if no log filtering is performed. But in this example, we will publish collected logs into Logstash and perform filtering to parse valuable message content.

The following command pulls Filebeat from the Elastic `docker` registry. This image can be used to create a Docker container that collects logs and publishes them to Logstash. You can find the latest Filebeat Docker image-pulling instructions on the `https://www.elastic.co/guide/en/beats/filebeat/current/running-on-docker.html` page:

```
docker pull docker.elastic.co/beats/filebeat:7.13.3
```

Then, we need to create a `filebeat.yml` file to instruct Filebeat to read logs from the `/usr/share/filebeat/logging_basic.log` file and publish them to Logstash:

```
filebeat.inputs:
- type: log
  paths:
    - /usr/share/filebeat/logging_basic.log
output.logstash:
  hosts: ["hostname:5044"]
```

The path we mentioned here needs to be mounted on the Docker container when it starts up. The following Docker `run` command creates a new Docker instance by mounting both the file we just created and the log files:

```
docker run --volume="<filebeat.yml_location>:/usr/share/
filebeat/filebeat.yml" --volume="<log_file_location>:/usr/
share/filebeat/logging_basic.log" docker.elastic.co/beats/
filebeat:7.13.3
```

In this command, two files are mounted from the host to the container. The first file is the configuration file used to configure Filebeat, and the second file is the `log` file that we need to collect. You need to update Logstash to collect logs from the TCP endpoint. The following configuration lets Logstash collect logs from the TCP endpoint:

```
input {
beats {
  port => 5044
  }
}
```

After running the Docker container, it continues to collects logs from the host computer and publishes logs to Logstash. Logstash parses incoming logs and prints onto the terminal console the output. Now, we can redirect these parsed logs into an analytics endpoint to analyze the logs or storage to store logs. In the next section, we will discuss how to use Elasticsearch to collect and store logs.

Using Elasticsearch to collect logs

Elasticsearch is a highly scalable and full-text search analytics engine. We can use Elasticsearch to store, search, and analyze a large amount of data. Elasticsearch uses the RESTful API in JSON format to interact with the Elasticsearch server.

Due to the strong capabilities of log aggregation and queries, Elasticsearch is the most popular log analytics and storage tool that is widely used in application monitoring. **Kibana** is the most popular interface that can be used with Elasticsearch to analyze and monitor logs that are published by the Logstash or Filebeat clients.

Elasticsearch also comes with an executable file that can be downloaded from its website (`www.elastic.co`). You can also use a Docker container of Elasticsearch to deploy Elasticsearch in a container environment. You can pull and run the latest Elasticsearch container by executing the following commands as given in the Elasticsearch Docker image documentation in `https://www.elastic.co/guide/en/ elasticsearch/reference/current/docker.html`:

```
docker pull docker.elastic.co/elasticsearch/
elasticsearch:7.13.3
docker run -p 9200:9200 -p 9300:9300 -e "discovery.type=single-
node" docker.elastic.co/elasticsearch/elasticsearch:7.13.3
```

Logstash configurations can be updated as follows to publish Logstash collected log messages to the Elasticsearch server. Here, we set the hostname to the Elasticsearch host and the port number to `9200`. If you are testing this setup locally, then you can set the hostname as `localhost`:

```
output {
  elasticsearch{
    hosts => "elasticsearch:9200"
    index => "order_management_system"
    document_type => "store_logs"
  }
}
```

When Filebeat reads log changes, it publishes log lines to Logstash. Logstash filters logs, generates messages, and publishes messages to Elasticsearch. Now we can use these log messages to analyze collected logs with the Kibana analytics tool.

Using Kibana to analyze Ballerina logs

Kibana provides an interface with which you can easily query data from Elasticsearch and visualize it. You can build and share dashboards that are useful for understanding how the system operates. Kibana visualization includes **Kibana Lens**, which is a drag-and-drop visualization tool, a time series visual builder that can represent time series data, and a geospatial analyzer tool that can analyze network failures and delay across different geolocations.

We can represent the whole logging observability process with the following diagram:

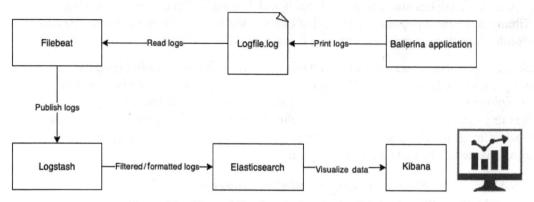

Figure 8.2 – Observability flow for logs in the ELK Stack

Here, we have started by logging in to a file on the file system and reading it with Filebeat. Next, we filtered/formatted logs with Logstash and published them to Elasticsearch. Finally, we can use Kibana to visualize data.

You can set up Kibana by downloading the Kibana distribution or by using Docker. The following is the command that you can use to pull Kibana from the Elastic Docker registry:

```
docker pull docker.elastic.co/kibana/kibana:7.13.3
```

You can find the Kibana Docker documentation on the https://www.elastic.co/guide/en/kibana/current/docker.html website. The Docker image can be executed as follows by exposing the default Kibana web interface port. You can access the Kibana interface through port 5601:

```
docker run --name kib01-test --net elastic -p 5601:5601 -e
"<elasticsearch_host>=http://es01-test:9200" docker.elastic.co/
kibana/kibana:7.13.3
```

You can change the default user interface ports and Elasticsearch port by changing the kibana.yml file in the Kibana config directory. You can mount these files into /usr/share/kibana/config/kibana.yml, the same as we did in the Filebeat configurations.

Once you set up Kibana, you can start creating an index for the logs that we have just created:

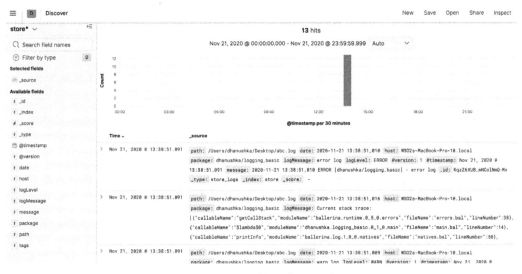

Figure 8.3 – List of logs for an indexer

As you can see, you can create a dashboard by manipulating data on the different axes. Kibana supports multiple types of graphs from simple line charts to much more complex multi-dimensional graphs, as shown here:

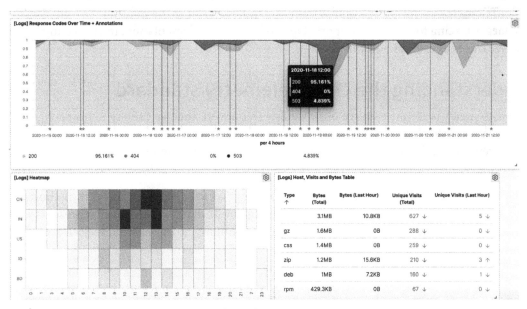

Figure 8.4 – Visualizing logs in the Kibana dashboard

You can expand the ability to find issues and monitor the system by adding meaningful logs with proper log levels. Visualizing errors and warning logs is very important in finding potential issues. You can check what the modules that are frequently failing are, time relations for failures, and so on. There is also some advancement in log analyzing methodologies to use machine learning techniques to identify errors in advance. You can easily integrate these services with Elasticsearch and create a much more reliable system.

So far, we have discussed collecting logs over a distributed system and visualizing them on a dashboard. But debugging an error in a distributed system is hard, as each request that comes into the system can go through multiple services.

In this section, we learned about the simplest method of observability, that is, logging. We also discussed logging messages with Ballerina's log functions, collecting logs with Logstash and Filebeat, storing logs with Elasticsearch, and visualizing logs with Kibana.

In the next section, we will discuss how to use traces on distributed system and analyze how requests go through different services.

Tracing with Ballerina

Traces are one of the important pillars of observability. This approach is important, especially in a distributed cloud system, where the requests go through different servers. Tracking a single message is complex just by using logs in a distributed system. **OpenTelemetry** is one of the popular tracing standards used in the industry. We will use the **Jaeger** tracing tool, which supports OpenTelemetry implementation, to collect and analyze traces as examples in this section.

Understanding the OpenTelemetry standard

OpenTelemetry is an open source observability framework that we can use to observe cloud native systems with traces. Traces can be used to track how a request flows through the system over different services. In this section, let's understand multiple terms used in tracing.

Span is the logical unit of work. A span has an operation life cycle with the starting time and ending time. A trace is an acyclic graph of spans with references. A trace represents the complete life cycle of multiple spans. A request coming into a Ballerina service can be considered a trace.

The initial span can be considered as the **root span**. The root span can be separated into multiple spans and travel through the code. If a service calls another service, the HTTP request wraps the span with a new **child** span and sends it. Each span that was generated has a timestamp that can be used to reconstruct the message flow. This is illustrated in the following diagram:

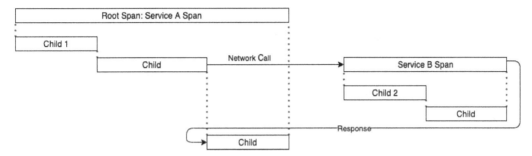

Figure 8.5 – Trace flow over the network

A service can generate multiple child spans while it is executing a set of instructions. These child spans can also generate child spans for each execution segment. Network calls wrap span data and send it to other services. Traces have inject functionality to add a trace ID to track the traces and extract functionality to extract trace data from an HTTP request. Usually, HTTP headers are used to send trace data over two services. In the next section, we will learn how to use Jaeger to implement a simple tracing scenario that contains multiple services.

Using Jaeger as a tracing platform

In this example, we use Jaeger as the tracing platform to collect and visualize traces over a distributed system. Jaeger is an open source distributed tracing platform developed by **Uber Technologies**. Jaeger is built on the OpenTracing and OpenTelemetry standards. The Jaeger backend was built with the **Go** language and you can use **Cassandra**, **Elasticsearch**, and **Kafka** as the storage backends.

Jaeger mainly provides support to monitor distributed transactions. You can easily visualize how requests are going through different services and how they are interacting with each other. You can use these analyses to optimize performance and perform root cause analysis in case of system failure.

We have already mentioned that Jaeger supports OpenTelemetry. In the latest release of Jaeger, they are moving on with OpenTelemetry rather than OpenTracing. Ballerina supports OpenTelemetry and for this example, we will use OpenTelemetry to collect traces. For that you need to get Jaeger from Docker Hub with the following Docker command.

Jaeger Docker images come in different flavors for different requirements. You can pull an all-in-one Docker image that contains all Jaeger services. Use the following command to pull Jaeger from Docker Hub:

```
docker pull jaegertracing/opentelemetry-all-in-one
```

Jaeger has the following components to collect, store, and preview data:

- **Jaeger client**: The Jaeger client library is a language-specific library that can be used to publish traces to the server. The Ballerina language has its own built-in libraries to publish traces to the Jaeger server. It also has built-in trace implementations of libraries such as network calls and databases. You can use these default publishing traces to track the program flow. If you need to add more traces, you can use the Ballerina tracing library to publish customer traces.

- **Jaeger agent**: This agent collects traces over **User Datagram Protocol** (UDP) by sitting on the same server. Agents collect traces from a given instance and send them as a batch to the Jaeger collector. In Kubernetes environments, the agent can be placed as a sidecar proxy.

- **Jaeger collector**: This collector collects traces from all the services and sends them to storage. The collector keeps traces in a queue and sends them to storage to speed up the trace collection process.

- **Jaeger storage**: Storage persists data on the storage backend. Jaeger supports **Cassandra**, **Badger**, and **Elasticsearch** as a storage backend. It also has in-memory storage to store data. But in production systems, you need to pick up a production-grade storage service since in-memory storage cannot handle a large number of traces.

- **Query**: The query service retrieves data from the storage.

- **Jaeger console**: This contains the user interface to view and analyze traces. In this interface, you can search and visualize the flow of traces in the system.

You can start running a pulled Jaeger Docker image with the following Docker command. This will start an all-in-one container that contains all the Jaeger components:

```
docker run -d -p 13133:13133 -p 16686:16686 -p 55680:55680
jaegertracing/opentelemetry-all-in-one
```

You can access the interface using `http://localhost:16686/` on a web browser. In the next section, let's discuss how to use Jaeger to monitor microservice applications.

Monitoring the Ballerina application with Jaeger

We will create the following sample application to demonstrate how Ballerina publishes traces and visualizes with Jaeger. For this example, we will continue to build the order management system with a series of HTTP calls to communicate with other services. Here, we are going to build the payment process scenario with the Ballerina language.

In this scenario, the client sends a request to the order service to continue the order to payment. The order service sends requests to the inventory service to calculate the total amount of the order. Once the amount to be paid has been determined, the order service sends the order to the payment service to continue with the payment.

The payment service performs the payment, sends an email notification to the customer, and sends the results back to the order service. Then, the order service sends requests to the inventory service to remove pending items. Finally, the order service sends another request to the delivery service to proceed with the delivery. This scenario can be represented with the following sequence diagram:

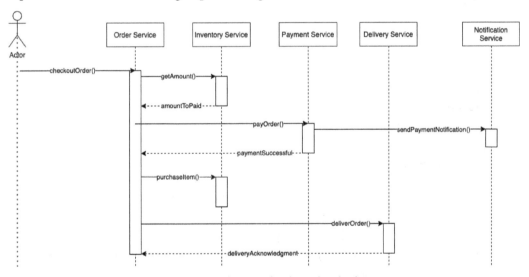

Figure 8.6 – Sequence diagram for the order checkout process

The order management code for this scenario can be implemented as follows:

```
OrderItemTable orderItems = check
  getOrderItemTableByOrderId(orderId);
if orderItems.length() == 0 {
    check sendError(caller, "No order items found");
} else {
    check getAmount(caller, orderItems, orderId);
}
```

This source code first gets the list of order items for a given order ID. Then, it sends the order item list to the `getAmount` function, which calls the inventory service to calculate the total amount that needs to be paid for the order:

```
function getAmount (http:Caller caller, OrderItemTable
    orderItems, string orderId) returns error?{
    io:println("Get amount");
    json payload = check inventoryEndpoint->post(string
        '/InventoryService/getAmount', check
        orderItems.toArray().cloneWithType(json));
    check payOrder(caller, orderItems, orderId, <float>
        check payload.totalPrice);
}
```

Once the total amount is calculated, it proceeds to the payment. The payment service performs the payment, and it sends the payment status back to the order service:

```
resource function get payOrder/[string orderId](http:Caller
    caller, http:Request req) returns error? {
    OrderItemTable orderItems = check
        getOrderItemTableByOrderId(orderId);
    if orderItems.length() == 0 {
        check sendError(caller, "No order items found");
    } else {
        check getAmount(caller, orderItems, orderId);
    }
}
```

If the payment process succeeds, the inventory service calls to remove pending orders for the given order ID. It also calls the delivery service to add orders to the items to be delivered list.

When you are starting any Ballerina server, you need to provide the `--observability-included` option to inform Ballerina to publish tracing to the Jaeger agent. Once you start up all of the servers, send a request to the order service to proceed with the payment. This generates traces for each data flow and publishes them to the Jaeger backend.

Now, we can access the Jaeger user interface through the default port `16686` and visualize all the traces collected for the request flow. In the Jaeger interface **Search** tab, you can find the **Service** dropdown, which includes all services that are interacting with the system. You can select **OrderService** and click the **Find Traces** button at the bottom of the screen.

Then, it displays all traces that are relevant to the order service, as shown here:

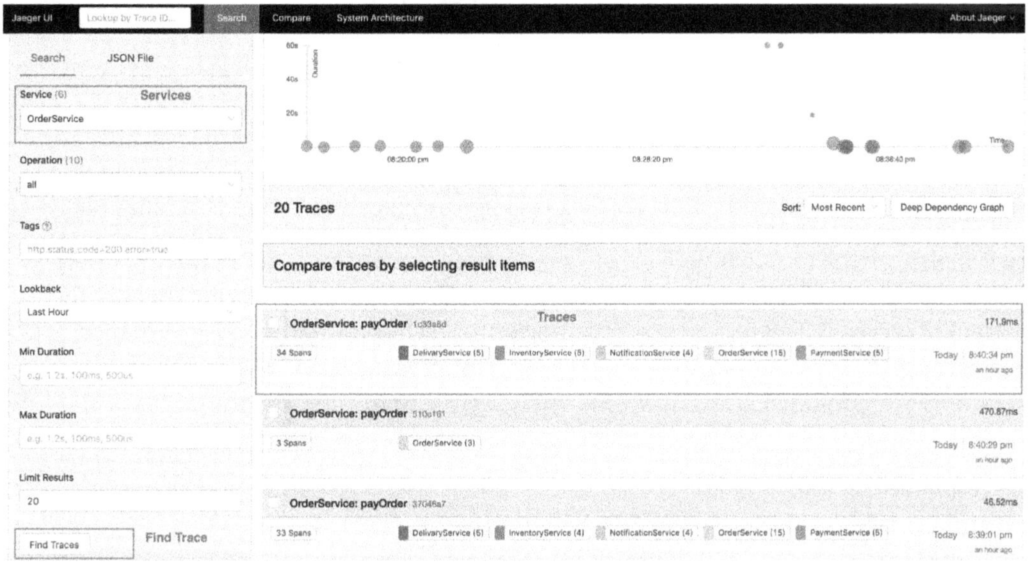

Figure 8.7 – Jaeger view of all traces

You can search for traces with different properties in this view. In each of the traces, you can find the number of spans generated in the request flow, how the request goes through each Ballerina service, the duration of the span, and the time taken for the generation of the request. Check the following example screenshot, which represents spans on the Jaeger interface:

Figure 8.8 – Span timeline view

Next, you can select the `payOrder` trace containing the span for the payment program flow to further analyze how the span is generated through the request flow.

In the preceding screenshot, you can find a set of spans that are generated and listed on the left panel. Each span's timeline is shown on the right side of the interface. In this view, you can check each of the spans and how long it took to execute each span. This information is important when the span is getting more time to complete its execution. We can easily identify that there might be a performance issue with that program flow.

You can also see the logs that we have put on the program by the dark vertical lines over the span timeline. You can also read these logs from the Jaeger interface and identify the program flow. Furthermore, you can drill down into each span and check information about each span. For example, the following screenshot is the span representation for an HTTP `POST` request that is sent from the order service to the inventory service:

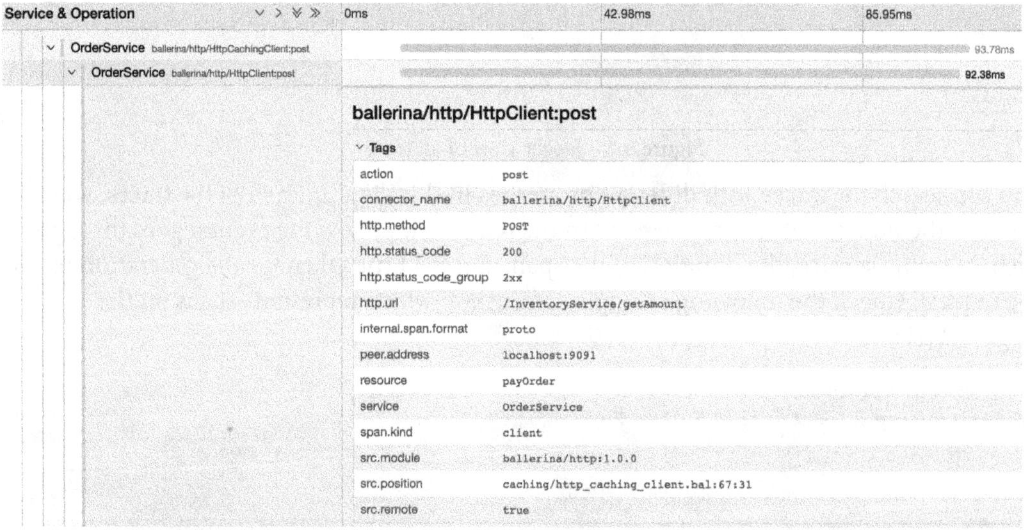

Figure 8.9 – Content of the span

In the preceding screenshot, you can find a list of tags that are attached to this particular span. It contains request details such as the HTTP method, URL, and hostname. These request details can be conveniently used to find what might go wrong in the case of failure. For example, the order service sends requests to the wrong service or wrong URL, so that information can be tracked here.

Jaeger also provides a graphic view to visualize how each request travels through each service. You can click on the **Trace Graph** option in the dropdown in the top-right corner of the interface, and you will get a trace graph as shown in the following screenshot:

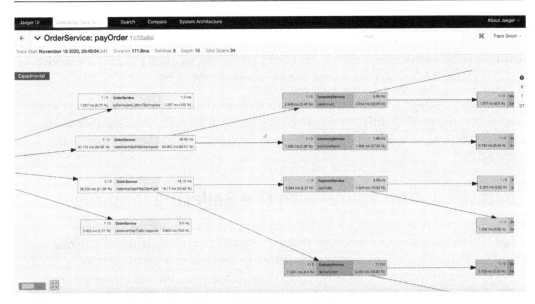

Figure 8.10 – Trace graph view

This view shows how different spans communicate with each other. Different services are visualized in different colors. This view shows how a span is distributed over the deployment among each service and how each span is connected.

A more generalized view of services can be accessed with the **System Architecture** tab from the top menu bar, as shown here:

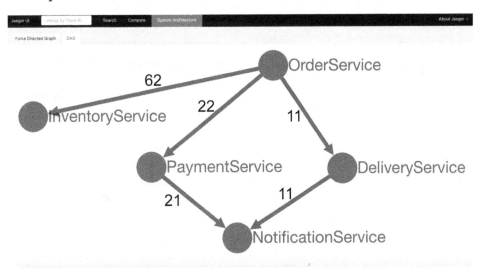

Figure 8.11 – System architecture diagram in Jaeger

This diagram visualizes how each service is connected and the direction of the request flow. Here, you can clearly see that the order service has called the inventory service, payment service, and delivery service. Also, the payment service and the delivery service have both called the notification service to send emails to the client.

So far, we have discussed viewing traces that are published by the Ballerina language itself. Ballerina libraries publish different types of informative properties with OpenTelemetry. In the next section, we will learn how to use a custom span and add tags to analyze the program flow.

Creating custom spans with the Ballerina language

Ballerina lets you create a new span with the `observe` library in the `ballerina` module. This library contains functions that are required to access spans and generate new spans. You can use the `startSpan` function to generate new spans and the `addTagToSpan` function to add new tags to the selected span. We will implement these functions by modifying the order management system that we just discussed.

For this example, we will modify the price calculation process in the order service to publish traces. In this service, it reads a list of order items that were sent from the order service and calculates the sum of all order items. We keep the `totalPrice` variable to keep the total price of calculated order items. Also, we can start creating a new span with the `startSpan` function with the name `GetAmountSpan`:

```
json[] orderJson = <json[]>orderDetails;
float totalPrice = 0;
error? result;
int getAmountSpan = check observe:startSpan("GetAmountSpan");
```

You can use the `getAmountSpan` reference to access this particular span. In the next step, we can loop through all of the order items that are sent by the order service and calculate the total price of the order:

```
foreach json orderItems in orderJson {
    int orderItemSpan = check observe:startSpan
      ("OrderItemSpan", (), getAmountSpan); // Line 2
    string inventoryItemId = <string> check
      orderItems.inventoryItemId;
    string orderItemId = <string> check
```

```
        orderItems.orderItemId;
    float quantity = <float> check orderItems.quantity;
    float fullPrice = check getTotalAmountForItem
        (inventoryItemId) * quantity;
        totalPrice += fullPrice;
    log:printInfo("Item id " + inventoryItemId +
        "fullPrice " + fullPrice.toString());
    check observe:addTagToSpan("Quantity of "+orderItemId,
        quantity.toString(), orderItemSpan); // Line 9
    result = observe:finishSpan(orderItemSpan); // Line 10
    if (result is error) { // Line 11
        log:printError("Error in finishing span", result);
    } // Line 13
}
```

In the loop, first we create another span that is the child span of GetAmountSpan. We can use the startSpan function with the root span reference to create a child span. All the data that is generated inside the loop is kept in this span. On *line 10*, you can see that this span is marked as finished. The finishSpan function finishes the execution timeline of this span. With each iteration of this loop, it generates a new span.

In *lines 2* to *5*, a new variable was created to access inventoryItemId, orderItemId, and quantity. To calculate the price, we need to get the value of each product and multiply it by the quantity. To get the price of each item, the getTotalAmountForItem function gets called, which returns a price for the given inventoryItemId. This function gets the price for the given item from the Products table on the database. By multiplying the price by the quantity, we can find the price for each order item. By summing up all of these prices, we can find the total price of the order.

On *line 9*, we added a tag to the current span. The addTagToSpan function takes three arguments to set the tag to the given span. The first argument is the key, the second argument is the value, and the third argument is the span reference that the tag needs to add. Here, we set orderItemId as the key, the price details as the value, and orderItemSpan as the span.

Once the tag details have been added to the span, we can finish the span. Ballerina provides the finishSpan function to mark a span as finished. After you finish the span, you cannot add details to that span again. From *lines 11* to *13*, there is an error check to log whether the finishSpan functionality generates an error.

Next, we will also add overall details to the `GetAmountSpan` span. The following code adds tags to the `GetAmountSpan` span after looping through the list of order items:

```
check observe:addTagToSpan("Total", totalPrice.toString(),
  getAmountSpan);
    result = observe:finishSpan(getAmountSpan);
    if result is error {
        log:printError("Error in finishing span", result);
    }
    http:Response res = new;
    json responseData = {
        totalPrice: totalPrice
    };
    res.setPayload(responseData);
    check caller->respond(res);
```

Here, we have also used the `addTagToSpan` function to add a new tag to the `GetAmountSpan` span. It adds a new tag with a key of `Total` and a value with the total price of the order. Then it closes the `GetAmountSpan` span with the `finishSpan` function and checks for any error.

Finally, it generates a JSON response and sends a response back with the total price to the order service. The order service can use these price details to proceed with the payment functionality.

When you start up an updated inventory service with tracing enabled, it publishes these new spans to the Jaeger backend. You can visualize this trace data as shown in the following screenshot with the Jaeger interface:

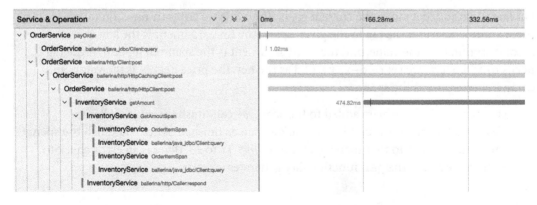

Figure 8.12 – Newly added spans

In this interface, you can see the span list containing the spans that we have added. It contains `GetAmountSpan` as the root span and two `OrderItemSpan` spans to represent two items in the order items. We can expand these spans to check the tags that we have added as shown in the following screenshot:

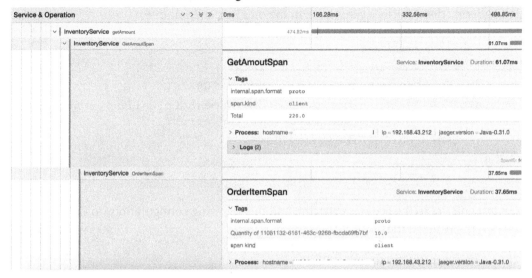

Figure 8.13 – Tags in the newly created span

You can add more spans and tags to these services depending on the observability requirements. You can use this information to debug the program and see how the program flows easily by tracking what the values given to the variables in each program segment are.

In this section, we learned about tracing, which is important in observing a microservices/distributed application. We discussed the OpenTelemetry protocol, which gives a standard definition for tracing. We have discussed the tracing features provided by the Ballerina language and using Jaeger to collect and visualize traces in Ballerina services. In the next section, we will discuss metrics, which is the third pillar of observability in a microservices platform. We will further discuss collecting metrics and visualizing them with different tools.

Collecting and visualizing metrics with Ballerina

Collecting metrics is a critical aspect of monitoring an application. It helps system engineers to identify the system status and how those applications have performed. In this section, we will learn how to expose Ballerina metrics to **Prometheus** and visualize those metrics with the **Grafana** dashboard. We will also discuss building our own custom metrics to observe important metrics on the system.

Exposing metrics from Ballerina

Ballerina has built-in support for exposing metrics. If you are creating a service, Ballerina exposes some useful metrics to the observability tools that can be used to analyze the performance and the load. For example, we can create the following simple HTTP service using Ballerina that returns a value back to the `caller` endpoint:

```
import ballerina/http;
import ballerinax/prometheus as _;
service /Customer on new http:Listener(9092) {
    resource function get getCustomerName(http:Caller
        caller, http:Request req) returns error? {
        check caller->respond("Tom");
    }
}
```

To expose metrics to the metrics collector, add the following configurations to the `Config.toml` file in the project directory:

```
[ballerina.observe]
metricsEnabled=true
metricsReporter="prometheus"
[ballerinax.prometheus]
port=9797
host="0.0.0.0"
```

These configurations expose Ballerina metrics over port `9797`. Same as the Jaeger sample, when you are running the application, make sure to use the `--observability-included` option with the `run` command. For example, to run Ballerina code with observability enabled, you should use the `bal run --observability-included` command. Once the server is ready, you can check the exposed metrics via the `http://localhost:9797/metrics` URL. Here, you need to change the hostname where the Ballerina server is hosted. This interface lists down metrics exposed by the Ballerina application. To analyze HTTP metrics, send a single request to the service that we have just created. Once you send a request, you'll get the following line on the metrics web view:

```
requests_total_value{listener_name="http",src_position="main.
bal:7:5",src_object_name="/Customer",entrypoint_service_
name="/Customer",protocol="http",src_resource_path="/
getCustomerName",entrypoint_resource_accessor="get",src_
service_resource="true",entrypoint_function_
```

```
module="user/package:0.1.0",http_url="/Customer/
getCustomerName",src_resource_accessor="get",src_
module="user/package:0.1.0",entrypoint_function_name="/
getCustomerName",http_method="GET",} 1.0
```

These metrics expose the number of requests that come into the service that we have just created. This is a key-value pair that contains the key as some details of the metrics and the value. If you send only one request, it shows as 1.0 at the end. If you send another request to the same service and check for the metrics, the number of requests will be shown as 2.0 at the end.

Other than the number of requests, you can also find the response time taken for each of the services with the following metrics:

```
response_time_seconds_value{listener_name="http",entrypoint_
function_name="/metrics",protocol="http",entrypoint_resource_
accessor="get",src_service_resource="true",src_resource_path="/
metrics",src_resource_accessor="get",entrypoint_service_
name="ballerinax_prometheus_svc_0",src_module="ballerinax/
prometheus:0.1.8",http_url="/metrics",entrypoint_
function_module="ballerinax/prometheus:0.1.8",src_object_
name="ballerinax_prometheus_svc_0",src_position="metric_
reporter.bal:62:13",http_method="GET",} 0.080726382
```

These collected values are temporal and we need a way to store this information in time series storage to analyze further. In the next section, we will use Prometheus to collect and visualize these metrics.

Collecting metrics with Prometheus

Prometheus is a popular metrics collection tool that can be used to collect Ballerina metrics. Prometheus can be downloaded from the https://prometheus.io/ download/ website as an executable .zip file. You can download Prometheus from Docker Hub with the following Docker command for Prometheus Docker deployments:

```
docker pull prom/prometheus
```

Before starting the server, we need to configure Prometheus to collect metrics from the given endpoint. Prometheus uses pull-based metrics collection techniques instead of push-based metrics collection. Therefore, you need to provide an interface that publishes data so that Prometheus can obtain data from it. To configure Prometheus to check the endpoints, we need to add the following job to the `scrap_config` section in the `prometheus.yml` config file:

```
- job_name: 'ballerina'
  static_configs:
  - targets: ['localhost:9797']
```

Since we exposed Ballerina metrics through port 9797, we also need to configure Prometheus to collect metrics from port 9797. Also, remember to set the right hostname where the Ballerina server was hosted. Once you start the Prometheus server, you can view the interface via port 9090. On this web page, you can check whether Ballerina correctly connected to the Prometheus server by clicking on the **Status | Targets** options on the menu bar:

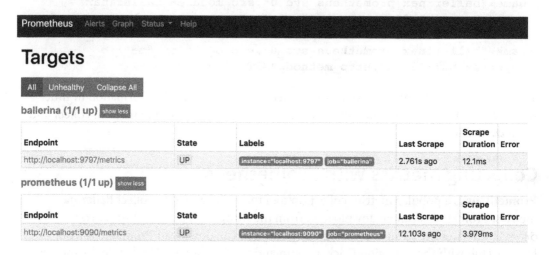

Figure 8.14 – List of targets on Prometheus

Now you can visualize different metrics with the Prometheus dashboard provided. For example, if you need to visualize the number of requests that have been sent over time, you can use the previous metric key in the home page text input to filter the metrics that collect a number of requests.

Then, click on the **Execute** button to list the metrics as a table. You can visualize the metrics over the timeline by clicking on the **Graph** tab in the same window. This view visualizes the time series view of the number of requests that come into the Ballerina server:

Figure 8.15 – Visualizing metrics with Prometheus

You can also visualize the response time by replacing the query with the response time metrics query.

However, you will find that this is not convenient to use as a dashboard due to the limited functionalities provided by Prometheus to visualize data. In the next section, we will discuss using Grafana to have more understanding of the metrics that we have collected using its dashboards.

Visualizing metrics with Grafana

The **Grafana** executable can be downloaded from the Grafana website at `https://grafana.com/grafana/download`, or you can use the Docker image to create a Grafana container. The following Docker command pulls the Grafana image from Docker Hub:

```
docker pull grafana/grafana
```

The Docker container can be started with the following Docker command that serves its interface through the default port `3000`:

```
docker run -d --name=grafana -p 3000:3000 grafana/grafana
```

Once you have downloaded and started the server, you can log in to its interface through the default port `3000`. In this interface, you can find features to create dashboards from the collected metrics.

To start visualizing metrics, first, you need to create a data source to instruct Grafana where to collect metrics. Since we use Prometheus to collect metrics, we will use Prometheus as the data source. To select the data source, click on the **Configuration |
Data Sources** options on the left-side menu bar:

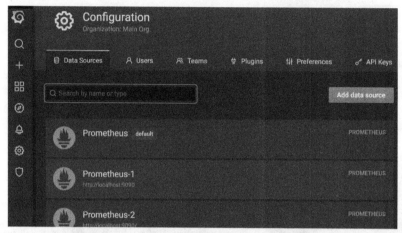

Figure 8.16 – The add data source interface

Then, click on the **Add data source** button and select **Prometheus** as the data source. In the next window, add the name of the data source. Also, set the URL as the Prometheus endpoint, which is `http://localhost:9090/` in this example:

Figure 8.17 – Setting up the data source

After setting up the data source, click on the **Save and task** button to complete adding a data source. Now, we can start creating dashboards to visualize collected metrics.

Click on **Create | Dashboard** on the left menu bar. Then, click on **Add new panel** to add a new panel to the dashboard. This operation will direct you to a workspace where you can create graphs:

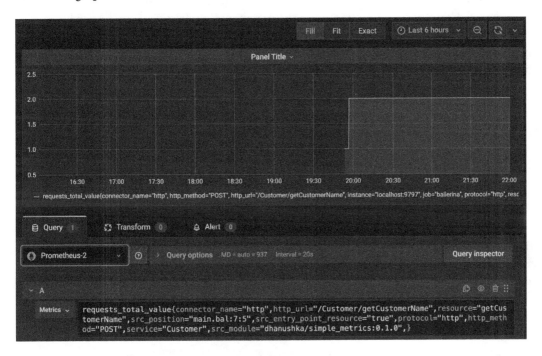

Figure 8.18 – Visualizing the request count with Grafana

For example, you can visualize the number of requests sent to the Ballerina service using the same filter that we used to filter the number of requests on the Prometheus interface. This query visualizes a time series graph on the interface that represents the number of requests that come into the system over time.

You can use different queries provided by Prometheus to filter and aggregate the collected metrics. Multiple graphs are also available for use with Grafana. Once you complete creating the graph, you can save the graph and even share it with another user.

Other than collecting Ballerina metrics, you can also use Prometheus plugins to collect metrics about the host computer or container. This information can be used to alert the infrastructure/operations team to identify when the system is about to crash or has already crashed. Having this information is valuable when you need to debug the system. Particularly when the system is running out of memory or a server suddenly crashes, we can identify failures in advance.

Creating custom metrics with Ballerina

Other than prebuilt metrics, you can define your own metrics with the Ballerina language. By using custom metrics, you can add more meaningful and application-specific observability to your code.

Ballerina provides two types of custom metrics implementations – counter-based and gauge-based. Counter-based metrics are only incremental metrics that can be used to measure incremental values. As an example, you can use counter-based metrics to store the number of requests served by a given API. Check the following example, which increments metrics when a new order is created:

```
observe:Counter registeredCounter = new ("total_orders",
  desc = "Total number of orders");
error? result = registeredCounter.register();
if result is error {
    io:println("Error in registering counter", result);
}
registeredCounter.increment();
```

The preceding code can be placed on the service to calculate the number of all order creations. Ballerina's `Counter` provides counter-based metrics that can be used to collect metrics that are always increasing. Each time this service gets triggered, it increases its value by 1 with the `registeredCounter.increment()` function. With the `increment` function, you can specify the value to increase as an input parameter. Having no arguments represents that the counter increases by 1.

After you run this code and send a request to this service, relevant metrics are published on the metrics list. If you send two requests to this server, it will print the following metrics on port `9797`:

```
total_orders_value 2.0
```

On the other hand, gauge-based metrics provide both increment and decrement functionality on metrics. This type of metric is important when it comes to metrics that can both increase and decrease. Computer processing power is a typical example of gauge-based metrics. This value can be increased or decreased. The following example shows collecting currently active order items when a user adds or removes items from the order:

```
observe:Gauge registeredGauge = new ("added_items", "Added
  items");
error? result = registeredGauge.register();
if result is error {
```

```
        io:println("Error in registering gauge", result);
}
registeredGauge.increment();
```

The preceding code block registers new metrics with the name of `added_items`.
When the user sends an order item added request, it increments the metrics by 1.
We can decrement the metrics by 1 when the order is removed from the system:

```
observe:Gauge registeredGauge = new ("added_items", "Added
    items");
error? result = registeredGauge.register();
if result is error {
    io:println("Error in registering gauge", result);
}
registeredGauge.decrement();
```

The preceding function is placed on the `deleteItem` function service's implementation,
and it decrements the value by 1 if an item deletion request comes to the server.

You can also find these metrics on the metrics view with the registered metrics name.
You can also visualize how the order items are placed over time on the Grafana dashboard.

Summary

In this chapter, we discussed monitoring cloud applications with the Ballerina language
by using tools that are commonly used in industry. We also discussed the importance
of observability and monitoring cloud native/distributed systems. We have discussed
different observability methods with reference to the three pillars of observability.

The first method we discussed was logging. This was the easiest way of collecting valuable
information from the system. We discussed using Filebeat to collect logs from an instance
or a container. Then, we discussed using Logstash to collect those logs and publish them
on Elasticsearch. We also discussed analyzing and visualizing logs with Kibana.

In distributed systems, it is difficult to track the requests flowing through different
services. Therefore, we learned how to use distributed tracing to collect and analyze how
requests flow among services. We used Jaeger to collect data, which was built on the
OpenTracing/OpenTelemetry standard. With a simple example, we discussed how to
collect traces and visualize them in the Jaeger dashboard.

In the final section of this chapter, we discussed metrics and how to use metrics published by Ballerina to monitor an application. We used Grafana to visualize the data in a dashboard, which gives important insight into system performance and load.

After reading this chapter, you should be able to build observable systems that are much easier to debug. Having more observability means fewer unknown bugs. With the tools and libraries that we have discussed in this chapter, you should be able to build highly maintainable cloud native/distributed systems with logs, metrics, and traces.

Rather than building everything from scratch, developers try to integrate existing solutions into the system. This makes the software development process much easier, faster, and error-free. In the next chapter, we will discuss integrating different services with the Ballerina language.

Questions

1. What are the best practices for monitoring cloud systems?
2. How can you secure sensitive data with monitoring?

Answers

1. These are the areas and best practices that we need to keep in mind while developing and monitoring a cloud application:

    a) Monitor trends and alerts on the cloud system. Create an alert system that notifies the infrastructure/operations team of any high consumption of resources.

    b) Use a single platform to monitor all of the deployments. Even though your application is deployed on multiple cloud vendor platforms, use a monitoring tool that lets you create a single monitoring platform that can be used to monitor the overall system.

    c) Make sure to cover all essential components within the application with monitoring tools.

2. When you are using monitoring applications, you might accidentally expose some sensitive data such as emails, passwords, and key files. Make sure to mask this sensitive data before sending it to monitoring applications. Also, use dynamic secrets that change periodically to avoid exposing API keys and any other sensitive data.

Further reading

- *Hands-On Microservices – Monitoring and Testing: A performance engineer's guide to the continuous testing and monitoring of microservices*, by D. Raj put, 2018, published by Packt Publishing available at `https://www.packtpub.com/product/hands-on-microservices-monitoring-and-testing/9781789133608`

- *Hands-On Microservices with Kubernetes: Build, deploy, and manage scalable microservices on Kubernetes*, by G. Sayfan, 2019, published by Packt Publishing available at `https://www.packtpub.com/product/hands-on-microservices-with-kubernetes/9781789805468`

9
Integrating Ballerina Cloud Native Applications

Cloud applications connect over a network and communicate with other services to perform tasks. This means that an application may need to connect with various other applications that offer different services. Your applications should have APIs exposed so that other applications can communicate with them. Therefore, we need a proper way of exposing our services to clients. When you are connecting to or exposing external services, you need to consider security, system performance, monetization, and more. Ballerina caters to these challenges with its own set of connectors and tools. In this chapter, we will learn about the different aspects of integrating services with the Ballerina language. This chapter will discuss the following topics:

- Fronting Ballerina services with an API gateway
- Building Ballerina integration flows with Choreo

By the end of this chapter, you should be able to expose Ballerina services through an API gateway and use Choreo as a low-code platform to build integration scenarios on a **Integration Platform as a Service(iPaaS)** cloud platform.

Technical requirements

In this chapter, we will look at exposing the Ballerina application as an API gateway and using the Choreo low-code platform. To expose Ballerina services, we will be using WSO2 API Microgateway as the API gateway. WSO2 API Microgateway installation instructions can be found in the *Setting up and using WSO2 API Microgateway* section of this chapter. When you are using Microgateway, check the Ballerina version that is supported and use the correct version syntax. To try out Choreo, you can get access to the Choreo platform at `https://console.choreo.dev/`.

You can find the code files for this chapter at `https://github.com/PacktPublishing/Cloud-Native-Applications-with-Ballerina/tree/master/Chapter09`.

The Code in Action video for the chapter can be found here: `https://bit.ly/2WuyqSw`

Fronting Ballerina services with an API gateway

Integrating services in a microservices application is essential as there are services that need to be connected. When client services are required to access a backend service, then the request should be routed to the target service. An API gateway is a component that is used to decouple internal services from external services. In this section, we will discuss building an API gateway with Ballerina and how to use the existing API gateway to handle requests.

Building an API gateway with Ballerina

An API gateway contains an API that exposes external services and applications to interact with. The API gateway provides a single entry point for client applications to connect with the backend application. The API gateway plays a major role in microservice architecture as it provides the following features:

- **Routing requests**: The API gateway endpoint needs to route requests to the corresponding services that can handle a particular request. An API gateway can identify a target service based on the request headers, URL, or message content. Then, requests can be routed to the target service. This routing can take the form of either just passing requests through or transforming and routing them. The API gateway should be capable of identifying requests and routing them to the corresponding service.

- **Monitoring the platform**: The API gateway is the ideal place to collect the metrics of incoming requests. By adding monitoring to the API gateway, you can identify and analyze the requests that have come into the system, successful requests, failed requests, performance bottlenecks, and bugs in the system.

- **Throttling requests**: The backend system might have a maximum traffic limit for the amount of traffic that it can handle at a given time. One of the tasks that can be done by the API gateway is limiting the number of requests that go to the service by applying rate-limiting policies. Rate-limiting policies limit the number of requests to be taken on at once. While it helps to maintain the overall health of the backend servers, it also helps to avoid **Distributed Denial of Service (DDoS)** attacks.

- **Security**: An API gateway provides security to internal backend services. It is common to perform certificate validation on the API gateway and send requests to target backend services. Thereafter, internal services trust the requests and perform operations such that each service does not have any certificate validation overhead. On the other hand, request authorization can also be performed in the API gateway against a user store. An API gateway can be configured to work with a user store and authorize each API call for each service.

- **Monetization**: You can charge a price based on the usage of your APIs for third-party client applications. The API gateway is the best place for you to measure usage and charge accordingly.

- **Caching**: For some APIs, responses can be the same for subsequent requests. Therefore, we can cache the response for a given request. If an incoming request is already in the cache, then it can respond using the cache instead of making another call to the backend. This can improve overall performance significantly and reduce the load on backend services.

You can use the Ballerina program as an API gateway that handles incoming requests and route them to the target service. Ballerina services can be used to expose backend APIs externally such that the client can access those services. In Ballerina services, we can route requests, monitor traffic, throttle requests, validate certificates, and apply authentication, authorization, monetization, and caching. For example, let's create a simple Ballerina gateway that passes a request through to a backend service.

The following sample Ballerina source code passes through a request that comes into the Ballerina service:

```
service / on new http:Listener(9090) {
    resource function 'default google(http:Request req)
            returns http:Response|http:InternalServerError|
            error {
        http:Client clientEP = check new ("https://google.com");
        http:Response clientResponse = check clientEP-
            >forward("/", req);
        return clientResponse;
    }
}
```

If a request comes to the specified resource, it forwards the request to the defined backend endpoint. The service starts on port 9090, and you can use your browser to access this resource function using http://localhost:9090/google. This request forwards you to the actual Google search home page.

Here we have given the resource function as default, which accepts all types of HTTP methods. The service context URL is / and the resource function has a context URL of https://google.com. Inside the resource function, there is a client endpoint definition that contains a URL of the Google home page. The forward function of the client endpoint forwards the request to the target endpoint. If the response message is a valid response, then it responds back with response received from Google with the response. In this scenario, the Ballerina service acts as a passthrough endpoint that routes the request to the target endpoint.

We can expand this scenario to route messages based on message content and headers, handle certificates, transform messages, cache responses, and so on. With Ballerina, you can easily access the message headers and content. Then, you can route the requests to a backend endpoint based on the request content or URL. Furthermore, you can transform message content, since Ballerina supports easily working with JSON, XML, and plain text message formats. You can easily apply message transformation logic to transform the incoming message and forward it. Also, you can perform message type conversion as well.

In some cases, you might need to integrate different types of protocols. The client application may support HTTP endpoints while the backend application accepts messages from message brokers. This type of integration can be implemented with Ballerina as it provides support for a variety of communication protocols. Certificate management and user management features provide a simple way of handling client requests and validating the client.

Building an API gateway from scratch is challenging as it is complex by nature. While you can do simple passthrough scenarios and message transformations with the Ballerina language, you need to have a proper API gateway to handle incoming requests to your backend application, as well as to handle other requirements concerning **Quality of Service (QoS)**. The API gateway should have the features that we discussed at the beginning of this section. One such prebuilt API gateway is **WSO2 API Microgateway**. WSO2 API Microgateway is an open source gateway that was built on top of the Ballerina language. It provides an API gateway to expose backend applications to external parties. In the next section, we will talk more about setting up WSO2 API Microgateway and how to customize it with the Ballerina language.

Setting up and using WSO2 API Microgateway

WSO2 API Microgateway can be configured on your computer by downloading and installing it or by using the Docker image. Microgateway comes with two separate components, which are the toolkit and the runtime. The toolkit contains tools to generate API gateway artifacts with the required configurations. The API gateway runtime, on the other hand, executes gateway artifacts generated by the toolkit and acts as the API gateway. Microgateway generates API gateway artifacts using Ballerina. These generated Ballerina artifacts form the program that runs on the API gateway runtime.

Based on the **Operating System (OS)**, you can download and install the Microgateway runtime and toolkit from `https://wso2.com/api-management/api-microgateway/`, the WSO2 API Microgateway download page. You can use the Docker image as your API gateway runtime for microservices applications as well. Once you have installed the API gateway runtime, you can use it with the `micro-gw` command. You can check your Microgateway version with the `micro-gw version` command and verify the installation.

To start creating a project, use the `micro-gw init <gw_project_location>` command. Replace the `gw_project_location` placeholder with the location of where you need to create the gateway project. This will create a set of directories and files in the location you have specified. In these files, you can find configurations and artifacts that can be used to configure the API gateway. Let's create a simple API gateway that will pass requests:

1. First, create a Ballerina service as the backend service. For this example, let's create a simple `Hello World` program that has a service context of `/hello` and a single resource that responds with `Hello World` and a resource URL of `/greeting`. See the following example `Hello World` Ballerina service:

    ```
    service /hello on new http:Listener(9091) {
        resource function get greeting() returns error|string
    {
            return "Hello, World!";
        }
    }
    ```

2. Generate the OpenAPI YAML definition for this service by executing the `bal openapi -i <path_to_service_bal_file>` command. For the Ballerina service port, use a port number other than `9090` and `9095`, since the API gateway also uses these default ports. For this example, let's export port `9091` for the `Hello World` service.

3. In the generated OpenAPI YAML file, set the base path with the `x-wso2-basePath: /hello` YAML property. The following is the sample YAML definition for the `Hello World` API that you can use to build the API gateway:

    ```
    ---
    openapi: 3.0.1
    info:
      title: Hello service
      version: 1.0.0
    servers:
      - url: http://localhost:9091/hello
    x-wso2-basePath: /hello
    paths:
      /greeting:
        get:
    ```

```
        operationId: operation1_get_sayHello
        responses:
          '200':     # status code
            description: Hello world
            content:
              application/json:
                schema:
                  example: Ok
  components: {}
  x-original-swagger-version: "2.0"
```

4. Once you have generated the OpenAPI spec YAML file for the API gateway, keep that file on the `<gw_project_location>/api_definitions` folder. This is the API definition that we will use to expose the API gateway.

5. The next step is to generate runtime artifacts for the API gateway with these YAML configurations. Execute the `micro-gw build <gw_project_location>` command to build API gateway artifacts. Once you have executed this command, it will generate a `jar` file in the `<gw_project_location>/target` directory. We will use this file to execute the API gateway runtime.

6. We can use this `jar` file and start running the API gateway. For this, you need to install the API gateway runtime.

7. Once you have downloaded and installed the API gateway runtime, you can verify the installation with the `gateway version` command. This command gives you the version of your WSO2 API Microgateway.

8. Now you can run the API gateway with the `gateway <jar_file_location>` command. `jar_file_location` is the location of the `jar` file that we generated with the API gateway toolkit.

    This will start the API gateway server on the default HTTP port, `9090`, and HTTPS port `9095`.

Now we can test whether the API gateway is functioning properly. Make sure to start the `Hello World` backend Ballerina service before trying to access the API gateway. Then, invoke the API gateway with the following `curl` command:

```
curl -X GET "https://localhost:9095/hello/greeting" -H "accept:
application/json" -H "Authorization:Bearer $TOKEN" -k
```

Here we invoke the backend service using the API gateway. The `curl` command invokes the API gateway and the API gateway forwards the request to the backend endpoint. The response generated from the backend endpoint is routed back to the client application. This will show you the `Hello World` response as the `curl` command output.

You can use the features granted by WSO2 API Microgateway to customize the API gateway functionalities. For example, you can apply throttling to a given API by adding the `x-wso2-throttling-tier: "5PerMin"` attribute to the OpenAPI YAML file definition. This will limit the rate of access for a given API to only 5 requests per minute. In the same way, you can set up security policies and observability for the API gateway.

Other than these configurations, you can have custom mediation logic in the API gateway. You can have custom Ballerina code interceptors that can handle incoming requests and validate requests and responses. In the next section, let's discuss using interceptors in API gateways.

Using interceptors in Microgateway with Ballerina

WSO2 API Microgateway provides support for adding interceptors for each request and response and engaging custom logic for each request that comes into the API gateway. Microgateway interceptors can be defined with both the Ballerina and Java languages. Interceptors can be executed in the request flow and the response flow. If you need to validate or transform the request or the response, then you can use an interceptor. In this section, let's go through a simple scenario where interceptors are used to validate the request and the response.

To have interceptors for Microgateway, you need to add a Ballerina source code file inside the `<gw_project_location>/interceptors` directory. Create a new Ballerina source file in the interceptor directory and add the following Ballerina source code to it:

```ballerina
import ballerina/http;
public function interceptRequest (http:Caller outboundEp,
  http:Request req) {
var foo = req.getQueryParamValue("foo");
if foo !== "bar" {
    http:Response res = new;
    res.statusCode = 400;
    json message = {"Error": "Invalid path parameter foo"};
    res.setPayload(message);
    var status = outboundEp->respond(res);
  } else {
```

```
        req.setPayload("From interceptor");
    }
}
public function interceptResponse (http:Caller outboundEp,
    http:Response res) {
    var payload = res.getTextPayload();
    if payload is error {
        res.statusCode = 500;
        json message = {"Error": "Unable to read the response"};
        res.setPayload(message);
    } else {
        res.setPayload(<@untainted>payload + " interceptor
            example");
    }
}
```

This Ballerina interceptor contains two functions that handle the request flow and the response flow. `interceptRequest` handles the request flow and `interceptResponse` handles the response flow. The `interceptRequest` interceptor tries to read the query parameter from the request. If there is a query parameter of `foo` and the value is `bar`, then it will forward the request to the backend endpoint. While it sends the request to the backend, it adds a custom payload to the request as well. If the `foo` query parameter does not contain `bar`, then it responds with an error message.

The `interceptResponse` interceptor works the same way as `interceptRequest` does. Rather than validating and transforming the request, though, this interceptor validates and transforms the response. The response interceptor tries to read the message payload as text content. If it cannot read the response as text content, it responds with an error message. If the response is a text message, then it appends another string to it and sends it back to the client.

> **Note**
>
> When you are coding interceptors with the Micro-Integrator, make sure to use the correct Ballerina version. The Ballerina source code syntax should be compatible with the Micro-Integrator version you are using. For example, Microgateway 3.2.0 supports Ballerina 1.2.x.

When the interceptor code is ready, we can add interceptors to the API gateway OpenAPI YAML definition. Open the OpenAPI YAML definition for the API gateway in the `<gw_project_location>/api_definitions` directory. In this YAML file, you can add interceptors for the whole service or each resource. You can add interceptors, using `x-wso2-request-interceptor: interceptRequest` for request interceptors and `x-wso2-response-interceptor: interceptResponse` for response interceptors. See the following updated YAML file that adds interceptors for the request and response flows:

```yaml
---
openapi: 3.0.1
info:
  title: Hello service
  version: 1.0.0
servers:
  - url: http://localhost:9091/hello
x-wso2-basePath: /hello
paths:
  /greeting:
    get:
      operationId: operation1_get_sayHello
      x-wso2-request-interceptor: interceptRequest
      x-wso2-response-interceptor: interceptResponse
      responses:
        '200':    # status code
          description: A JSON array of user names
          content:
            application/json:
              schema:
                example: Ok
components: {}
x-original-swagger-version: "2.0"
```

Here the `interceptRequest` and `interceptResponse` interceptors are placed under the `paths[0]./greeting.get` location with keys of `x-wso2-request-interceptor` and `x-wso2-response-interceptor`. You can keep these configurations on the root level to have interceptors at the API level.

Once you have set everything up, generate the artifacts with the `micro-gw build <gw_project_location>` command. Then, execute the API gateway runtime with the updated `jar` file. This will start the API gateway and you can invoke the `Hello World` service using the previous `curl` command. But when you try this, it will give you the request interceptor error message, saying that there is no query parameter. Next, try setting up the path parameter as follows:

```
curl -X GET "https://localhost:9095/hello/greeting?foo=bar" -H
 "accept: application/json" -H "Authorization:Bearer $TOKEN" -k
```

This command invokes the backend API through the API gateway and sends the response along with the concatenated text from the API gateway response interceptor.

The API gateway is a complex component that contains lots of features. Building an API gateway from scratch is not easy or practical. Therefore, we can use prebuilt API gateways and configure them according to our requirements. WSO2 API Microgateway is one such solution that you can use as an API gateway for your applications. As it supports writing interceptors in the Ballerina language, you can easily implement interceptor logic.

In this section, we focused on building an API gateway for the Ballerina language. First, we discussed building an API gateway from scratch. But building an API gateway from scratch has a lot of practical challenges. Therefore, we have discussed using WSO2 API Microgateway as an API gateway. In the *Setting up and using WSO2 API Microgateway* section, we discussed how to configure WSO2 API Microgateway and how to execute it. Furthermore, we discussed adding interceptors to the API gateway and how to transform and validate requests and responses with Ballerina code.

In the next section, we will discuss another brand-new platform provided by WSO2 to build integrations with the Ballerina language. Choreo is an **Integration Platform as a Service (iPaaS)** that you can use to build integrations easily with a low-code editor. We will discuss how to use Choreo and how to build a simple integration scenario on the cloud.

Building Ballerina integration flows with Choreo

Ballerina is a programming language that you can use to build different kinds of programs. Ballerina applications are meant to be developed in a standalone computer by a programmer with an IDE. When you want to deploy a Ballerina application, you need to create the underlying infrastructure to run the Ballerina application. Choreo is a platform that internally uses Ballerina to run applications on the cloud. Rather than developing and deploying Ballerina applications on your infrastructure, Choreo manages all the underlying infrastructure for you, allowing you to focus just on the business use case. You can build an application using the visual designer in the browser and some Ballerina program statements. In this section, we will learn about what the Choreo platform is and how to use it to build cloud applications.

Introduction to Choreo low-code development

Choreo is a cloud service that helps you easily develop integration services. While on-premises deployments are moving to the cloud, the Choreo platform provides an easy way of building applications directly on the cloud. If you are building your cloud application on an **Infrastructure as a Service** (**IaaS**) platform, then you need to manage all the different layers that your application needs to run on. For example, you need to install and manage Kubernetes, Docker, Helm, observability tools, and so on. Also, you need to configure the security and deployment pipelines of the cloud application.

Choreo is a low-code iPaaS platform that you can use to implement services and integrations in a cloud environment. You can visually create the flow of your program without writing any code; Choreo automatically generates the code you need behind the scenes. Starting with simple code blocks, you can implement complex scenarios on the Choreo platform. Choreo internally uses Ballerina as the programming language to run applications. Choreo converts a visual program flow to source code and deploys it on the cloud. Internally, Choreo uses containers to host Ballerina applications and exposes them externally by using an API gateway. You can invoke these services over an internet connection.

Choreo provides the following key features to users to build cloud native applications easily with low maintenance costs:

- **Low-code/no-code support**: On the Choreo platform, you don't need to use code to define integration flows. The Choreo integration design dashboard provides an interface that you can use to build an integration flow in just a few clicks. You can build simple integration scenarios with no code. Optionally, you can use custom Ballerina code to build complex integration scenarios.

- **Testing**: On the Choreo platform, you can write test cases for each integration flow you are designing. You can define test cases for each deployment and test the validity of the integration flows that you are building on Choreo.

- **Observability**: You can build APIs on the Choreo cloud and observe the metrics of the system. This is important for identifying performance bottlenecks and possible failures in a system.

You can create an account on the Choreo cloud by logging in to `https://console.choreo.dev/`. Create a new account on it to start developing cloud applications. Once you have logged in to the Choreo console, the home page of Choreo will be displayed:

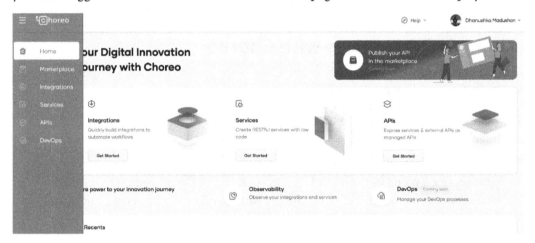

Figure 9.1 – Choreo console home page

In *Figure 9.1*, you can see the main navigation bar on the left panel of the console. In this panel, there are multiple links that you can use to build your cloud application. The following are the features provided by the Choreo platform to help you build an application:

- **Integrations**: You can build integration scenarios if you need to integrate two or more endpoints. By using the visual editor, you can design the program flow and integrate different services together. You can start designing your application with a trigger and implement the integration flow.

- **Services**: Services let you create and expose web services. Just like integrations, you can design the business logic of services with a low-code editor and run applications.

- **APIs**: Once you have created your services, you can expose them as an API with an API gateway. APIs let you expose services with an API gateway. You can apply different policies in the API gateway to handle incoming requests.

- **Remote Apps**: This feature lets you develop a Ballerina application on a local computer and observe it from the Choreo platform. You can create a Ballerina project with Choreo or set up an existing Ballerina project to work with the Choreo platform.

- **Observability**: Observability lets you observe and monitor Choreo applications. With observability support, you can find app status data, failures, logs, and other information that is required when debugging a Choreo application. It provides various graphs as log views to analyze system data.

These are the main features provided by the Choreo platform to create cloud applications. Let's discuss further how to build services on the Choreo platform. The next section will cover how to build a simple service on the Choreo platform.

Building HTTP services with Choreo

You can follow these steps to create a simple Choreo application that has a single API resource that responds with a payload:

1. Click on the **Services** link in the left navigation panel of the Choreo console to start building an HTTP service.

2. This will show you a new page where it will ask you to create your first service. Give your service a name in the input box and click on the **Create** button to create the new service (you can also click on the **Try out samples** button to list all available sample services).

3. Once you have clicked on the **Create** button, you will be redirected to a page where you can design the service. A screenshot of the Choreo service design window follows:

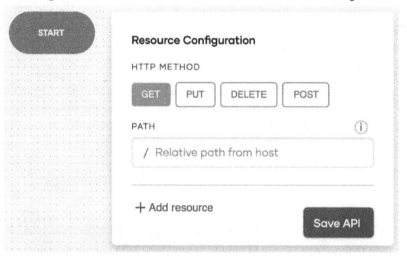

Figure 9.2 – Choreo service design page

On this page, you can select the HTTP method and the context path for the service. Let's select the **GET** method for **HTTP Method** and set the path as /hello. You can add more resources to the service by clicking on the **Add Resource** button.

4. Once you have set the resource information, click on the **Save API** button to save the service resource configurations. Then, it will generate another pop-up window with a set of components that you can add to your service definition. The following is a screenshot of the pop-up window:

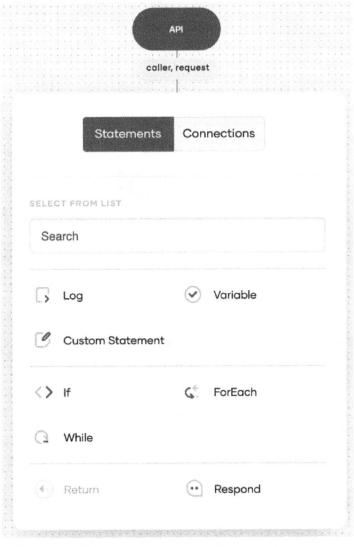

Figure 9.3 – Choreo components to build the service business logic

In the list, you can add logs, variables, if conditions, and so on.

5.   For instance, let's add a **Log** component to the service to print logs. Once you have clicked on it, it will show you another pop-up window that asks for logging configurations. Here, it asks for the type of log, which is the log level. In Choreo, you can select either **Info, Debug, Warn**, or **Error**. The **Info** log level is used to log normal information about program flow. The **Error** log level is used to log the abnormal behavior of the application. Let's select the **Info** log level. For the expression, let's give a string value of `Hello world`. Here, you need to make sure to have surrounded the string value with double-quote marks to mark this value as a string value.

6.   Then, click on the **Save** button to save your changes. This will add a new box with a label of **Log** to the service flow chart diagram.

7.   Let's add another component to this service definition to send a result back to the client. Click on the plus button between the **Log** component and the end of the flow chart (**END**). See the following screenshot for the location of the plus button:

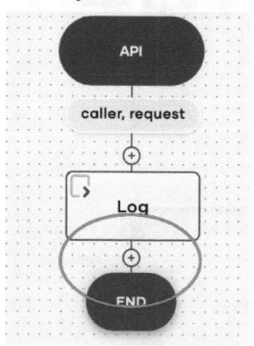

Figure 9.4 – Plus button to add a new component

Once you have clicked on the plus button, it will give you a component list so that you can add to the service definition.

8. This time, add the **Respond** component instead of the **Log** component. This will open a window asking you about the **Respond** component properties. Here, let's give a payload and a status code to send to the client. Let's set **Respond Expression** as `{"result": "success"}` and set **Status Code** as `200`.

9. Click **Save** to save your changes. Once you have saved changes, the flow chart diagram will be updated accordingly.

As we described previously, the Choreo platform uses Ballerina as the programming language to convert these flow chart diagrams into an executable program. You can see the Ballerina program code generated for this service by clicking on the <> button on the right side of the service editor window. See the following screenshot showing the Ballerina code generated for the service that we have created:

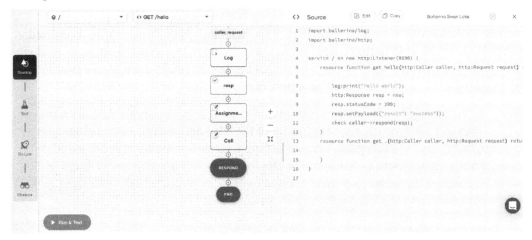

Figure 9.5 – Ballerina code editor with the Choreo low-code editor

On the right-hand side of this window, you can see the Ballerina code for the service implementation. All the components that we have added in the flow chart diagram are reflected in this source view. You can directly edit the Ballerina source code in the right panel and deploy the changes. Click on the **Edit** button at the top of the source editor window to change the Ballerina code. Once you are finished with the editing, you can click on the **Save** button to update the changes. This will also update the flow chart diagram according to the Ballerina source code changes.

If there are scenarios where you do not have a graphical component that you can use to implement your logic, you can use Ballerina statements to add the component to the service flow chart diagram. In the same way that you would add a new component to the flow chart, select **Other** to add a custom Ballerina statement to the service. All Ballerina source code editing windows on the Choreo platform give you auto-suggestions and syntax highlighting to easily work with the Ballerina language.

Once you have completed the implementation of your service, you can deploy the service to test it. To deploy the service, click on the **Run & Test** button at the bottom left of the service design page. This will open up a new panel named **Run & Test** at the bottom of the service design page and show you some deployment logs. Just as if you were running some Ballerina source code on your computer, it will give Ballerina compiler logs and a message saying the Ballerina application started on port 8090. In this panel, you can find the URL where you can access this service from. The URL is in the following format:

```
https://<service_name>-<user's_name>-<environment>.choreo.dev
```

With this URL, you can access the service that we have created. For example, `https://choreotest-dhanushkamadushan-test.choreo.dev/hello` is the URL for me to access this Ballerina service. When you invoke this service, it will respond to you with a JSON message and a response code of 200. Also, you can find the logs generated by this service in the **Run & Test** panel. Whenever you invoke this API, it will print the INFO log that we added to the Ballerina program. Once the testing is completed, you can click on the **Stop** button (the black square button) in the **Run & Test** panel to stop running the test environment. By clicking on this button, you will undeploy the running Ballerina artifacts and stop running the application.

Once you have tested the application, you can push changes to the deployment with the deployment view. Click on the **Go Live** button in the left-hand navigation panel to open the deployment page. On this page, you can deploy your service on the Choreo cloud. To start deploying your service, click on the **Deploy** button. This will trigger a build in Choreo and deploy the service you have created on the Choreo cloud. The following are the steps that will happen while deploying a service:

1. **Initialize**: Setting up and checking out the code
2. **Build**: Compiling and packaging
3. **Expose**: Creating an API and deploying to a gateway to expose the app
4. **Deploy**: Generating deployment artifacts and starting the app

Deployment logs can be seen in the bottom log panel. Once deployed, you will see a URL that can be used to access services. For this example, `s://choreotest-dhanushkamadushan.dv.choreoapis.dev/[ServiceName]/1.0.0` is the URL to access the service.

If you need to stop running the service, then you can click on the **Stop** button in the top right of the page.

In this example, we will build a service on the Choreo cloud. In the next section, let's discuss building an integration scenario where we connect multiple parties together to perform some tasks.

Integrating services with Choreo

In the same way that we created services on the Choreo platform, we can build integration scenarios as well with Choreo. You can integrate different services with each other and build complex business use cases. Choreo supports multiple connectors that you can use to integrate services. You can use Ballerina connectors on the Choreo platform as well. Services and integrations are pretty much the same on the Choreo platform, but an integration starts with a trigger while a service starts with an API. When you are building an integration flow on Choreo, you need to start with a trigger that starts the integration flow. Just like you created services on the Choreo platform, you can implement integrations scenarios by adding a program flow controller and Ballerina connectors.

To start creating an integration flow, follow these steps:

1. Click on **Integrations** in the left-hand navigation panel. This will take you to a page where you can create new integrations. Click on the **Create** button located in the top-right corner. It will take you to a **Create Integration** page that asks you for the name of your project. Give a name for the integration project and click the **Create** button to create an integration.

2. Once you have created the integration, you will get the same design page as the service design page. Here, you need to start your integration with a trigger. From a given list of triggers, you can select a trigger and start building an application.

    For example, let's select the GitHub trigger to start our integration flow.

3. Once you have selected **GitHub**, a **Configure GitHub Trigger** pop-up window will open and ask you to authenticate your GitHub account with the Choreo platform. Give authorization for the Choreo app.

    Once that is done, Choreo will show you your username in the pop-up window.

4. Then, select the GitHub repository name that you need to monitor. Select the event that will trigger the integration flow. Here, we selected **issues** as the event. Then, select an action for the event.

    For this example, we selected the **opened** event. Then, click **Save** to save the trigger information. The following is a screenshot of the GitHub trigger configurations:

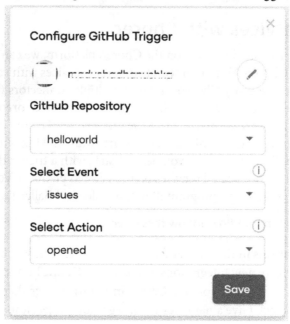

Figure 9.6 – GitHub trigger configurations

5. For the next step, let's connect the Gmail connector for this integration flow to send emails. In the component selection window, shift to the **API Calls** tab to connect Choreo with third-party applications using APIs.

    This will list many client connectors that you can use to connect your Choreo application to. For this example, let's click on the **Mail by Choreo** connector. This will give you a new **Mail by Choreo Connection** pop-up window to configure the email details.

6. In the pop-up window, for **Recipient**, give the email addresses you need to send emails to. Give a subject name in the **Subject** field and put the email message content in the **Message Body** fields. All of them accept string values as input. Save your changes to finish adding the email component.

7. Just as you deploy services, you can deploy integrations as well. Go to the **Go Live** page and deploy the application.

You can test the integration by creating an issue in the GitHub repository you specified. The Choreo integration will send an email to the email address you specified.

On the Choreo platform, you can test your integration flow with test cases. Just as you would write test cases for a Ballerina application, you can design flows to test your Choreo application as well. You can use both the visual designer and the Ballerina code editor window to create test cases. You can configure Choreo to run test cases before each deployment to test and make sure that the design has no issues.

Choreo provides an iPaaS platform to implement services and integrations. Choreo low-code development lets you create services and integrations with less code. Optionally, you can use the Ballerina code editor to edit your flow. The Choreo platform will automatically convert Ballerina source code to a visual representation and vice versa. SaaS platforms let you build integration scenarios simply without buying IaaS systems, installing tons of software, and having to manage it all. Choreo will manage all of your complex dependencies and let you focus on building the services and integrations required for your business.

Summary

In this chapter, we focused on the different aspects of building integrations with the Ballerina language. We discussed different techniques and tools that we can use with Ballerina to build integration flows. In the first section, we discussed exposing services externally with an API gateway. We discussed how to build a simple gateway with Ballerina. Due to the complexity and security compliance concerns of API gateways, we introduced WSO2 API Microgateway to expose our services to external client applications. With WSO2 API Microgateway, you can expose services with OpenAPI definitions and apply different policies such as throttling and security. Furthermore, we discussed using Microgateway interceptors with the Ballerina language to perform operations on request and response flows.

Then, we discussed using the Choreo platform to build integration workflows. Choreo provides a SaaS platform via your web browser. You can build integration flows and services with the Choreo platform and the Choreo low-code development experience. You can use Ballerina on the Choreo platform to build complicated integration solutions. Rather than building everything from scratch, you can use the Choreo platform, which will handle all of your infrastructure requirements. Choreo also provides a set of built-in connectors that help you to integrate with a lot of third-party applications/services easily. In this chapter, you have learned the importance of API gateways to integrate services and have learned how to use the Choreo platform to simplify the integration process with an iPaaS platform.

In the next chapter, we will be discussing building CI/CD pipelines with the Ballerina language. Deploying an application manually is hard when it comes to a microservices architecture, as there can be hundreds of independent services. In the next chapter, we will discuss how to automate the deployment of a Ballerina microservices application along with testing and Ballerina Central.

Questions

1.   How do you deploy WSO2 API Microgateway on a Kubernetes cluster?

2.   What are the observability tools provided with the Choreo platform?

Answers

1.   You can deploy Microgateway on a Kubernetes cluster with the `micro-gw` toolkit command. First, you need to generate build artifacts by creating the Microgateway project with the OpenAPI specs.

     Then, add the following configurations to the `<project_home>/conf/deployment-config.yaml` file:

```
[kubernetes]
  [kubernetes.kubernetesDeployment]
    enable = true
    name = 'hello_service'
    tag = 'v1'
    replicas = '2'
    buildImage = true
    push = 'true'
    imagePullPolicy = 'Always'
    registry = 'index.docker.io/<DOCKER_USERNAME>'
    baseImage = 'wso2/wso2micro-gw:latest'
    [kubernetes.kubernetesDeployment.livenessProbe]
      enable = true
      initialDelaySeconds = '20'
      periodSeconds = '20'
    [kubernetes.kubernetesDeployment.readinessProbe]
      enable = true
      initialDelaySeconds = '30'
```

```
        periodSeconds = '30'
    [kubernetes.kubernetesServiceHttps]
        enable = true
        name = 'httpsService'
        serviceType = 'NodePort'
    [kubernetes.kubernetesServiceHttp]
        enable = true
        name = 'httpService'
        serviceType = 'NodePort'
    [kubernetes.kubernetesConfigMap]
      enable = true
      ballerinaConf = '<MGW_TOOLKIT_HOME>/resources/conf/
        micro-gw.conf'
```

Make sure to replace the DOCKER_USERNAME placeholder with your Docker Hub username and MGW_TOOLKIT_HOME with the toolkit location. Also, set the DOCKER_USERNAME and DOCKER_PASSWORD environment variables in the terminal. Then, build the project with the micro-gw build <project_name> -d <path_to_deployment-config.toml_file> command. This will generate Docker artifacts and push them to your Docker registry. Then, you can deploy the Docker image in the Kubernetes environment. You can find the Kubernetes artifacts to deploy Microgateway in the <project_home>/target directory.

2. The Choreo platform supports observability with application metrics, request tracing, and logs. Choreo lets you collect metrics from an application and visualize them using graphs. It provides multiple graphs to help you easily analyze metrics along with logs on the Choreo dashboard. You can filter logs and perform root cause analysis with the **Root Cause Analysis** view.

10
Building a CI/CD Pipeline for Ballerina Applications

Cloud applications are generally larger in size and complex in terms of deployment. Regardless of whether the system is complex or simple, automating the system has multiple advantages. Software that has been started on a small scale can be scaled up to complex code that can be distributed over several services.

Having a manual process to deploy an application can become a nightmare when it comes to such a large system. On the other hand, a single adjustment to the system could break down the entire production in a few milliseconds. **Continuous Integration** (**CI**) and **Continuous Delivery** (**CD**) pipe systems in **DevOps** culture offer an ideal way to maintain larger cloud applications.

In this chapter, we will discuss the features provided by the Ballerina language along with other popular tools to implement CI/CD pipelines for cloud native applications. We will discuss freely available CI/CD pipeline development tools that we can use to implement a deployment pipeline with the Ballerina application. We will improve this pipeline with the Ballerina test framework and modularize the development process with Ballerina Central.

The following are the topics that you will learn about in this chapter:

- Writing automated tests in the Ballerina language
- Deploying a Ballerina application into the cloud with cloud deployment tools
- Learning about Ballerina Central and modularizing Ballerina components

By the end of this chapter, you should understand the basics of using automated deployment tools along with Ballerina. You will learn about popular tools that can be used with Ballerina to build an easily maintainable system.

Technical requirements

The samples given here use GitHub as the **Source Control Management** (**SCM**) tool and you can change them according to your requirements. Therefore, you need to create an account in GitHub to try out these samples. You can use other SCM cloud platforms such as GitLab or Bitbucket. But to try out GitHub Actions, you must have a GitHub account.

You can find all the code used in this chapter at `https://github.com/ PacktPublishing/Cloud-Native-Applications-with-Ballerina/tree/ master/Chapter10`.

The Code in Action video for the chapter can be found here: `https://bit.ly/2THhszi`

Testing Ballerina applications

It is common for programmers to spend more time testing applications than writing programs. Each time you write simple code, you might run the application and check whether the application is functioning in the way that is expected. In this scenario, you might be focused on providing the expected functionality for the feature you are adding. *What if the code you have added breaks the functionality of another feature that you haven't checked?* This is where you need to use automated testing to make sure that your changes do not break other functionalities. Testing is an extensive subject that has different test methods to test different types of applications. In this section, we will focus more on writing tests for Ballerina applications that are running on the cloud.

Testing cloud native applications

Testing is an essential phase in the software development process. Once you develop the code, you should be able to add new features at a later point in time without breaking the previous functionality.

Users who perform tests on a system are known as **testers**. Tests can be either automated or manual. Manual testing might be easy for testing a small-scale system. But when the system grows, it is hard to do manual testing even with a group of testers. This is one of the main reasons to have automated testing instead of manual testing. Manual testing also might not be accurate due to human errors.

Automated testing, on the other hand, is much more reliable than manual testing. But also keep in mind that in manual testing, testers can perform random tests as well, which might be tricky to implement in automated testing. The best thing to do is to automate tests as much as possible while doing some manual acceptance tests as well.

Software testing can be separated into the following types:

- **Unit testing**: A unit is a small component of your program. For example, this might be a `util` function that outputs a simple calculation. Unit testing is done to test whether these small components are working fine and provide the expected outcome. This is the easiest type of test case that you can implement since smaller components have well-defined functionalities and fewer dependencies.

- **Integration testing**: Unit tests check the functionality of all units. However, there might be unexpected issues when all units work together. In cloud native applications, it is common to have different, small services working together to perform certain tasks. Since each service depends on others, the integration test is important to ensure all the services interact with each other in an expected way.

- **System testing**: System testing is used to test the overall functionality of the system. Use cases defined in requirement analysis should be satisfied by the system. The system test should verify the overall system functionalities and performance.

- **Acceptance testing**: This test guarantees that the product can be delivered. This test should make sure that the system meets all the business requirements and is production-ready. Acceptance testing is conducted in two stages: **alpha** and **beta testing** stages.

- **Non-functional testing**: As its name suggests, non-functional testing is testing non-functional requirements of the system, such as performance, usability, security, and reliability.

- **Performance testing**: Performance testing is used to test the system performance. The system should be able to handle the expected amount of traffic without failing. Performance testing includes **load testing**, **stress testing**, **endurance testing**, and **spike testing**.

- **Security testing**: Security is a key aspect of any application. There are tools available to surf through code to find possible vulnerabilities.

- **Usability testing**: Usability testing is performed to ensure that the interfaces on the application are easily used by the end user. There are different monitoring tools available to check how users navigate through the interface. With this data, designers can identify how the user interacts with the system and how to make it more user-friendly.

- **Compatibility testing**: Compatibility testing is used to test whether different platforms are compatible with the system. For example, a backend server can be tested with different types and versions of the operating system. Frontend applications can be tested on different browsers.

Testing can be represented as a pyramid to show the complexity of these tests:

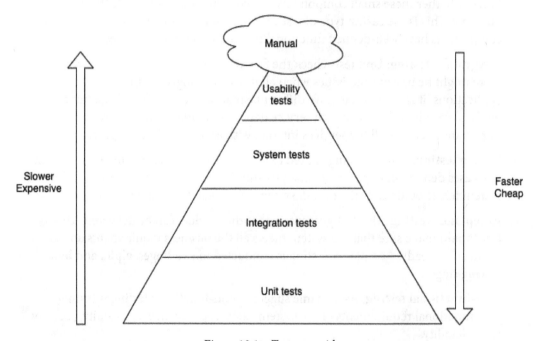

Figure 10.1 – Test pyramid

As you can see, **unit tests** are the easiest to implement and faster in getting feedback on the system status. On the other hand, usability testing is much more complex to implement and requires a longer time to get the system status. The higher up in the pyramid, the harder it becomes to automate the system.

The final goal of having these different types of testing is to have more reliable software that can be easily used by the end user. Ballerina provides a built-in testing framework to automate testing. You can easily write unit tests with different features, such as **functional mocking**. Integration tests can also be performed over a cloud application by sending requests to different services and validating responses. In the next section, we will learn about writing a simple test with the Ballerina testing framework.

Writing a simple test in the Ballerina language

To start adding tests to the Ballerina program, you need to create a new directory inside your project home directory as `tests`. Inside this directory, you can put all the tests that you are going to implement. For this example, we will create a `main_test.bal` file inside the `tests` directory. The full path to the test file is `<project_home>/tests/main_tests.bal`. This test is written in the Ballerina default module, which is in the root of the package. You can also write test cases for each module. To add test cases to a module, create a `tests` directory in the module home and add test cases to that directory. For example, to add a test case with the `module_tests.bal` filename, create a file in `<project_home>/modules/<module_name>/tests/module_tests.bal`.

The simplest test you can implement with Ballerina is as follows; you can add the following content to the `main_test.bal` file:

```
import ballerina/test;
@test:Config
function testAssertEquals () {
    test:assertTrue(true, msg = "Failed!");
}
```

The `@test:Config` annotation at the beginning of the test file indicates that this is a test case with no additional configurations. You can have any name given to each test case as the function name. However, you can use the Ballerina naming convention to name test case functions. The function name always begins with the word `test` and appends what this test case does.

Inside the test case, we can implement the tests we need to perform. In the previous example, we have a simple assertion that checks whether the first input argument for the `assertTrue` function is `true`. If the first argument is `false`, then the Ballerina testing framework marks this as a failed test case.

You can add more test cases to the test file with multiple assertions on each of the test cases. You can have any number of assertions in a given test case. If any assertion fails, then the Ballerina compiler will mark the test case as a failed test case.

You can run tests by using the `bal test` command. This command will run all of the test cases that belong to a given package, including tests defined in the module. Ballerina also lets you run a single test function with the `bal test --tests <test_function>` command. This function only runs the function with the given name. For example, the command to run the `testAssertEquals` test function is `bal test --tests testAssertEquals`.

If you need to execute tests only for a particular module, you can use the `bal test --tests <package_name>.<module_name>:*` command. For example, if the package name is `basic_module`, then the command is `bal test --tests testing_basic.basic_module:*`. If you need to execute only a single test case on the module, then you can replace the `*` character with the test function name.

When you run this command, Ballerina compiles the project and runs the test. In the terminal, you can see that there is one test that gets executed and it passes. If you try to change the first argument of the `assertTrue` function to `false` and try to run the test cases, you will see it failing and displaying the stack where it failed. In the next section, we will discuss further the different assertion functions that can be used in the Ballerina test suite.

Writing test functions with the Ballerina test framework

In the previous example, we discussed using assert functions to check the value of a variable and mark the given test as passed or failed based on the assertion status. We can use different types of assert functions to compare variables. Consider the following `sum` function in the `main.bal` file, which takes two integers and returns the sum as the result:

```
public function sum(int a, int b) returns int{
    return a + b;
}
```

We can add a test case to test the functionality of the `sum` function, as follows:

```
@test:Config { }
function testAssertSum() {
    test:assertEquals(sum(3, 2) , 5, msg =
        "Sum does not equal");
}
```

This test function checks whether the sum of two values is returned correctly from the sum function that we have defined. Here, the assertEquals function compares whether two variables are equal. This is similar to the == equality operator, which checks deep value equality.

The first argument for the assertEquals function is the actual result from the sum function, which is the sum(3, 2) function call. The second argument is the expected result, which is 5 in this case. The third argument is an error message that needs to be printed if two values are not equal. For example, if you change the expected value to 6 in the preceding example, it will print an error message that says Sum does not equal. You can use the assertEquals function in any case where you need to pass the test case if two values are equal.

The opposite of the assertEquals function is the assertNotEquals function. This test function gets passed if values are not equal after a deep value equality check. For example, the following assert function checks whether the return value of the sum of 3 and 2 is not equal to 6:

```
test:assertNotEquals(sum(3, 2), 6, msg = "Sum is equal");
```

This function also gets passed, since the return value of the sum function is not equal to 6.

Ballerina also provides an assert function to perform a reference equality test. The assertExactEquals function compares whether the two given variables are exactly equal. Other than checking the value, it also checks for type equality as well. This function is much similar to ===, which references value equality. Check the following example, which compares two integer values:

```
test:assertExactEquals(5.0, 5.0,
  msg = "Sum does not exact equal");
```

This assert function gets passed since both values and types are similar. If you have changed the value of one of the arguments or the data type, this assertion fails.

The negation of assertExactEquals is assertNotExactEquals. The assertNotExactEquals function checks the reference value for inequality. If two values are not equal, then it passes the test. Check the following example:

```
test:assertNotExactEquals(4, 5.0, msg = "Sum is exact equal");
```

This function argument has two different values of two different types. The first argument is an integer and the second argument is a float. Therefore, this assertion gets passed.

There are two functions available in the Ballerina testing framework to assert Boolean values. The `assertTrue` function, which we discussed earlier, is used to check for a Boolean. The Boolean should be `true` for the test to be marked as passed. The `assertFalse` function also checks for a Boolean and the Boolean should be `false` to mark the test as passed. Check the following example, which uses both of these functions:

```
test:assertTrue(true, "Value is false");
test:assertFalse(false, "Value is true");
```

Ballerina also provides the `assertFail` function, which can be called when the program flow reaches a place where the test should have failed. This assert function can be used in situations where the program flows into an exception. We will discuss the `assertFail` function's practical usage later in this chapter.

Ballerina test reports

In previous examples, the test running status was only printed on the terminal as a list of test cases. The Ballerina test framework also offers the test status as an HTML document. To generate a test report in HTML format, you can execute the `bal test --test-report` command. The test report can be found in the `<project_home>/target/report/index.html` file. The following screenshot shows us the test report generated by the Ballerina testing framework:

Figure 10.2 – Test report generated by Ballerina

This test report visualizes all the modules that ran tests and the number of passed and failed test cases. You can select a particular module and check the failed test cases as well. If any test case has failed, the page will show the reason as well.

You can also find the test result in JSON format on the `<project_home>/ target/ report/test_results.json` file. This JSON file is human-readable and you are able to use it with your own API to analyze the test running status.

Other than just the test execution result, you can also take a code coverage report to analyze how much code is covered by the test cases. This report can be generated by executing the `bal test --code-coverage` command. This command generates a JSON file that contains code coverage details. You can generate an HTML report by using the `bal test --test-report --code-coverage` command, the same as with the previous test report. The following screenshot is the result of an HTML web page generated by this command:

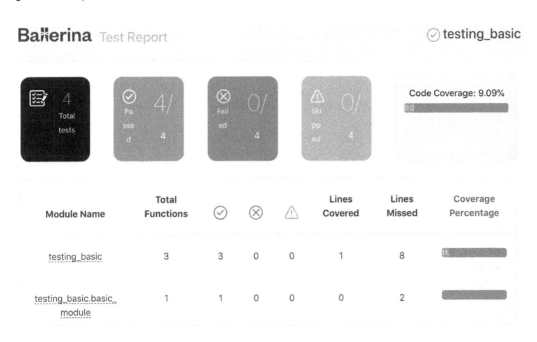

Figure 10.3 – Coverage test results

As you can see, the preceding report is the same as the previous report (*Figure 10.2*) and you can check all the test cases executed as previously. **Code Coverage** shows the number of lines that interacted with the test versus the number of code lines in total. Additionally, you can check the code that is covered by checking each of the files, as follows:

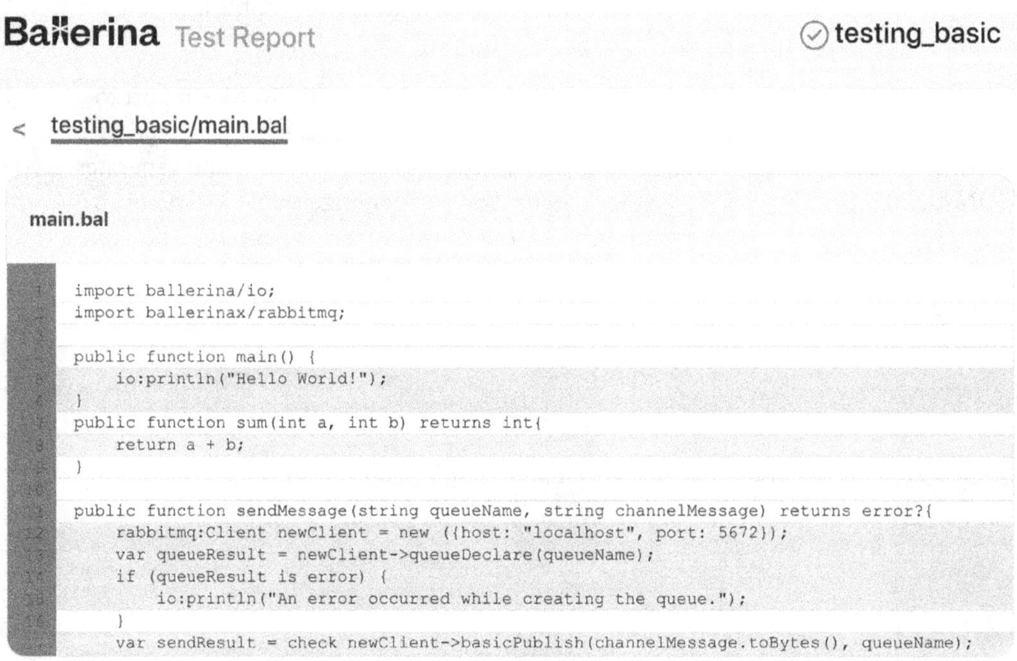

Ballerina Test Report   ⊘ testing_basic

< testing_basic/main.bal

```
main.bal

     import ballerina/io;
     import ballerinax/rabbitmq;

     public function main() {
         io:println("Hello World!");
     }
     public function sum(int a, int b) returns int{
         return a + b;
     }

     public function sendMessage(string queueName, string channelMessage) returns error?{
         rabbitmq:Client newClient = new ({host: "localhost", port: 5672});
         var queueResult = newClient->queueDeclare(queueName);
         if (queueResult is error) {
             io:println("An error occurred while creating the queue.");
         }
         var sendResult = check newClient->basicPublish(channelMessage.toBytes(), queueName);
```

Figure 10.4 – Test coverage report over the code

As you can see, in this view, you can check which line of code is executed and which line of code still needs to be covered with test cases. Having an idea about code coverage is an important factor in any type of software development. More code coverage makes the program more reliable. With this view, you can add test cases to cover code lines as much as possible.

We discussed running multiple test cases sequentially, but sometimes we need to execute a set of instructions before and after running test cases. The most common example is running the same test cases with different configurations provided by different configuration files. In Ballerina, we could have **life cycle hooks** that can trigger at each phase of the test cases. In the next section, we will discuss the Ballerina testing life cycle further.

Understanding Ballerina's testing life cycle

The Ballerina test framework contains a life cycle of how tests should be run. You can add functionalities to each of these life cycles to implement test cases. A set of test cases that are grouped with a single Ballerina test file is known as a **test suite**. A test suite can be used to group several test cases together to run as a single unit. There may be several test cases in a single test suite.

Each of these test suites can have a function that triggers before running test cases and after running test cases. These functions can be used to configure the testing infrastructure before running each test case. For example, if this particular set of tests requires a config file to run the test cases, you can use the `before` function to copy configuration files to the desired location. Similarly, you can delete these files after executing tests in the `after` function.

Also, each of these test cases can have `before test` and `after test` functions that trigger before and after running the test cases. Same as the test suite, you can configure the environment to run a given `before test case` function and clear the changes with the `after test case` function. We can represent this life cycle in the following diagram:

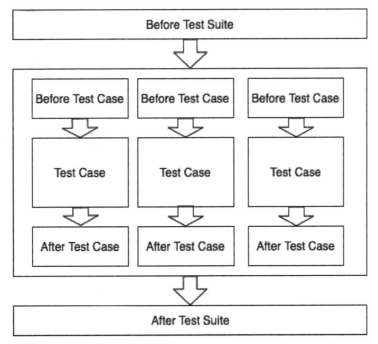

Figure 10.5 – Ballerina test life cycle

You can use the Ballerina `@test:BeforeSuite` annotation and `@test:AfterSuite` `{}` annotation to access test suite life cycles. Each test file can have one `BeforeSuite` function and one `AfterSuite` function. Check the following example code, which copies a file from one location to another before running test cases on the current test suite:

```
@test:BeforeSuite
function beforeSuiteFunc() {
    io:println("Coping config files");
    file:Error? copyDirResults =
        file:copy("/usr/data/conf.toml",
            "/usr/test/conf.toml", file:REPLACE_EXISTING);
    if copyDirResults is () {
        io:println("bar.txt file is copied to new path ");
    }
}
```

Similarly, you can implement the `AfterSuite` function to delete the copied file from the filesystem after running test cases in the test suite. The following example deletes the file we created in the `BeforeSuite` function:

```
@test:AfterSuite {}
function afterSuiteFunc() {
    io:println("Removing config files");
    file:Error? removeResults =
        file:remove("/usr/test/conf.toml");
    if removeResults is () {
        io:println("File removed");
    }
}
```

Similarly, you can implement the `before` test case and the `after` test case function for each test case. Here, we need to use Ballerina before and after using the test annotation to reference the function that we are going to execute before and after the test case. The following example defines the test function along with the references to the `before` and `after` test case functions:

```
@test:Config {
    before: beforeFunction,
    after: afterFunction
}
```

```
function testFunction() {
    io:println("I'm in test function!");
    test:assertTrue(true, msg = "Failed!");
}
```

The next step is to implement beforeFunction and afterFunction definitions, as follows:

```
function beforeFunction() {
    io:println("I'm the before function!");
}
function afterFunction() {
    io:println("I'm the after function!");
}
```

Once you run this test, you can see that beforeFunction gets executed first, then testFunction is executed, and finally, afterFunction is executed. Here, we will just print a message, but you can use the beforeFunction and afterFunction functions to create the initial conditions to set up the test environment.

You can also set the order of the test execution with the dependsOn option provided by the Ballerina platform. Check the following example, which has dependencies on each test execution:

```
function testFunction1() {
    io:println("This is function 1");
    test:assertTrue(true, msg = "Failed!");
}
@test:Config { dependsOn: [testFunction1] }
function testFunction2() {
    io:println("This is function 2");
    test:assertTrue(true, msg = "Failed!");
}
```

In this example, the testFunction2 test function depends on the testFunction1 function. If you try to execute testFunction2, you will see that testFunction1 is executed first and then testFunction2 gets executed. This behavior is important when you have test cases that need to be executed in sequential order. If one test case depends on another test case, we can use dependsOn to specify the test dependencies.

Grouping Ballerina tests

The life cycle that we discussed earlier is rigid and does not allow us to run a particular set of tests together. With Ballerina test grouping, you can execute a particular set of tests that are grouped together. You can configure a group and name it in the definition of the test function, as follows:

```
@test:Config { groups: ["GroupA"] }
function testFunctionGroupA() {
    io:println("Testing group A");
    test:assertTrue(true, msg = "Failed!");
}
```

Here, we named this function to belong to GroupA. You can have any number of groups for a given test case. For example, check the following example, which belongs to both GroupA and GroupB:

```
@test:Config { groups: ["GroupA", "GroupB"] }
function testFunctionGroupAB() {
    io:println("Testing group A and B");
    test:assertTrue(true, msg = "Failed!");
}
```

If you run this example with the normal `bal test <module_name>` command, you can see that it executes all of the test cases. This command executes all the test cases belonging to all the groups. If you need to run a particular group of test cases, you can use the `bal test --groups <group_1>,<group_2>` command to execute a given set of test groups. For example, to execute only GroupA over the entire project, you can use the `bal test --groups GroupA` command. This command executes all of the GroupA test cases over the Ballerina project.

You can also execute multiple groups by running the `bal test --groups GroupA,GroupB` command. This command executes both test cases belonging to GroupA and GroupB. If there are test cases that do not belong to any of the groups, those tests do not get executed if you run tests with groups.

You can use the `--disable-groups` option to execute the tests that do not belong to the given set of test groups. This is the opposite of the command that was discussed earlier. For example, if you don't need to execute GroupA tests, then you can execute the `bal test --disable-groups GroupA` command.

You can also list down all available test groups with the `bal test --list-groups` command. You can select the available test groups from the list and execute those groups.

As with the test suite and test cases, groups can also have before and after functions that get executed before and after running test groups. The BeforeGroups and AfterGroups annotations can be used as follows to execute functions before and after the GroupA test group:

```
@test:BeforeGroups { value:["GroupA"] }
function beforeGroupB() {
    io:println("Before running GroupA");
}
@test:AfterGroups { value:["GroupA"] }
function afterGroupA() {
    io:println("After running GroupA");
}
```

So far, we have discussed testing the Ballerina application and different life cycle states in the Ballerina test suite. When you are building complex software, there can be multiple layers of software that interact with each layer. Testing components in each layer is much more complicated than testing a simple util function. In this case, you need to create mock functions and objects to replicate functionalities and perform the test. In the next section, we will learn how to mock functions and run tests in a Ballerina application.

Mocking functions with Ballerina

Testing a simple function that has no underlying function calls is as easy as directly calling the function and doing the assertion. But if there are function calls, such as database calls and HTTP calls, within the function that we need to test, then we need to isolate those function calls and return a mock value.

This is an important fact in testing applications. We can use this mocking capability in unit testing to test some particular functionality only. If we can isolate all of the other underlying functions, we can easily write a test case to test that particular functionality.

Check the following checkValidOrder function that we described earlier in *Chapter 5*, *Accessing Data in Microservice Architecture*, for the order management system that validates the order deliverability against inventory availability:

```
function checkValidOrder(jdbc:Client jdbcClient,
    OrderItemTable orderItems) returns boolean|error{
        foreach OrderItem item in orderItems {
            int orderQuantity = item.quantity;
            int inventoryQuantity = check
```

```
                         getAvailableProductQuantity(jdbcClient,
                         item.inventoryItemId);
        if orderQuantity > inventoryQuantity {
            return false;
        }
    }
    return true;
}
```

The checkValidOrder function calls the getAvailableProductQuantity function to get data from the database by giving inventoryItemId as an input argument. In unit testing, if we can mock getAvailableProductQuantity, we can easily write a test case to check the functionality of the checkValidOrder function.

The first step is to define the mock function that we are going to mock. You can do this with Mock annotations. Here, you need to provide the function name that you are going to mock. Since we need to mock the getAvailableProductQuantity function, we can set the functionName parameter as getAvailableProductQuantity. If you need to mock a function in another module, then you can use moduleName to specify the module name. The following code mocks the getAvailableProductQuantity function with the availableProductQuantityMockFn function:

```
@test:Mock { functionName: "getAvailableProductQuantity" }
test:MockFunction availableProductQuantityMockFn = new();
```

Here, availableProductQuantityMockFn is the function that we are going to use as the mock function. We will use this to refer to whenever we need to mock the getAvailableProductQuantity function. Since we need to provide the jdbcClient object to handle the database, we can use the h2 database as follows instead of using the **MySQL** database:

```
jdbc:Client jdbcClient = check new ("jdbc:h2:file:./target/
sample1");
```

A test can be implemented as follows that checks whether the checkValidOrder function operates correctly:

```
@test:Config {}
function testCheckValidOrder() {
    test:when(availableProductQuantityMockFn).thenReturn(20);
    OrderItemTable = table [
```

```
                {orderItemId: "3234",
                orderId: "38403294",
                quantity: 10,
                inventoryItemId: "834209"}
        ];
        boolean|error status = checkValidOrder(jdbcClient,
            orderItemTable);
        if status is boolean {
            test:assertTrue(status, msg = "Cannot fulfil order");
        } else {
            test:assertFail(msg = "Error returned" +
                status.toString());
        }
    }
```

We can specify having a mock function instead of the default getAvailableProductQuantity function in *line 3*. Here, we define the return value of the mocked getAvailableProductQuantity function as 20.

We need the OrderItemTable record to call the checkValidOrder function. This record contains each of the order items in the order. Here, we can set the quantity as a hardcoded value of 10. When we call the checkValidOrder function, it goes through a loop and checks whether the order items can be fulfilled by inventory. Since we have hardcoded the value to 20, it returns as true. Since the return value is true, the assertion is passed, and the test case also passes.

In this section, we have learned about different methods of testing provided by Ballerina to run automated tests for Ballerina applications. You can mock objects just like how we mock functions with the Ballerina language. You can use Ballerina to write integration tests as well.

If you are creating user interfaces, then make sure to use appropriate testing tools such as Selenium or Cypress to test interfaces as well. Always write tests to cover your code to avoid any failures due to newly added features. And always automate the system to run these tests before the software release and code merge. This will make your effort much more productive and failure-proof.

In the next section, we will discuss creating a delivery pipeline for the Ballerina application. We will also discuss how to add automated testing for a delivery pipeline.

Automating a cloud native application's delivery process

Manually deploying a simple server application with limited capabilities might be easy. But eventually, the application will increase in size and complexity. The main aspect of a cloud native application is to design a fine-grained system rather than a coarse-grained system. When the system becomes more complex, the number of components on the system will also increase.

In microservice architecture, the number of services increases with the complexity of the application. When the number of services increases, it makes another overhead of performing deployments. Each component should be deployed with a well-defined set of rules. Performing this manually for larger microservice applications is a nightmare. To make things easy, we can use automated delivery pipelines to perform the deployment with predefined rules. In this section, we will focus on building a fully automated delivery pipeline for cloud native applications with different tools that are freely available to use.

CI/CD pipeline in a cloud native application

CI and CD are fundamental concepts for building any cloud native application. Those two terms are closely associated with each other and are widely used in cloud native development. CI is all about maintaining your code base. In the early days, developers used **Subversion (SVN)** as a version controlling system. Later, developers gradually moved into the Git version controlling system. In this section, we will learn about a common deployment pattern for cloud native applications that you can use along with Ballerina applications as well.

With the introduction of a Git-based version controlling system, versioning has been revolutionized. Git is much faster than SVN and convenient for creating branches. Developers can have their own repository on a local computer and a central repository on a remote computer. There are multiple free cloud versioning systems available, such as **GitHub**, **GitLab**, and **Bitbucket**, which are based on Git. Developers can collaborate with each other to create some features by using these cloud platforms.

When developers need to add new features to the system, first they need to create a new branch from the current master branch (or the branch that will be used to be released). The team can work together to complete features by adding code changes to this branch. Once a feature is deliverable, they can merge the code into the master branch.

When developers need to add some changes to a particular branch, they need to verify that the changes do not break the system. In this case, they can execute automated tests to check whether the new feature is good to merge. On the other hand, when developers need to release new features into the system, it is easy to automate the deployment since cloud applications can be large in size.

CI is building the pipeline to deploy changes rapidly to deployment. The deployment environment might be a testing, staging, or production environment. Developers can easily deploy code whenever there is a change that should be deployed.

Testing is a major aspect of the deployment pipeline. Each time you need to perform a deployment, it is good to perform tests and make sure the application does not break the system.

The combination of Docker and Kubernetes is a great one that can be used to build and maintain a complex cloud native application. Since Docker isolates most of the dependency issues, it simplifies the deployment effort significantly. Kubernetes acts as the container orchestration platform that has been used to handle the state of the overall system. We can combine the SCM tools, Docker, and Kubernetes with the deployment pipeline, as follows:

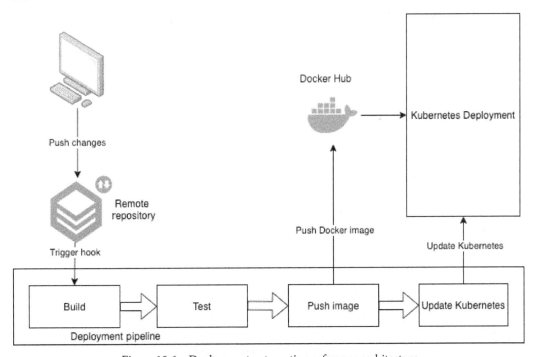

Figure 10.6 – Deployment automation reference architecture

As you can see, when developers are building applications on their local computers and they need to merge changes with the feature branch, they can push changes to the remote repository. Once the code has been reviewed, it merges into the main branch.

Once changes are merged, remote repositories can trigger a hook to the deployment pipeline. Deployment pipelines contain instructions for building, testing, running, and deploying the program. If one of these steps fails, the next step should not be taken.

The first step of the deployment pipeline is for building the program. The program must be compiled properly to complete this step. The next step is to run tests and validate whether the program works as expected. In the deployment step, it deploys artifacts into the target deployment.

As you learned earlier, containers are the ideal tool that we can use to deploy changes into a production or test environment. Here you can package all of your artifacts and dependencies into a Docker image and publish them on Docker Hub. This image can be pulled from deployment to update the system with the latest updates.

If you are using Kubernetes to manage Docker containers, then you can perform Kubernetes operations after pushing changes into Docker Hub. By applying new Kubernetes artifacts from the new build, you can deploy Kubernetes changes along with the deployment pipeline.

There are multiple tools available to implement the deployment pipeline. **Jenkins**, **Wercker**, **CircleCI**, and **GitHub Actions** are some popular tools that we can use to implement deployment pipelines. In the upcoming sections, we will explore how to use GitHub Actions to build the Ballerina delivery pipeline.

Using GitHub Actions with Ballerina

GitHub Actions can be triggered by events. These events can be creating pull requests, joining new contributors, creating issues, merging pull requests, and so on. GitHub Actions can have different workflows triggered for each event. A GitHub workflow consists of multiple actions. When an event occurs, a sequence of actions gets triggered. Each of these actions can do different things, such as creating issues, labeling issues, running tests, or deploying code.

Workflow is a generalized concept that can be used to do various things with GitHub events. We can use the workflow to implement a CI/CD pipeline as well.

To start with GitHub Actions, create a Git repository on GitHub to store the source code. For this example, we will use the simple `Hello World` service that we learned about in *Chapter 3, Building Cloud Native Applications with Ballerina*. Here, we have given the project name as `cicd_samples`. To create Kubernetes deployment artifacts, we will add the following `Cloud.toml` file to the project home directory. These artifacts contain an image name definition along with the Kubernetes deployment policies:

```
[container.image]
repository="dhanushka"
name="cicd_samples"
tag="v0.1.0"
```

```
[cloud.deployment]
min_memory="100Mi"
max_memory="256Mi"
min_cpu="1000m"
max_cpu="1500m"
```

Once we have defined the cloud configurations, we can perform the deployment manually and test the application. Performing this manually is a headache in Agile development methods as the program changes rapidly. Therefore, let's use these artifacts and build an automated pipeline to deploy the Ballerina application automatically. In the next section, we will discuss how to use GitHub Actions to build this deployment pipeline.

Setting up GitHub Actions

As we have discussed in the previous section, we will use GitHub Actions to define our build pipeline. You can follow these steps to build a simple deployment pipeline with GitHub Actions for a Ballerina program:

1. Create a GitHub repository and push the source code into the remote repository.

2. Next, navigate to the **Actions** tab on the GitHub repository page to start creating GitHub actions.

3. On the GitHub Actions page, click on the **Set up this workflow** button to create a new GitHub action, as shown in the following screenshot:

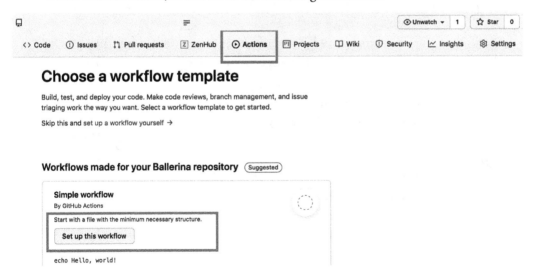

Figure 10.7 – GitHub Actions page

4.   Once you click on that button, you will be redirected to a page where you can edit a `.yml` file. In this file, you can specify what the steps are that GitHub Actions needs to take. The following is the `.yml` file content that builds the Ballerina project and publishes Docker images to Docker Hub:

```yaml
name: Ballerina example
on: [push]
jobs:
  build:
    runs-on: ubuntu-latest
    steps:
      - name: Checkout
        uses: actions/checkout@v1

      - name: Ballerina Build
        uses: ballerina-platform/ballerina-action@master
        with:
          args:
            build --cloud=docker

      - name: Log in to Docker Hub
        uses: docker/login-
          action@f054a8b539a109f9f41c372932f1ae047eff08c9
        with:
          username: ${{ secrets.docker_username }}
          password: ${{ secrets.docker_password }}

      - name: Build and push Docker images
        uses: docker/build-push-
          action@ad44023a93711e3deb337508980b4b5e9bcdc5dc
        with:
          context: target/docker/CICDSample
          push: true
          tags: user/cicd_sample:latest
```

From the preceding code, the first option, `name`, is to define a name that is given to this particular action. GitHub lists down these actions on the **Actions** tab on the GitHub website.

The next option, `on`, is to mark that these actions should be run each time code is pushed to any branch in the repository. Optionally, you can set these events to be triggered for each pull request, or issue creation, and so on. Since this application only needs to be built and published when code is updated, we will use `push` as the trigger event.

The next option, `jobs`, represents a set of instructions that are needed to be executed for each action. In GitHub Actions, you can have single or multiple jobs. These jobs are run in parallel by default. If you need to perform multiple tasks for a given action, you can set up multiple jobs.

Here, we set up a single job, which is the `build` job. The first option, `runs-on`, in the `build` job is to specify the operating system of the VM on which we are going to run this job. GitHub Actions supports executing jobs on **Windows, Linux**, and **macOS** operating systems. Here, we used the latest **Ubuntu** instance to run this job. You can check the available operating systems and versions on the GitHub Actions website and select the appropriate one.

The next option is `steps`, which is used to define a list of instructions that the deployment needs to follow to run this job. This `.yml` file contains two jobs. The first job is to check out the code from the GitHub repository and the other job is to build the project with Ballerina. The first step instructs GitHub Actions to check out the code from the repository. The `name` option is the name given to the action that is shown on the GitHub **Actions** page when it gets executed. The `use` option is used to point out the action that needs to be performed in the job. To check out the repository, we have used the `actions/checkout@v1` action.

The second action is set to build the Ballerina project with the Ballerina actions. Here, we set up the Ballerina actions to build the Ballerina program. The `args` option represents the command that we need to execute to perform the `build` operation with the Ballerina program. You can set any argument according to your requirements. For example, if you only need to execute tests, then you can use the `test` option as an argument. Here, we provide `build --cloud=docker` to build with Docker artifacts.

5.   Since we need to push the Docker image we have generated to Docker Hub, we can use GitHub Actions for Docker Hub. First, you need to log in to Docker Hub with your Docker Hub username and password. The `Log in to Docker Hub` build step is to log in to Docker Hub. You need to provide the Docker Hub username and password. To keep these values a secret, we use **GitHub secrets** to maintain these values. Here, we only set up a reference for the username and the password.

6.   The next step is to publish Docker image artifacts to Docker Hub with the `Build and push Docker images` step. In this step, we can define the Dockerfile location in the `context` attribute. The `push` attribute allows you to publish the Docker image to Docker Hub with the defined tags.

Click on the **Start commit** button. This will pop up another view and ask you to provide the commit description. Provide a description and click on the **Commit new file** button to add this file to the repository. Once you perform the commit, then you can find the committed file in the `.github/workflow/` directory on the repository.

Now, to add a new secret to the GitHub repository, we'll use the following steps:

1.   Navigate to the **Settings | Secret** option on the GitHub repository home page, as shown in this screenshot:

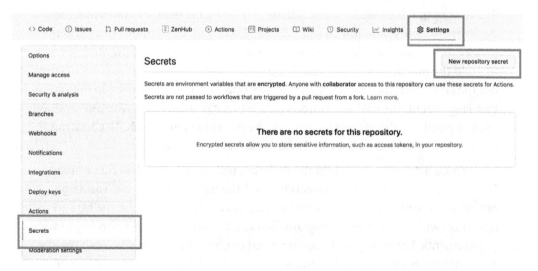

Figure 10.8 – Adding secrets to GitHub

2.   Now click on the **New repository secret** button to add a new secret. Set the name as `docker_username`. For **value**, add the username of the Docker login. Do the same for the password as well.

When all is set up, GitHub Actions executes this job every time someone pushes code to the repository. When new code is pushed, it creates Docker image artifacts and publishes them on Docker Hub. Now, you can use this Docker image on Docker Hub to create a deployment pipeline for your application.

In this section, we learned about GitHub Actions, which is a GitHub solution for easily building a deployment pipeline for source code hosted in the GitHub repository. Other than using GitHub Actions, you can use **Jenkins**, **CircleCI**, **Wercker**, or **Travis CI** to create CI/CD pipelines.

The applications that you need to build can be huge in size. Handling a large application as a single code base is not a good practice as it is harder to maintain. Also, software is not a single unit, and it can be divided into multiple library components. In the next section, we will discuss how to use Ballerina Central to organize Ballerina code with libraries.

Building and deploying applications with Ballerina Central

Building applications in a single code base makes code bases larger and bulky. A widely used programming practice is to break reusable components up into multiple libraries and reuse them when needed. As mentioned in *Chapter 1*, *Introduction to Cloud Native*, the first factor in the 12-factor app is to maintain code in a single repository and move dependent libraries to separate repositories. When an application is built, it resolves dependencies and produces executables. In the upcoming sections, we will discuss how to create a Ballerina package, how to push packages to Ballerina Central, and how to use packages from Ballerina Central.

Introduction to Ballerina Central

Ballerina uses Ballerina Central as the central registry to manage its dependencies. If you are developing an application and you have found reusable segments of code, you can publish those packages on Ballerina Central. Not only can you reuse these packages, but other developers can also use your packages to build their applications.

The Ballerina Central concept is the same as the **npm Registry** and the **Maven Central Repository**. In Ballerina Central, developers can search for packages and use them in their applications. The idea of a central library plays a key role in the development of an automated deployment pipeline. The reuse of components makes production more agile. Different teams in an organization can work with different components and publish them as reusable components in the central registry.

Next, we'll see how we can create a reusable Ballerina package, publish it in Ballerina Central, and reuse it in the Ballerina application. To do this, you first need to create a Ballerina Central account on the `https://central.ballerina.io/` website. You can easily create a Ballerina Central account with either a GitHub account or your **Gmail** account. Once you log in to the Ballerina account, you can go to the dashboard and manage the packages and the organization and invite members.

Building and publishing packages in Ballerina Central

For this example, we will build a simple package to generate an **XML** payload with a given Ballerina record. In the order management system that we discussed in *Chapter 1, Introduction to Cloud Native*, there is a notification service that sends notifications to the customer via email. This service takes details such as the customer name, mobile number, and order details and sends invoices to the customer. The `invoice_util` package reads customer data and order data to generate invoices in HTML format.

We will use the following steps to build packages in Ballerina Central:

1. We can start building the project by creating a new project with the `bal new invoice_util --template lib` command. This will create a Ballerina project with default `Ballerina.toml`, `Package.md`, and `invoice_util.bal` files.

2. Then, create `product.bal` inside the project directory. This record type contains the data structure that can be used to store a list of customer-selected products:

    ```
    public type Product record {|
        string productName;
        int quantity;
        float unitPrice;
    |};
    ```

    In the preceding code, `productName` is the product name, `quantity` is the number of products that the customer purchased, and `unitPrice` is the price of a product.

3. After that, we create the `invoice.bal` file inside the project directory. This record type is used to send invoice data to the `util` function:

    ```
    import ballerina/time;
    public type Invoice record {
        string customerName;
        string customerAddress;
        string customerBillingAddress;
    ```

```
    string customerMobileNumber;
    time:Utc invoiceDate;
    float discount;
    table<Product> productList;
};
```

Here, `customerName`, `customerAddress`, `customerBillingAddress`, and `customerMobileNumber` are the customer details. `invoiceDate` is the date the invoice is generated. For each invoice, there could be `discount`. The list of products is stored in `productList` as a table of products.

4. The next step is to create the function that generates HTML output from the given invoice. Open the `invoice_util.bal` file inside the module's home directory.

This file contains a function that generates HTML code based on the given list of products. The `foreach` loop goes through the given list of products and calculates the total value by multiplying the quantity by the unit price. Add the following content to the `invoice_util.bal` file:

```
function generateProductlist(table<Product>
    productList) returns xml{
    xml output;
    foreach var product in productList {
        float total = <float>product.quantity *
          product.unitPrice;
        output =  xml '<td>${product.productName}</td>'
        + xml '<td>${product.quantity}</td>'
        + xml '<td>${product.unitPrice}</td>'
        + xml '<td>${total}</td></tr>';
    }
    return output;
}
```

5. The `generateInvoice` function uses the `generateProductList` function to add a list of items to the final HTML document. This function converts the invoice data to the Ballerina `string` type and then adds it to the XML result. The XML returned from this function is generated from the data given in the `invoice` variable in the function input argument. In the following code, we have used the XML template laterals to generate the XML document from the XML template:

```
public function generateInvoice(Invoice invoice) returns
    xml {
    string time = time:utcToString(invoice.invoiceDate);
    xml invoiceOutput = xml '<div>
    <div><b>Prime Mart</b></div>
    <div>Name: ${invoice.customerName}</div>
    <div>Address: ${invoice.customerAddress}</div>
    <div>Billing Address: ${invoice.
        customerBillingAddress}</div>
    <div>Mobile: ${invoice.customerMobileNumber}</div>
    <div>Billing Date: ${time}</div>
    <div><table>
    <tr>
        <th>Product Name</th>
        <th>Quantity</th>
        <th>Per price</th>
        <th>Total</th></tr><tr>
        ${generateProductlist(invoice.productList)}
    </tr></table></div>
    </div>';
    return invoiceOutput;
}
```

6. The next step is to push your library into Ballerina Central. To do this, first, you need to make sure you define the organization name and package version in the `Ballerina.toml` file in your project. Define the organization name and the package version in the `Ballerina.toml` file:

```
[package]
org = "dhanushka"
name = "invoice_util"
version = "0.1.0"
```

7. Then, we give a description of your module in the `Package.md` file that is located inside the project home directory. This description goes to Ballerina Central and previews it along with your module name. This helps another developer to understand what your module does. Now, we can build the project with the `bal build -c` command. Finally, you can push the module into Ballerina Central with the `bal push` command:

```
bal push <package-name>
```

8. Then, you can log in to the Ballerina Central dashboard and see your package listed under the **My Packages** section. There, you can see a list of the different versions that you pushed into Ballerina Central. Check the following screenshot of the pushed package libraries on the Ballerina Central **Overview** page:

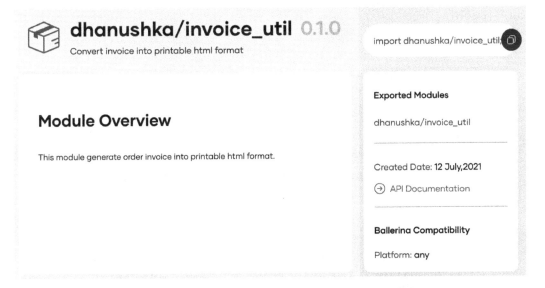

Figure 10.9 – Pushed module versions in Ballerina Central

Ballerina Central only allows you to push the latest version of the package. You can't push changes to the same version multiple times. Once you've made a new change to the package, you need to update the version in the `Ballerina.toml` file and build the project before you push changes to Ballerina Central. You can use the same `bal push` command to push updates to Ballerina Central.

Also, make sure you have the correct organization name in the `Ballerina.toml` file. From the Ballerina Central dashboard, you can create and manage organizations. Optionally, you can set up an organization in Ballerina Central and use it in the `Ballerina.toml` code instead of the default organization name that is inferred from your computer's username.

Pushing into Ballerina Central can be automated with GitHub Actions if you are using GitHub as the SCM tool. In the next section, we will learn about using GitHub Actions for pushing the package into Ballerina Central automatically when any changes merge into the source code.

Publishing packages to Ballerina Central with GitHub Actions

GitHub Actions can be easily integrated with the Ballerina code to push package changes into Ballerina Central. The first step is to create a new GitHub action on GitHub with the following Actions definition:

```
name: Ballerina example
on: [push]
jobs:
  build:
    runs-on: ubuntu-latest
    steps:
      - name: Checkout
        uses: actions/checkout@v1
      - name: Ballerina Build
        uses: ballerina-platform/ballerina-action@master
        with:
          args:
            build -c
      - name: Ballerina Push
        uses: ballerina-platform/ballerina-action@master
        env:
          BALLERINA_CENTRAL_ACCESS_TOKEN: ${{ secrets.
            BALLERINATOKEN }}
        with:
          args:
            push
```

Here, we set this action to be triggered on a push event. The first step is used to check out the source code from GitHub. The second step builds the Ballerina package along with tests. In the third step, it pushes packages into Ballerina Central. Here, we can specify the push command as push.

For authentication purposes, we need to provide a key to Ballerina Central to authorize the push command. You can find the access token generated for you on the Ballerina Central dashboard **Token** page, as shown in the following screenshot:

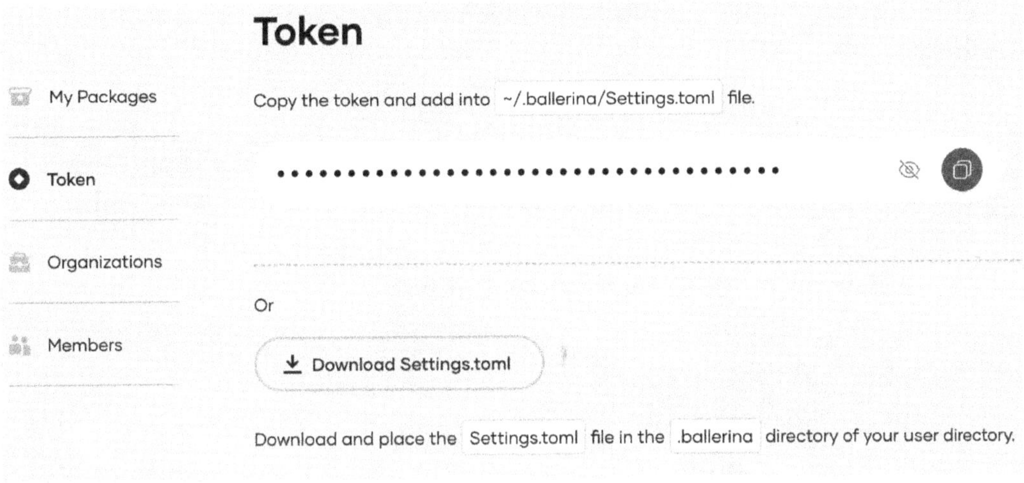

Token

🛒 My Packages	Copy the token and add into ~/.ballerina/Settings.toml file.
● Token	••••••••••••••••••••••••••••••••••••••• 👁̶ 📋
👥 Organizations	
	Or
👫 Members	⬇ Download Settings.toml

Download and place the Settings.toml file in the .ballerina directory of your user directory.

Figure 10.10 – Ballerina Central access token

You can provide this token in the GitHub Actions Docker push command and publish the source into Ballerina Central. Once you set up GitHub Actions, whenever you change the code on GitHub, it will automatically publish the package on Ballerina Central.

So far, we have created a package that formats a record to HTML content and publishes it on Ballerina Central. In the next section, we will learn about using that package in a Ballerina program to generate HTML output from a record.

Using Ballerina Central packages

Once you push the package into Ballerina Central, you can reuse it in another Ballerina project. We will execute the following steps to use the invoice_util package in our Ballerina project:

1. First, create a new Ballerina project with the bal new test_central command.

2. Add the following dependency in the Dependencies.toml file. Create a new Dependencies.toml file if one does not exist already:

```
[dependencies]
"dhanushka/invoice_util" = "0.1.0"
```

Make sure to set the correct organization name and version according to the package you have published.

3. Then, pull the package into the local computer by executing the `bal pull dhanushka/invoice_util:0.1.0` command. This command pulls libraries from Ballerina Central and caches them inside your local computer. When your project is built, it reads the content from the local library and generates artifacts. The generalized version of Ballerina's `pull` command is as follows:

```
bal pull <org-name>/<package-name>[:<version>]
```

If you don't specify the version, it will pull the latest available version from Central. It is not compulsory to pull libraries from Central since Ballerina automatically pulls packages at compile time.

4. Now, add the following code to the `main.bal` file:

```ballerina
import ballerina/io;
import ballerina/time;
import dhanushka/invoice_util;


public function main() {
    table<invoice_util:Product> products = table [
        {productName: "SD card reader", quantity: 1,
            unitPrice: 20},
        {productName: "USB cable", quantity: 2,
            unitPrice: 1}
    ];
    time:Utc invoiceDate = time:utcNow();
    invoice_util:Invoice invoice = {customerName: "John",
    customerAddress: "New York",
    customerBillingAddress: "California",
    customerMobileNumber: "0123456",
    discount: 0,
    productList: products,
    invoiceDate: invoiceDate};
    xml invoiceHTML = invoice_util:generateInvoice
        (invoice);
    io:println(invoiceHTML);
}
```

The preceding code imports the newly created package and uses the `Product` type to create a table of products. Then, by using the `Invoice` type, we create an `invoice` variable and send it to the `generateInvoice` function in the `invoice_util` package. This function returns an **XML** type that contains HTML content for the relevant invoice.

5. Finally, we can run Ballerina code with the `bal run` command. The output gives the HTML version of the `invoice` object as follows:

```
<div>
    <div>
        <b>Prime Mart</b>
    </div>
    <div>Name: John</div>
    <div>Address: New York</div>
    <div>Billing Address: California</div>
    <div>Mobile: 0123456</div>
    <div>Billing Date: Tue, 28 Mar 2017 23:42:45 -
        0500</div>
    <div>
        <table>
            <tr>
                <th>Product Name</th>
                <th>Quantity</th>
                <th>Per price</th>
                <th>Total</th>
            </tr>
            <tr>
                <td>USB cable</td>
                <td>2</td>
                <td>1.0</td>
                <td>2.00</td>
            </tr>
        </table>
    </div>
</div>
```

You can implement different libraries that can be reusable as separate Ballerina packages. Ballerina Central is an ideal place for developers to share open source content with other developers.

Summary

The automated deployment process is important when building a large application. Cloud native applications are built with small, deployable components. Therefore, the automated deployment process is a must in cloud native architecture. When developers need to deploy some code changes into a production system or testing environment, these code changes should be properly tested before deployment.

In the testing section, we discussed the Ballerina testing framework and writing automated testing. We also learned about mocking functions, different assert functions, and Ballerina's testing life cycle.

We also discussed widely used pipeline deployment tools such as GitHub Actions. GitHub Actions is an ideal platform if you are using GitHub as the source code management platform. GitHub Actions provides multiple plugins to automate the build and deployment pipeline. Ballerina Central plays a key role in building scalable applications by separating a large Ballerina project into smaller pieces. You can separate smaller components into packages and combine those packages in the deployment stage.

We have gone through different aspects of building cloud native applications with the Ballerina language throughout this book. This book focused on building cloud applications that are scalable, resilient, and maintainable. In each chapter of this book, we covered these aspects along with underlying theories and practical examples with the Ballerina language. Starting with simple code samples, we have gone through much more complex designs with many freely available third-party tools.

In the first section of this book, we learned about cloud native applications and the basics of the Ballerina programming language. The second section helped us to learn about building microservices applications with the Ballerina language. In the third section of this book, we grasped the general and advanced concepts that we need to know in cloud native application development.

We covered different aspects of building cloud native applications that are run on the cloud. We also used the order management system example, which we discussed as a sample implementation, throughout the book. With the sample that we have discussed, you have understood the basic building blocks of cloud native applications. You can use these blocks to create much more complicated and advanced cloud solutions.

Questions

1. What are alpha and beta testing?

2. What are the different methods of testing?

3. What are the benefits of CI and CD?

4. What is the role of the **Quality Assurance** (**QA**) team in DevOps culture?

Further reading

- *Learning DevOps: Continuously Deliver Better Software*, by Joakim Verona, Michael Duffy, and Paul Swartout, published by Packt Publishing, available at `https://www.packtpub.com/product/learning-devops-continuously-deliver-better-software/9781787126619`

- *Kubernetes – A Complete DevOps Cookbook*, by Murat Karslioglu, published by Packt Publishing, available at `https://www.packtpub.com/product/kubernetes-a-complete-devops-cookbook/9781838828042`

Answers

1. Both alpha and beta testing are the final stages of the testing process. Alpha tests are done by testers within the developer organization. But beta testing is performed by the end user. Both tests are considered acceptance tests.

2. The testing method can be separated into two parts as black box testing and white box testing. Black box testing does not require knowledge about the system implementation. Black box testing plays a key role in usability testing. Assume a scenario where users enter an invalid username and password. The user interface should inform the user that this is not the correct combination of username and password. White box testing, on the other hand, requires an internal implementation to perform the test. For example, unit testing checks the internal implementation of a program.

3. The CI/CD pipeline unifies code changes into a single repository and automates the whole deployment and delivery pipeline. Any change can be deployed to the target platform with a single click or after a code push. Automation makes deployment easy in complex cloud applications. Agile development methods push changes frequently to production. Therefore, automated CI/CD pipelines simplify the whole process.

4. By automating the testing process, we might think that the QA role will disappear and automation will do everything. But the QA team is still responsible for handling the overall execution of CI/CD and making sure there are no issues in the deployment process. Also, there are tests that cannot be automated. The QA team should take responsibility for delivering a reliable system to the end user by maintaining the quality of the final outcome.

`Packt.com`

Subscribe to our online digital library for full access to over 7,000 books and videos, as well as industry leading tools to help you plan your personal development and advance your career. For more information, please visit our website.

Why subscribe?

- Spend less time learning and more time coding with practical eBooks and Videos from over 4,000 industry professionals

- Improve your learning with Skill Plans built especially for you

- Get a free eBook or video every month

- Fully searchable for easy access to vital information

- Copy and paste, print, and bookmark content

Did you know that Packt offers eBook versions of every book published, with PDF and ePub files available? You can upgrade to the eBook version at `packt.com` and as a print book customer, you are entitled to a discount on the eBook copy. Get in touch with us at `customercare@packtpub.com` for more details.

At `www.packt.com`, you can also read a collection of free technical articles, sign up for a range of free newsletters, and receive exclusive discounts and offers on Packt books and eBooks.

Other Books You May Enjoy

If you enjoyed this book, you may be interested in these other books by Packt:

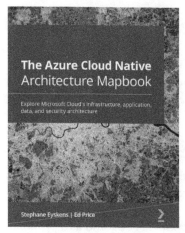

The Azure Cloud Native Architecture Mapbook

Stephane Eyskens

ISBN: 978-1-80056-232-5

- Gain overarching architectural knowledge of the Microsoft Azure cloud platform
- Explore the possibilities of building a full Azure solution by considering different architectural perspectives
- Implement best practices for architecting and deploying Azure infrastructure
- Review different patterns for building a distributed application with ecosystem frameworks and solutions
- Get to grips with cloud native concepts using containerized workloads
- Work with AKS (Azure Kubernetes Service) and use it with service mesh technologies to design a microservices hosting platform

Cloud Native with Kubernetes

Alexander Raul

ISBN: 978-1-83882-307-8

- Set up Kubernetes and configure its authentication

- Deploy your applications to Kubernetes

- Configure and provide storage to Kubernetes applications

- Expose Kubernetes applications outside the cluster

- Control where and how applications are run on Kubernetes

- Set up observability for Kubernetes

- Build a continuous integration and continuous deployment (CI/CD) pipeline for Kubernetes

- Extend Kubernetes with service meshes, serverless, and more

Packt is searching for authors like you

If you're interested in becoming an author for Packt, please visit `authors.packtpub.com` and apply today. We have worked with thousands of developers and tech professionals, just like you, to help them share their insight with the global tech community. You can make a general application, apply for a specific hot topic that we are recruiting an author for, or submit your own idea.

Share Your Thoughts

Now you've finished *Cloud Native Applications with Ballerina*, we'd love to hear your thoughts! Scan the QR code below to go straight to the Amazon review page for this book and share your feedback or leave a review on the site that you purchased it from.

https://packt.link/r/1800200633

Your review is important to us and the tech community and will help us make sure we're delivering excellent quality content.

Index

www.ingramcontent.com/pod-product-compliance
Lightning Source LLC
Chambersburg PA
CBHW081454050326
40690CB00015B/2796